United States Environmental Protection Agency	Office of Water (4601)	EPA 815-R-06-009 November 2005

MEMBRANE FILTRATION GUIDANCE MANUAL

Note on the Membrane Filtration Guidance Manual

Purpose

The purpose of this guidance manual is to provide technical information on the use of membrane filtration and application of the technology for compliance with the Long Term 2 Enhanced Surface Water Treatment Rule, which would require certain systems to provide additional treatment for *Cryptosporidium*. The requirements of this rule, as they relate to membrane filtration, are summarized in Chapter 1 of this manual.

This guidance is not a substitute for applicable legal requirements, nor is it a regulation itself. Thus, it does not impose legally binding requirements on any party, including US EPA, States, or the regulated community. Interested parties are free to raise questions and objections to the guidance and the appropriateness of using it in a particular situation. Although this manual covers many aspects of implementing membrane filtration, it is not intended to be a comprehensive resource on the subject. The mention of trade names or commercial products does not constitute endorsement or recommendation for use.

Authorship

This manual was developed under the direction of US EPA's Office of Water, and was prepared by Malcolm Pirnie, Inc., Separation Processes, Inc., and The Cadmus Group, Inc. Questions concerning this document should be addressed to:

Steve Allgeier
United States Environmental Protection Agency
Office of Ground Water and Drinking Water
26 West Martin Luther King Dr.
Cincinnati, OH 45268
Voice: (513) 569-7131
Fax: (513) 569-7191
Email: allgeier.steve@epa.gov

Acknowledgements

This document was prepared by the United States Environmental Protection Agency, Office of Ground Water and Drinking Water, Standards and Risk Management Division, Technical Support Center. The Work Assignment Manager was Steven Allgeier, and the Contract Project Officer was Jane Holtorf.

Technical consultants played a significant role in the development of this document. The work was conducted jointly by Malcolm Pirnie, Inc and Separation Processes, Inc. under contract with The Cadmus Group, Inc., as administered by Maureen Donnelly. The primary contributors to the document included:

- Steven Allgeier (US EPA)
- Brent Alspach (Malcolm Pirnie)
- James Vickers (Separation Processes)

Other contributors included Steve Alt (Separation Processes), Janet Cherry (The Cadmus Group), Maureen Donnelly (The Cadmus Group), Christopher Hill (Malcolm Pirnie), and Sheryl Patrick (Separation Processes).

Additional support for this work was provided by a panel of peer reviewers, including: Pierre Côté (ZENON Environmental), Scott Freeman (Black & Veatch), Joseph Jacangelo (MWH), Larry Landsness (Wisconsin Department of Natural Resources), James Lozier (CH2M Hill), Charles Liu (Pall Corporation), David Paulson (GE Osmonics / David Paulson Consulting Services), Richard Sakaji (California Department of Health Services), and James Schaefer (Pall Corporation / Reiss Environmental).

Contents

Appendices

List of Tables

List of Figures

Acronyms

AFT	Alternate Filtration Technology
ALCR	Air-Liquid Conversion Ratio
ANSI	American National Standards Institute
APHA	American Public Health Association
ASTM	American Society for Testing and Materials
AWWA	American Water Works Association
AWWARF	American Water Works Association Research Foundation
BOD	Biochemical Oxygen Demand
CA	Cellulose Acetate
CAM	Correlated Airflow Measurement
CFR	Code of Federal Regulations
CIP	Clean-In-Place
CL	Control Limit
CSTR	Continuous Stirred Tank Reactor
CT	(Disinfectant Residual) Concentration (mg/L) × (Contact) Time (minutes)
DBP	Disinfection Byproduct
DOC	Dissolved Organic Carbon
ED	Electrodialysis
EDR	Electrodialysis Reversal
ETV	Environmental Technology Verification
FR	Federal Register
GWR	Ground Water Rule
GWUDI	Ground Water Under the Direct Influence (of Surface Water)

HAA	Haloacetic Acid
HAA5	(sum of five) Haloacetic Acids
HFF	Hollow Fine Fiber
HPC	Heterotrophic Plate Count
IESWTR	Interim Enhanced Surface Water Treatment Rule
IMS	Integrated Membrane System
ISO	International Organization for Standardization
IVP	Integrity Verification Program
LCL	Lower Control Limit
LED	Light Emitting Diode
LRC	Log Removal Credit
LRV	Log Removal Value
LT1ESWTR	Long Term 1 Enhanced Surface Water Treatment Rule
LT2ESWTR	Long Term 2 Enhanced Surface Water Treatment Rule
MCF	Membrane Cartridge Filtration
MCL	Maximum Contaminant Level
MF	Microfiltration
MWCO	Molecular Weight Cutoff
NDP	Net Driving Pressure
NDPT	Non-Destructive Performance Test
NF	Nanofiltration
NSF	National Sanitation Foundation
OEM	Original Equipment Manufacturer
PA	Polyamide
PAC	Powdered Activated Carbon

PAN	Polyacrylonitrile
PES	Polyethersulfone
PFR	Plug Flow Reactor
PP	Polypropylene
PS	Polysulfone
PVDF	Polyvinylidene Fluoride
QA	Quality Assurance
QC	Quality Control
QCRV	Quality Control Release Value
RO	Reverse Osmosis
SDI	Silt Density Index
SEM	Scanning Electron Microscopy
SSDR	Stock Solution Delivery Rate
SUVA	Specific Ultraviolet (Light) Absorbance
SWTR	Surface Water Treatment Rule
TCEQ	Texas Commission on Environmental Quality
TCF	Temperature Correction Factor
TCPP	Total Challenge Particulate Population
TDS	Total Dissolved Solids
TMP	Transmembrane Pressure
TOC	Total Organic Carbon
TSS	Total Suspended Solids
TTHM	Total Trihalomethanes
UCL	Upper Control Limit
UF	Ultrafiltration

USEPA	United States Environmental Protection Agency
UV	Ultraviolet (Light)
VCF	Volumetric Concentration Factor

Symbols Used in Equations

Symbol	Description	Units	
		English	**Metric**
A_1	Membrane area based on the inside of a hollow fiber	ft^2	m^2
A_2	Membrane area based on the outside of a hollow fiber	ft^2	m^2
A_{20}	Membrane area required for design at a reference temperature of 20 °C	ft^2	m^2
A_d	Design membrane area	ft^2	m^2
A_m	Membrane area	ft^2	m^2
ALCR	Air-liquid conversion ratio	-	-
BP	Backpressure	psi	kPa dynes/cm^2
BP_{max}	Maximum backpressure	psi	kPa dynes/cm^2
C	Coefficient of discharge	-	-
C_f	Influent feed water concentration	-	mg/L number/L
$(C_f)_1$	Influent feed water concentration associated with the first stage of a multi-stage membrane filtration process	-	mg/L number/L
$C_{f\text{-}max}$	Maximum feed concentration (in a challenge test)	-	mg/L number/L
$C_{f\text{-}min}$	Minimum feed concentration (in a challenge test)	-	mg/L number/L
C_m	Concentration maintained on the feed side of the membrane	-	mg/L number/L
$(C_m)_i$	Concentration maintained on the feed side of the membrane associated with the second or subsequent stage (i.e., stage "i") of a multi-stage membrane filtration process	-	mg/L number/L

Symbol	Description	Units	
		English	**Metric**
$C_m(t-1)$	Concentration on the feed side of the membrane immediately after the previous backwash	-	mg/L number/L
$C_m(x)$	Concentration maintained on the feed side of the membrane as a function of position with the membrane unit	-	mg/L number/L
C_p	Filtrate concentration	-	mg/L number/L
C_{test}	Feed concentration of the challenge particulate in a challenge test	-	mg/L number/L
C_{ss}	Concentration of challenge particulate in stock solution	-	mg/L number/L
D_{am}	Diffusion coefficient for air through a saturated semi-permeable membrane	-	cm^2/s
D_{aw}	Diffusion coefficient for air in a water matrix	-	cm^2/s
D_{base}	Baseline pressure decay	psi/min	kPa/min
d_{cap}	Capillary diameter	in	cm mm µm
d_{fect}	Defect diameter	in ft	cm mm µm
d_{fiber}	Fiber diameter	in ft	cm mm µm
d_{res}	Integrity test resolution requirement	in	cm mm µm
DL	Detection limit	-	-
DOC	Dissolved organic carbon	-	mg/L
e	Specific roughness	in	cm mm

Symbol	Description	Units		
		English	**Metric**	
f	Friction factor	-	-	
g	Gravitational constant	$\dfrac{lbm - ft}{lbf - s^2}$	$\dfrac{kg - m}{s^2}$	
H	Henry's constant for air-water system	$\dfrac{mol}{psi - m^3}$		
J	Flux	gfd	Lmh	
J_{20}	Normalized flux at 20 ^{o}C (common for MF, UF, and MCF systems)	gfd	Lmh	
J_{25}	Normalized flux at 25 ^{o}C (common for NF and RO systems)	gfd	Lmh	
J_T	Flux at temperature T	gfd	Lmh	
K	Flow resistance coefficient	-	-	
K_{air}	Resistance coefficient of air	-	-	
K_{water}	Resistance coefficient of water	-	-	
L	Defect length	in ft	cm mm	
LRC	Log removal credit	-	-	
LRV	Log removal value	-	-	
LRV_t	Target log removal value in a challenge test	-	-	
LRV_{DIT}	Log removal value that can be verified by the direct integrity test (i.e., method sensitivity)	-	-	
$LRV_{C\text{-}Test}$	Log removal value demonstrated during challenge testing	-	-	
M	Specific flux	gfd/psi	Lmh/kPa	
M_{20}	Specific flux normalized at 20 ^{o}C	gfd/psi	Lmh/kPa	
M_{25}	Specific flux normalized at 25 ^{o}C	gfd/psi	Lmh/kPa	
NDP	Net driving pressure [1]	psi	kPa	
P	Pressure	psi	kPa	

Symbol	Description	Units	
		English	**Metric**
P_{atm}	Atmospheric pressure [2]	psia	kPa dynes/cm^2 atm
P_{bp}	Bubble point pressure	psi	kPa
P_c	Concentrate pressure	psi	kPa
P_f	Feed pressure	psi	kPa
P_p	Filtrate pressure	psi	kPa
P_{test}	Pressure at which a pressure-based direct integrity test is conducted	psi	kPa dynes/cm^2
ΔP	Transmembrane differential pressure	psi	kPa dynes/cm^2
ΔP_{eff}	Effective integrity test pressure	psi	kPa dynes/cm^2
ΔP_{test}	Rate of pressure decay / loss	psi/min	kPa/min
Q_{air}	Flow of air	ft^3/s ft^3/min	cm^3/s cm^3/min L/min
Q_b	Backwash flow	gpm	L/min
Q_{breach}	Flow from an integrity breach associated with the smallest integrity test response that can be reliably measured	ft^3/s ft^3/min	cm^3/s cm^3/min L/min
Q_c	Concentrate (i.e., bleed or reject) flow	gpm	L/min
Q_{diff}	Diffusive flow of air through the water matrix in the pores of a fully-wetted membrane	ft^3/min	L/min cm^3/min
Q_f	Feed flow	gpm	L/min
Q_p	Filtrate flow	gpm gpd MGD	L/hr L/min
$Q_p(x)$	Filtrate flow as a function of position within a membrane unit	gpm gpd MGD	L/hr L/min

Symbol	Description	Units English	Units Metric
Q_{water}	Flow of water	ft^3/s ft^3/min	cm^3/s cm^3/min L/min
R	Recovery	%	%
Re	Reynolds number	-	-
$R(x)$	Recovery as a function of position within a membrane unit	%	%
R_f	Foulant layer resistance	$\dfrac{psi}{gfd-cp}$	$\dfrac{bar-hr-m^2}{L-cp}$
R_{gas}	Universal gas constant	$\dfrac{L-psia}{mol-K}$	
R_m	Intrinsic membrane resistance	$\dfrac{psi}{gfd-cp}$	$\dfrac{bar-hr-m^2}{L-cp}$
R_t	Total membrane resistance	$\dfrac{psi}{gfd-cp}$	$\dfrac{bar-hr-m^2}{L-cp}$
r_1	Inner radius of a hollow fiber	in	mm
r_2	Outer radius of a hollow fiber	in	mm
SF	Safety factor	-	-
SSDR	Stock solution delivery rate	gpm	L/min
SUVA	Specific ultraviolet (light) absorbance	-	L/mg-m
T	Temperature	^oF	^oC
t	Filtration cycle time	min	min
t_b	Backwash duration	min	min
t_f	Filtration cycle duration	min	min
T_{min}	Minimum challenge test duration	min	min
TCF	Temperature correction factor	-	-
TCPP	Total challenge particulate population	number lbs	number g
TDS_c	Total dissolved solids concentration in the concentrate	-	mg/L

Symbol	Description	Units	
		English	**Metric**
TDS_f	Total dissolved solids concentration in the feed	-	mg/L
TDS_p	Total dissolved solids concentration in the filtrate	-	mg/L
TMP	Transmembrane pressure	psi	kPa dynes/cm^2
TMP_{20}	Transmembrane pressure at 20 $^\circ$C	psi	kPa
TMP_{25}	Transmembrane pressure at 25 $^\circ$C	psi	kPa
TMP_T	Transmembrane pressure at temperature T	psi	kPa
U	Membrane-specific temperature correction factor constant (manufacturer-supplied)	-	1/K
UCL	Upper control limit: (for airflow) (for pressure decay rate)	- psi/min	L/min kPa/min
UV_{254}	Ultraviolet absorbance at 254 nm	-	1/m 1/cm
VCF	Volumetric concentration factor	-	-
VCF(x)	Volumetric concentration factor as a function of position within a membrane unit	-	-
VCF(t)	Volumetric concentration factor as a function of time	-	-
VCF_{avg}	Average volumetric concentration factor in a membrane unit	-	-
VCF_{max}	Maximum volumetric concentration factor in a membrane unit	-	-
V_{eq}	System volume required to attain equilibrium feed concentration during a challenge test	gal	L
V_{hold}	Unfiltered test solution volume remaining in the system at the end of a challenge test (i.e., the hold-up volume)	gal	L
V_r	Total recirculation loop volume	gal	L
V_{ss}	Challenge particulate stock solution volume	gal	L
V_{sys}	Volume of pressurized air in a membrane system during a pressure- or vacuum-decay test	ft^3	L
V_{test}	Minimum challenge test solution volume	gal	L

Symbol	Description	Units	
		English	Metric
x	Position in a membrane unit in the direction of flow	ft	m
x_{max}	End position in a membrane unit in the direction of flow	ft	m
Y	Net expansion factor for compressible flow through a pipe to a larger area	-	-
z	Membrane thickness	in	cm
ϵ	Membrane porosity	-	-
θ	Liquid-membrane contact (i.e., "wetting") angle	degrees	degrees
κ	Pore shape correction factor	-	-
σ	Surface tension	-	dynes/cm
$\Delta\pi$	Transmembrane osmotic pressure differential	psi	kPa dynes/cm^2
ρ_{air}	Density of air	lb/ft^3	g/m^3
ρ_w	Density of water	lb/ft^3	g/cm^3
μ	Viscosity of water or air (as specified)	lbs/ft-s	cp
μ_{20}	Viscosity of water at 20 ^{o}C	lbs/ft-s	cp
μ_{air}	Viscosity of air	lbs/ft-s	cp
μ_T	Viscosity of water or air at temperature T (as specified)	lbs/ft-s	cp
μ_w	Viscosity of water	lbs/ft-s	cp
τ	Volumetric turnover time	min	min

1 For simplicity, equations applicable to both MF/UF and NF/RO systems use the term "TMP" throughout this document; as these equations are applied to NF/RO systems, the term "TMP" should be considered synonymous with "NDP"

2 Atmospheric pressure (P_{atm}) is expressed as absolute pressure (i.e., psia); note that this pressure varies with elevation

Glossary

Anion – a negatively charged ion resulting from the disassociation of salts, acids, or bases in aqueous solution

Anti-Scalant – a chemical agent added to water to inhibit the precipitation or crystallization of salt compounds; also referred to as a "scale-inhibitor"

Anti-Telescoping Device – a rigid structure firmly attached to each end of a spiral-wound nanofiltration (NF) or reserve osmosis (RO) membrane module that prevents telescoping, unwinding, or other undesirable movement of the membrane module

Array – a description of a nanofiltration (NF) or reverse osmosis (RO) membrane system based upon the ratio of the number of pressure vessels in each stage that operate in parallel (e.g., a 24:12 (absolute) two-stage array or a 2:1 (relative) two-stage array

Asymmetric – having a varying consistency throughout (e.g., a membrane that varies in density or porosity across its structure)

Backwash – 1) the intermittent waste stream from a microfiltration (MF) or ultrafiltration (UF) membrane system; also, 2) a term for a cleaning operation that typically involves periodic reverse flow to remove foulants accumulated at the membrane surface

Biofouling – membrane fouling (and associated decreases in flux) that is attributable to the deposition and growth of microorganisms on the membrane surface and/or the adsorptive fouling of secretions from microorganisms

Bleed – the continuous waste stream from a microfiltration (MF) or ultrafiltration (UF) system operated in a crossflow hydraulic configuration

Boundary Layer – thin layer of water at the surface of a semi-permeable membrane containing the rejected contaminants from the filtrate (i.e., permeate) flow in higher concentrations than the bulk feed/brine stream (called concentration polarization), affecting the osmotic pressure and salt passage

Brackish Water – saline water in which the dissolved solids content generally falls between that of drinking water and seawater

Breach – see Integrity Breach

Brine – a saline solution with a concentration of dissolved solids exceeding that of seawater (i.e., approximately 35,000 mg/L)

Brine Seal – a rubber seal around the circumference of a spiral-wound module between the module and the interior pressure vessel wall that separates the feed water from the concentrate stream, preventing the bypass of feed between the module and the inside of the pressure vessel wall

Bubble Point – the amount of applied air pressure required to evacuate the largest pores of a fully-wetted porous membrane

Cation – a positively charged ion resulting from the dissociation of salts, acids, or bases in aqueous solution

Cartridge – a term commonly used to describe a disposable filter element; included under the term "module" for the purposes of the LT2ESWTR

Challenge Particulate – the target organism or acceptable surrogate used to determine the log removal value (LRV) during a challenge test

Challenge Test – a study conducted to determine the removal efficiency (i.e., log removal value (LRV)) of a membrane material for a particular organism, particulate, or surrogate

Clean-In Place (CIP) – the periodic application of a chemical solution or (series of solutions) to a membrane unit for the intended purpose of removing accumulated foulants and thus restoring permeability and resistance to baseline levels; commonly used term for in-situ chemical cleaning

Colloid – type of particulate matter ranging in size from approximately 2 - 1,000 nm in diameter that does not settle out rapidly

Compaction – the compression or densification of a membrane as a result of exposure to applied pressure over a period of time, typically resulting in decreased productivity

Composite – made from different materials (e.g., a membrane manufactured from two or more different materials in distinct layers)

Concentrate – the continuous waste stream (typically consisting of concentrated dissolved solids) from a membrane process, usually in association with nanofiltration (NF) and reverse osmosis (RO) processes; in some cases also used to describe a continuous bleed stream of concentrated suspended solids wasted from microfiltration (MF) and ultrafiltration (UF) systems operated in a crossflow (or feed-and-bleed) hydraulic configuration

Concentrate Staging – a configuration of spiral-wound nanofiltration (NF) and reverse osmosis (RO) membrane systems in which the concentrate from each stage of a multi-stage system becomes the feed for the subsequent stage

Concentration Polarization – a phenomenon that occurs when dissolved and/or colloidal materials concentrate on or near the membrane surface in the boundary layer

Conductivity – a measure of the ability of an aqueous solution to conduct an electric charge; related to the amount of total dissolved solids (TDS)

Control Limit (CL) – a response from an integrity test, which, if exceeded, indicates a potential problem with the membrane filtration system and triggers a response; synonymous with "upper control limit" (UCL) as used in the *Membrane Filtration Guidance Manual* to distinguish from additional voluntary or State-mandated "lower control limits" (LCLs)

Crossflow – 1) the application of water at high velocity tangential to the surface of a membrane to maintain contaminants in suspension; also, 2) suspension mode hydraulic configuration that is typically associated with spiral-wound nanofiltration (NF) and reverse osmosis (RO) systems and a few hollow-fiber microfiltration (MF) and ultrafiltration (UF) systems

Dalton – a unit of mass equal to $1/12^{th}$ the mass of a carbon-12 atom (i.e., one atomic mass unit (amu)); typically used as a unit of measure for the molecular weight cutoff (MWCO) of a ultrafiltration (UF), nanofiltration (NF), or reverse osmosis (RO) membrane

Dead End Filtration – term commonly used to describe the deposition mode hydraulic configuration of membrane filtration systems; also synonymous with "direct filtration"

Deposition Mode – a hydraulic configuration of membrane filtration systems in which contaminants removed from the feed water accumulate at the membrane surface (and in microfiltration (MF)/ultrafiltration (UF) systems are subsequently removed via backwashing)

Desalination – the process of removing dissolved salts from water

Diagnostic Test – precise direct integrity tests that are specifically used to isolate any integrity breaches that may be initially detected via other means, such as coarser direct integrity tests that simply indicate the presence or absence of a breach within a membrane unit

Differential Pressure – pressure drop across a membrane module or unit from the feed inlet to concentrate outlet (as distinguished from transmembrane pressure (TMP), which represents the pressure drop across the membrane barrier)

Direct Filtration – (as used with respect to membrane filtration) term commonly used describe the deposition mode hydraulic configuration of membrane filtration systems; also synonymous with "dead end filtration"

Direct Integrity Test – (as defined under the LT2ESWTR) a physical test applied to a membrane unit in order to identify and/or isolate integrity breaches

Electrodialysis (ED) – a process in which ions are transferred through ion-selective membranes by means of an electromotive force from a less concentrated solution to a more concentrated solution

Electrodialysis Reversal (EDR) – a variation of the electrodialysis process in which the polarity of the electrodes is periodically reversed on a prescribed time cycle, thus changing the direction of ion movement, in order to reduce scaling

Element – a term commonly used to describe an encased spiral-wound membrane module; included under the term "module" for the purposes of the LT2ESWTR

Feed-and-Bleed Mode – a term used to describe a variation of the suspension mode hydraulic configuration of membrane filtration systems in which a portion of the crossflow stream is wasted (i.e., bled) rather than recirculated

Feed Channel Spacer – a plastic mesh spacer that separates the various leaves in a spiral-wound module, providing a uniform channel for feed water to reach the membrane surface and promoting turbulence in order to minimize the formation of a boundary layer at the membrane surface

Feed Water – the influent stream to a water treatment process

Filtrate – the water produced from a filtration process; typically used to describe the water produced by porous membranes such those used in membrane cartridge filtration (MCF), microfiltration (MF), and ultrafiltration (UF) process, although used in the context of the LT2ESWTR to describe the water produced from all membrane filtration processes, including nanofiltration (NF) and reverse osmosis (RO)

Flux – the throughput of a pressure-driven membrane filtration system expressed as flow per unit of membrane area (e.g., gallons per square foot per day (gfd) or liters per hour per square meter (Lmh))

Foulant – any substance that causes fouling

Fouling – the gradual accumulation of contaminants on a membrane surface or within a porous membrane structure that inhibits the passage of water, thus decreasing productivity

Heterogeneous – composed of a combination of different materials (e.g., composite and some asymmetric membranes)

Hollow-Fiber Module – a configuration in which hollow-fiber membranes are bundled longitudinally and either encased in a pressure vessel or submerged in a basin; typically associated with microfiltration (MF) and ultrafiltration (UF) membrane processes

Hollow-Fine-Fiber (HFF) Module – a relatively uncommon configuration in which very small diameter (i.e., approximately 50 μm (inside diameter)) semi-permeable hollow-fiber membranes are bundled in a "U" shape and potted into a pressure vessel; typically associated with reverse osmosis (RO) membrane processes

Homogeneous – composed of the same material throughout (e.g., symmetric and some asymmetric membranes)

Hydraulic Configuration – the pattern of flow through a membrane process by which the feed contaminants are removed or concentrated (e.g., crossflow, dead-end, etc.)

Hydrophilic – the water attracting property of membrane material

Hydrophobic – the water repelling property of membrane material

Indirect Integrity Monitoring – (as defined under the LT2ESWTR) monitoring some aspect of filtrate water quality that is indicative of the removal of particulate matter

Integrity Breach – one or more leaks in a membrane filtration system that could result in the contamination of the filtrate with unfiltered feed water

Irreversible Fouling – any membrane fouling that is permanent and cannot be removed by either backwashing (if applicable) or chemical cleaning

Leaf – a sandwich arrangement of flat sheet, semi-permeable membranes placed back-to-back and separated by a fabric spacer (i.e., permeate carrier) in a spiral-wound module

Log Removal Value (LRV) – filtration removal efficiency for a target organism, particulate, or surrogate expressed as $\log_{10}$ (i.e., $\log_{10}$(feed concentration) – $\log_{10}$(filtrate concentration))

Lower Control Limit (LCL) – a control limit (CL) that is not mandated by the LT2ESWTR but which is instead voluntarily implemented or which may be required by the State at its discretion

Lumen – the center or bore of a hollow-fiber membrane

Membrane Filtration – (as defined under the LT2ESWTR) a pressure- or vacuum-driven separation process in which particulate matter larger than 1 μm is rejected by an engineered barrier, primarily through a size-exclusion mechanism and which has a measurable removal efficiency of a target organism that can be verified through the application of a direct integrity test; includes common membrane classifications microfiltration (MF), ultrafiltration (UF), nanofiltration (NF), and reverse osmosis (RO), as well as any "membrane cartridge filtration" (MCF) device that satisfies this definition

Membrane Cartridge Filtration (MCF) – any cartridge filtration devices that meet the definition of membrane filtration as specified under the LT2ESWTR

Membrane Softening – semi-permeable membrane treatment process designed to selectively remove hardness (i.e., calcium, magnesium, and certain other multivalent cations) but allow significant passage of monovalent ions; typically used to describe the application of nanofiltration (NF) for hardness removal

Membrane Unit – (as defined under the LT2ESWTR) a group of membrane modules that share common valving which allows the unit to be isolated from the rest of the system for the purpose of integrity testing or other maintenance

Molecular Weight Cutoff (MWCO) – a measure of the removal characteristic of a membrane in terms of atomic weight (or mass), as opposed to pore size; typically measured in terms of Daltons

Microfiltration (MF) – a pressure-driven membrane filtration process that typically employs hollow-fiber membranes with a pore size range of approximately $0.1 - 0.2$ μm (nominally 0.1 μm)

Module – (as defined under the LT2ESWTR) the smallest component of a membrane unit in which a specific membrane surface area is housed in a device with a filtrate outlet structure; used in the *Membrane Filtration Guidance Manual* to refer to all types of membrane configurations, including terms such as "element" or "cartridge" that are commonly used in the membrane treatment industry

Nanofiltration (NF) – a pressure-driven membrane separation process that employs the principles of reverse osmosis to remove dissolved contaminants from water; typically applied for membrane softening or the removal of dissolved organic contaminants

Net Driving Pressure (NDP) – the pressure available to force water through a semi-permeable nanofiltration (NF) or reverse osmosis (RO) membrane, defined as the average feed side pressure (i.e., the average of the feed and concentrate pressures) less the filtrate side backpressure and the osmotic pressure of the system

Non-Destructive Performance Test (NDPT) – a physical quality control test typically conducted by a manufacturer to characterize some aspect of process performance without damaging or altering the membrane or membrane module

Normalization – the process of evaluating membrane system performance at a given set of reference conditions (e.g., at standard temperature, per unit pressure, etc.), allowing the direct comparison and trending of day-to-day performance independent of changes to the actual system operating conditions

Osmosis – the passage of a solvent (e.g., water) through a semi-permeable membrane from a solution of lower concentration to a solution of higher concentration so as to equalize the concentrations on either side of the membrane

Osmotic Pressure – the amount of pressure that must be applied to stop the natural process of osmosis

Particle Counter – an instrument used to count the number of discrete particles in a solution and classify them according to size

Permeability – the ability of a membrane barrier to allow the passage or diffusion of a substance (i.e., a gas, a liquid, or solute)

Permeate – the water that passes through a nanofiltration (NF) or reverse osmosis (RO) membrane; synonymous with the term filtrate, which is used in the context of the LT2ESWTR

Permeate Staging – a configuration of spiral-wound nanofiltration (NF) and reverse osmosis (RO) membrane systems in which the permeate (or filtrate) from each stage of a multi-stage system becomes the feed for the subsequent stage

Permeate Tube – the perforated tube in the center of a spiral-wound module that collects permeate (or filtrate) and transports it out of the membrane module

Permeate (Filtrate) Carrier – the fabric spacer in between two sheets of membrane material in one leaf of a spiral-wound module, serving to transfer the water that permeates through the membrane(s) (i.e., the filtrate) to a perforated central collector tube (i.e., the permeate tube)

Plate-and-Frame Module – a relatively uncommon configuration consisting of a series of flat sheet membranes separated by alternating filtrate spacers and feed/concentrate spacers; used with electrodialysis reversal (EDR) membrane systems

Plugging – the physical blockage of the feed side flow passages of a membrane or membrane module (e.g., a blockage in the lumen of an hollow-fiber module operated in inside-out mode or in the spacer of a spiral-wound module)

Pore Size – the size of the openings in a porous membrane expressed either as nominal (average) or the absolute (maximum), typically in terms of microns

Porosity – for a membrane material, the ratio of the volume of voids to the total volume

Post-Treatment – any treatment applied to the filtrate of a membrane process in order to meet given water quality objectives

Pretreatment – any treatment applied to the feed water to a membrane process to achieve desired water quality objectives and/or protect the membranes from damage or fouling

Productivity – the amount of filtered water that can be produced from a membrane module, filtration unit, or system over a period of time, accounting for the use of filtrate in backwash and chemical cleaning operations, as well as otherwise productive time that a unit or system is off-line for routine maintenance processes such as backwashing, chemical cleaning, integrity testing, or repair

Quality Control Release Value (QCRV) – a minimum quality standard of a non-destructive performance test (NDPT) established by the manufacturer for membrane module production that ensures that the module will attain the targeted log removal value (LRV) demonstrated during challenge testing in compliance with the LT2ESWTR

Rack – in a nanofiltration (NF) or reverse osmosis (RO) spiral-wound membrane filtration system, a group of pressure vessels that share common valving and which can be isolated as a group for testing, cleaning, or repair; synonymous with the terms train and skid; included under the term "unit" for the purposes of the LT2ESWTR

Recovery – the volumetric percent of feed water that is converted to filtrate in the treatment process over the course of an uninterrupted (i.e., by chemical cleaning or a solids removal process such as backwashing) operating cycle (excluding losses that occur due to the use of filtrate in backwashing or cleaning operations)

Reject – a continuous waste stream from a membrane system; used synonymously with the term concentrate for nanofiltration (NF) and reverse osmosis (RO) membrane systems

Rejection – the prevention of feed water constituents from passing through a semi-permeable membrane; typically used in association with dissolved solids rather than particulate matter

Resistance – the measurement of the degree to which the flow of water is impeded by a membrane material or fouling

Resolution – (as defined under the LT2ESWTR) the size of the smallest integrity breach that contributes to a response from a direct integrity test; also applicable to some indirect integrity monitoring methods

Reverse Osmosis (RO) – 1) the reverse of the natural osmosis process – i.e., the passage of a solvent (e.g., water) through a semi-permeable membrane from a solution of higher concentration to a solution of lower concentration against the concentration gradient, achieved by applying pressure greater than the osmotic pressure to the more concentrated solution; also, 2) the pressure-driven membrane separation process that employs the principles of reverse osmosis to remove dissolved contaminants from water

Salinity – amount of salt in a solution; usually used in association with salt solutions in excess of 1,000 mg/L and synonymously with the term total dissolved solids (TDS)

Salt Passage – the transport of a salt through a semi-permeable membrane; typically expressed either as a percentage or as mass of salt per unit of membrane area per unit time

Salt Rejection – the amount of salt in the feed water that is rejected by a semi-permeable membrane, expressed as a percentage; also referred to as "solids rejection"

Scale Inhibitor – a chemical agent added to water to inhibit the precipitation or crystallization of salt compounds; also referred to as "anti-scalant"

Scaling – the precipitation or crystallization of salts on a surface (e.g., on the feed side of a membrane)

Semi-Permeable – the property of a membrane barrier that allows it to selectively pass certain molecules in a solution while restricting the passage of others

Sensitivity – the maximum log removal value (LRV) that can be reliably verified by a direct integrity test; also applicable to some continuous indirect integrity monitoring methods

Skid – in a nanofiltration (NF) or reverse osmosis (RO) spiral-wound membrane filtration system, a group of pressure vessels that share common valving and which can be isolated as a group for testing, cleaning, or repair; synonymous with the terms train and rack; included under the term "unit" for the purposes of the LT2ESWTR

Softening – the removal of hardness (i.e., divalent metal ions, primarily calcium and magnesium) from water

Spacer – the material that separates the semi-permeable membrane layers and creates flow passages in a spiral-wound module; also called feed water spacer or brine spacer

Specific Flux – membrane flux normalized for pressure and temperature

Spiral-Wound Module – a configuration in which sheets of a semi-permeable membrane, a porous support matrix, and a spacer are wrapped around a central filtrate collector tube; typically associated with nanofiltration (NF) and reverse osmosis (RO) membrane processes

Stage – a group of membrane units operating in parallel

Suspension Mode – a hydraulic configuration of membrane filtration systems in which contaminants are maintained in suspension through the application of an external force, typically either air or water tangential to the membrane surface

Surrogate – a challenge particulate that is a substitute for the target microorganism of interest and which is removed to an equivalent or lesser extent by a membrane filtration device

Symmetric – having the same consistency throughout

Telescoping – the physical deformation of a spiral-wound membrane module due to high differential pressure in which the membrane, support, and spacer layers are displaced axially (i.e., in the direction of the feed flow) from the center, causing membrane fracture and element failure

Train – in a nanofiltration (NF) or reverse osmosis (RO) spiral-wound membrane filtration system, a group of pressure vessels that share common valving and which can be isolated as a group for testing, cleaning, or repair; synonymous with the terms rack and skid; included under the term "unit" for the purposes of the LT2ESWTR

Transmembrane Pressure (TMP) – the difference in pressure from the feed (or feed-concentrate average, if applicable) to the filtrate across a membrane barrier

Tubular Module – a relatively uncommon configuration (in drinking water applications) similar to that of a hollow-fiber module but utilizing membranes of much larger diameter; may be associated with either microfiltration (MF)/ultrafiltration (UF) or nanofiltration (NF)/reverse osmosis (RO) membrane processes

Turbidimeter – an instrument used to measure turbidity via the scattering of a light beam through a solution that contains suspended particulate matter

Volumetric Concentration Factor (VCF) – the suspended solids concentration on the high pressure side of the membrane relative to the that of the ambient feed water

Ultrafiltration (UF) – a pressure-driven membrane filtration process that typically employs hollow-fiber membranes with a pore size range of approximately 0.01 – 0.05 μm (nominally 0.01 μm)

Upper Control Limit (UCL) – a control limit (CL) for a membrane filtration system that is required by the LT2ESWTR; used in the *Membrane Filtration Guidance Manual* to distinguish CLs mandated by the rule from additional "lower control limits" (LCLs) that are either voluntarily implemented or which may be required by the State at its discretion

1.0 Introduction

Currently, the most common form of drinking water treatment for surface water sources involves the chemical/physical removal of particulate matter by coagulation, flocculation, sedimentation, and filtration processes, along with disinfection to inactivate any remaining pathogenic microorganisms. Filtration remains the cornerstone of drinking water treatment, conventionally in the form of granular media depth filters. Although granular media filters can produce high quality water, they represent a probabilistic rather than an absolute barrier; consequently, pathogens can still pass through the filters and pose a health risk. The disinfection process provides an additional measure of public health protection by inactivating these microorganisms. However, some microorganisms, such as *Cryptosporidium*, are resistant to common primary disinfection practices such as chlorination and chloramination. Furthermore, drinking water regulations have established maximum contaminant levels (MCLs) for disinfection byproducts (DBPs) that may create incentive for drinking water utilities to minimize the application of some disinfectants. As a result of the concern over chlorine-resistant microorganisms and DBP formation, the drinking water industry is increasingly utilizing alternative treatment technologies in an effort to balance the often-competing objectives of disinfection and DBP control. One such alternative technology that has gained broad acceptance is membrane filtration.

Although the use of membrane processes has increased rapidly in recent years, the application of membranes for water treatment extends back several decades. Reverse osmosis (RO) membranes have been used for the desalination of water since the 1960s, with more widespread use of nanofiltration (NF) for softening and the removal of total organic carbon (TOC) dating to the late 1980s. However, the commercialization of backwashable hollow-fiber microfiltration (MF) and ultrafiltration (UF) membrane processes for the removal of particulate matter (i.e., turbidity and microorganisms) in the early 1990s has had the most profound impact on the use, acceptance, and regulation of all types of membrane processes for drinking water treatment.

USEPA developed this guidance manual in support of the Long Term 2 Enhanced Surface Water Treatment Rule (LT2ESWTR), which has identified membrane filtration (i.e., MF, UF, NF, RO, and membrane cartridge filtration (MCF)) as one category of treatment technology in a "toolbox" of options that may be used to achieve the required level of *Cryptosporidium* treatment. Although the LT2ESWTR only regulates membrane filtration in terms of its application for compliance with the requirements of the rule, the concepts and guidance provided in this manual may also be relevant to the Interim Enhanced Surface Water Treatment Rule (IESWTR), the Long Term 1 Enhanced Surface Water Treatment Rule (LT1ESWTR), the Ground Water Rule (GWR), and other regulations, at the discretion of the State[1].

---

1 For the purposes of this manual, the term "State" refers to the State or primacy agency that is responsible for enforcement of drinking water standards.

1.1 Regulating Membranes for the Drinking Water Industry

Driven by the need to protect public health from waterborne pathogens, USEPA has progressively developed regulations that require higher standards for filtered water quality to prevent the passage of infectious pathogens through the treatment process and into the finished drinking water supply. In 1986, the Surface Water Treatment Rule (SWTR) required surface water systems to provide treatment equivalent to 3 log *Giardia* and 4 log virus reduction via a combination of removal and inactivation. USEPA estimated conventional filtration plants (i.e., including coagulation and sedimentation) meeting the filter effluent turbidity requirements provided a minimum of 2.5 log *Giardia* and 2 log virus reduction, while direct filtration plants (i.e., without sedimentation) provided 2 log *Giardia* and 1 log virus removal. The IESWTR (FR 63(241):69518) focused on filter effluent turbidity control, setting a combined filter effluent turbidity limit of 0.3 NTU (40 CFR 141.173) and requiring individual filter turbidity monitoring (40 CFR 141.174) for surface water systems serving at least 10,000 people. The rule also introduced a requirement for 2 log *Cryptosporidium* removal. The IESWTR additionally required that turbidity monitoring be conducted "continuously" (i.e., every 15 minutes) on individual filters. Based on available data, USEPA determined that conventional and direct filtration plants meeting the IESWTR filter effluent turbidity requirements provided a minimum 2 log removal of *Cryptosporidium*. The LT1ESWTR (FR 67(9):1812) promulgated the requirements of the IESWTR, with some modifications, for all surface water systems serving less than 10,000 people.

Under the existing surface water treatment rules, log removal credits for alternative filtration technologies (AFTs), such as membrane processes and bag and cartridge filters, were not explicitly addressed, but instead covered under a special State primacy requirement. For compliance with the SWTR, IESWTR, and LT1ESWTR, many States grant removal credits to membrane processes based on the guidelines for AFTs in the *Guidance Manual for Compliance With the Filtration and Disinfection Requirements for Public Water Systems Using Surface Water Sources* (commonly called the SWTR Guidance Manual) (USEPA 1991). However, there is significant variability in the manner in which States regulate membrane processes, as summarized for MF/UF systems in the USEPA report *Low-Pressure Membrane Filtration for Pathogen Removal: Application, Implementation, and Regulatory Issues* (2001).

The LT2ESWTR builds on the previous surface water treatment rules by requiring additional treatment for those systems with elevated influent *Cryptosporidium* levels. The rule identifies a number of "toolbox" technologies that may be employed to achieve additional *Cryptosporidium* treatment requirements. The range of removal or inactivation credits allocated to each of the various toolbox options under the rule varies based on the capabilities of the particular treatment technology. Because the various types of membrane filtration processes represent one of these toolbox alternatives, utilities have the option of using membrane filtration for compliance with the rule requirements as a distinct technology rather than simply as a general AFT. Consequently, USEPA has developed specific regulatory requirements and associated guidance, as contained in this manual, for membrane filtration processes used for compliance with the LT2ESWTR.

The regulatory framework established under the LT2ESWTR applies only to membrane filtration processes used to achieve *Cryptosporidium* removal for rule compliance. Thus, the LT2ESWTR does not supersede or conflict with any prior surface water treatment requirements, including the allowance of State primacy for regulating membrane filtration processes as AFTs for *Giardia* and virus removal. The LT2ESWTR regulatory framework could be employed for other applications of membrane filtration (e.g., for the removal of *Giardia*, viruses, or other pathogens), albeit solely at the discretion of the State.

1.2   Overview of the Long Term 2 Enhanced Surface Water Treatment Rule

Under the LT2ESWTR, systems that use either a surface water or ground water under the direct influence of surface water (GWUDI) source (collectively referred to as surface water systems) are required to conduct source water monitoring to determine average *Cryptosporidium* concentrations. Based on the average *Cryptosporidium* concentration, a system is classified in one of four possible Bins, as shown in Table 1.1 (40 CFR 141.711(a)). The Bin assignment dictates the supplemental level of *Cryptosporidium* treatment required in addition to the existing requirements of the SWTR, IESWTR, and LT1ESWTR. Utilities may comply with additional treatment requirements by implementing one or more management or treatment techniques from a toolbox of options that includes membrane filtration. Guidance for the use of membrane filtration for compliance with the LT2ESWTR is provided in this manual. A separate guidance manual has also been developed for the use of ultraviolet (UV) disinfection for LT2ESWTR compliance (*Ultraviolet Disinfection Guidance Manual*). Guidance for the use of all other toolbox options is given in the *Long Term 2 Enhanced Surface Water Treatment Rule Toolbox Guidance Manual*.

Table 1.1 LT2ESWTR Additional Treatment Requirements for Filtered Systems

LT2ESWTR Category		Type of Existing Filtration			
Bin	*Cryptosporidium* Concentration *(oocysts/L)*	Conventional Filtration[1]	Direct Filtration	Slow Sand or Diatomaceous Earth Filtration	Alternative Filtration Technologies[7]
1	< 0.075	No additional treatment	No additional treatment	No additional treatment	No additional treatment
2	≥ 0.075 and < 1.0	1 log[2]	1.5 log[2]	1 log[2]	As determined by the State[2,4]
3	≥ 1.0 and < 3.0	2 log[3]	2.5 log[3]	2 log[3]	As determined by the State[3,5]
4	≥ 3.0	2.5 log[3]	3 log[3]	2.5 log[3]	As determined by the State[3,6]

1 Applies to a treatment train using separate, sequential, unit processes for coagulation/flocculation, clarification, and granular media filtration; clarification includes any solid/liquid separation process following coagulation where accumulated solids are removed during this separate component of the treatment system.
2 Public water systems may use any technology or combination of technologies from the toolbox
3 Public water systems must achieve at least 1 log of the required treatment using bag filters, bank filtration, cartridge filters, chlorine dioxide, membrane filtration, ozone, and/or UV disinfection
4 Total *Cryptosporidium* removal and inactivation must be at least 4 log
5 Total *Cryptosporidium* removal and inactivation must be at least 5 log
6 Total *Cryptosporidium* removal and inactivation must be at least 5.5 log
7 Includes membrane filtration

1.3 Requirements for Membrane Filtration Under the LT2ESWTR

In order to receive removal credit for *Cryptosporidium* under the LT2ESWTR, a membrane filtration system must meet the following three criteria:

1. The process must comply with the definition of membrane filtration as stipulated by the rule.

2. The removal efficiency of a membrane filtration process must be established through a product-specific challenge test and direct integrity testing.

3. The membrane filtration system must undergo periodic direct integrity testing and continuous indirect integrity monitoring during operation.

The rule does not prescribe a specific removal credit for membrane filtration processes. Instead, removal credit is based on system performance as determined by challenge testing and verified by direct integrity testing. Thus, the maximum removal credit that a membrane filtration process may receive is the lower value of either (40 CFR 141.719(b)(1)):

- The removal efficiency demonstrated during challenge testing; **OR**

- The maximum log removal value that can be verified by the direct integrity test used to monitor the membrane filtration process

Based on this framework, a membrane filtration process could potentially meet the Bin 4 *Cryptosporidium* treatment requirements, as shown in Table 1.1. Additionally, if a membrane filtration system has been previously approved for 5.5 log *Cryptosporidium* removal by the State, the utility would not be required to conduct source monitoring under the LT2ESWTR (40 CFR 141.701(d)). These systems would ***not*** be subject to the requirements for membrane filtration systems under the LT2ESWTR.

These primary elements of the regulatory requirements for membrane filtration under the LT2ESWTR, including the definition of membrane filtration, as well as challenge testing, direct integrity testing, and continuous indirect integrity monitoring, are summarized in the following sections.

1.3.1 Definition of a Membrane Filtration Process

For the purposes of compliance with the LT2ESWTR, membrane filtration is defined as a pressure- or vacuum-driven separation process in which particulate matter larger than 1 µm is rejected by an engineered barrier primarily through a size exclusion mechanism and which has a measurable removal efficiency of a target organism that can be verified through the application of a direct integrity test (40 CFR 141.2). This definition is intended to include the common membrane technology classifications: MF, UF, NF, and RO. In addition, any cartridge filtration device that meets the definition of membrane filtration and which can be subject to direct integrity testing in accordance with rule requirements would also be eligible for *Cryptosporidium* removal credit as a membrane filtration process under the LT2ESWTR (40 CFR 141.719(b)(1)). In this guidance manual, such processes are called membrane cartridge filtration. Filtration processes that are reliant on mechanisms such as adhesion to filter media or accumulation of a fouling layer to remove particulate matter are excluded from the definition of membrane filtration.

1.3.2 Challenge Testing

Since there are no uniform design criteria that can be used to ensure the removal efficiency of a membrane process, challenge testing is required to demonstrate the ability of a membrane process to remove a specific target organism. The removal efficiency demonstrated during challenge testing establishes the *maximum* removal credit that a membrane process would be eligible to receive, provided that this value is less than or equal to the maximum log removal value that can be verified by the direct integrity test (40 CFR 141.719(b)(1)), as described in the following section. The LT2ESWTR only requires product-specific challenge testing; once the removal efficiency has been demonstrated, additional testing is not required unless the product is significantly modified. Data from challenge studies conducted prior to promulgation of this regulation can be considered in lieu of additional testing at the discretion of the State (40 CFR 141.719(b)(2)). However, the prior testing must have been conducted in a manner that demonstrates removal efficiency for *Cryptosporidium* equivalent to or greater than the treatment credit awarded to the process. The rule requirements for challenge testing, as well as associated guidance, are discussed in Chapter 3.

1.3.3 Direct Integrity Testing

While challenge testing can demonstrate the ability of an integral membrane process to remove the target organism, integrity breaches can develop in the membrane during routine operation that could allow the passage of microorganisms. In order to verify the removal efficiency of a membrane process during operation, direct integrity testing is required for all membrane filtration processes used to comply with the LT2ESWTR (40 CFR 141.719(b)(3)). A direct integrity test is defined as a physical test applied to a membrane unit in order to identify and isolate integrity breaches. The rule does not mandate the use of a specific type of direct integrity test, but rather performance criteria that any direct integrity test must meet. These criteria include requirements for resolution, sensitivity, and frequency (40 CFR 141.719(b)(3)):

- **Resolution**: The direct integrity test must be applied in a manner such that a 3 μm breach contributes to the response from the test.

- **Sensitivity**: The direct integrity test must be capable of verifying the log removal value awarded to the membrane process by the State.

- **Frequency**: The direct integrity test must be applied at a frequency of at least once per day, although less frequent testing may be permitting by the State at its discretion if appropriate safety factors are incorporated.

A control limit must also be established for a direct integrity test, representing a threshold response which, if exceeded, indicates a potential integrity problem and triggers subsequent corrective action. For the purposes of LT2ESWTR compliance, this threshold response must be indicative of an integral membrane unit capable of achieving the *Cryptosporidium* removal credit awarded by the State. The criteria for direct integrity testing are discussed in detail in Chapter 4, along with guidance describing how these criteria apply to commonly used direct integrity tests.

1.3.4 Continuous Indirect Integrity Monitoring

Because currently available direct integrity test methods require the membrane unit to be temporarily taken out of service, or are either too costly or infeasible to apply continuously, direct testing is only conducted periodically. Thus, in the absence of a continuous direct integrity test that meets the resolution and sensitivity requirements of the LT2ESWTR, continuous indirect integrity monitoring is required (40 CFR 141.719(b)(4)). Although the indirect monitoring methods are typically not as sensitive as direct tests for detecting a loss of membrane integrity, the indirect methods do provide some measure of performance assessment between applications of direct testing. For the purposes of the LT2ESWTR, indirect integrity monitoring is defined as monitoring some filtrate water parameter that is indicative of the removal of particulate matter, and "continuous" is defined as monitoring at a frequency of no less than once every 15 minutes (40 CFR 141.719(b)(4)(ii)). Although turbidity monitoring is specified as the default method of continuous indirect integrity monitoring under the rule, other methods, such as particle counting or particle monitoring, may be used in lieu of turbidity monitoring at the discretion of the State (40 CFR 141.719(b)(4)(i)). For any indirect method used, a control limit must be established that is indicative of acceptable performance. Monitoring results exceeding the control limit for a period of more than 15 minutes must trigger immediate direct integrity testing (40 CFR 141.719(b)(4)(iv)). The requirements and associated guidance for continuous indirect integrity monitoring are detailed in Chapter 5.

1.4 Considering Existing Membrane Facilities Under the LT2ESWTR

As shown in Table 1.1, the LT2ESWTR only requires additional treatment measures for those drinking water systems with source water *Cryptosporidium* levels greater than or equal to 0.075 oocysts/L – Bins 2, 3, or 4. Utilities with existing membrane filtration facilities will be affected by the LT2ESWTR in one of five ways. These cases are summarized as follows:

- **Case 1**: The utility has previously been awarded 5.5 log *Cryptosporidium* treatment credit via a combination of physical removal (which may include membrane filtration) and chemical inactivation. In this case the utility is not required to conduct source water monitoring (40 CFR 141.701(d)(1)).

- **Case 2**: The utility conducts source water monitoring for *Cryptosporidium* and determines that its source water is in Bin 1 (i.e., concentrations less than 0.075 oocysts/L). In this case, the system may continue to operate under the previous surface water treatment rules (i.e., the SWTR and either the IESWTR or the LT1ESWTR) as administered by the State. No additional action is required under the LT2ESWTR.

- **Case 3**: The utility conducts source water monitoring for *Cryptosporidium* and determines that its source water is in either Bins 2, 3, or 4 (i.e., concentrations greater than or equal to 0.075 oocysts/L). The utility then successfully demonstrates to the State that its membrane filtration system can achieve ***all*** of the total required *Cryptosporidium* treatment credit, as listed in the right hand column (i.e., under "Type of Existing Filtration – Alternative Filtration Technologies") of Table 1.1, up to a maximum of 5.5 log credit. In this case, the utility would be required to meet all requirements for membrane filtration specified in the LT2ESWTR to achieve *Cryptosporidium* removal credit.

- **Case 4**: The utility conducts source water monitoring for *Cryptosporidium* and determines that its source water is in either Bins 2, 3, or 4 (i.e., concentrations greater than or equal to 0.075 oocysts/L). The utility then successfully demonstrates to the State that its membrane filtration system can achieve ***part*** of the required *Cryptosporidium* treatment credit, as listed in the right hand column (i.e., under "Type of Existing Filtration – Alternative Filtration Technologies") of Table 1.1. In this case the utility would be required to use other toolbox options to obtain the balance of *Cryptosporidium* treatment credit required under the rule. Furthermore, the utility would be required to meet all requirements for membrane filtration specified in the LT2ESWTR to achieve *Cryptosporidium* removal credit.

- **Case 5**: The utility conducts source water monitoring for *Cryptosporidium* and determines that its source water is in either Bins 2, 3 or 4 (i.e., concentrations greater than or equal to 0.075 oocysts/L). However, the utility opts not to use its membrane filtration system for the purposes of LT2ESWTR compliance. In this case the utility would be required use other toolbox options to obtain all of the *Cryptosporidium* treatment credit required under the rule, and the membrane regulatory framework under the LT2ESWTR would not apply to the utility's membrane filtration system.

Under the LT2ESWTR, the regulatory basis for a membrane filtration process to receive treatment credit for *Cryptosporidium* is the demonstration of removal efficiency through challenge testing and the verification of membrane system integrity through routine direct integrity testing and continuous indirect integrity monitoring. These criteria form the basis for the potential for membrane filtration systems to be awarded up to a maximum of 5.5 log *Cryptosporidium* removal credit for complying with the requirements of the previously promulgated surface water treatment rules in combination with any additional *Cryptosporidium* treatment credit that may be required under the LT2ESWTR.

With respect to the challenge testing requirements of the rule, the two most likely options available to utilities with existing membrane filtration systems are to grandfather data generated during the pilot testing conducted as part of the permitting process (if applicable) or to use data from challenge testing conducted by the membrane manufacturer in its effort to qualify its product(s) for *Cryptosporidium* removal credit under the LT2ESWTR (since challenge testing is required on a product-specific and not a site-specific basis). Challenge testing, including recommendations for grandfathering data, is described in detail in Chapter 3.

Existing membrane facilities will also have to meet the direct integrity testing and continuous indirect integrity monitoring requirements of the LT2ESWTR to qualify for treatment credit under the rule. This may necessitate that some facilities implement new integrity verification practices, since State requirements vary widely, and some may not require direct integrity testing at all (USEPA 2001). However, this is not anticipated to be problematic for many existing facilities, since most membrane filtration systems applied to surface water are equipped with the ability to conduct some form of direct integrity testing. In addition, although States may not have explicit indirect integrity monitoring requirements, turbidity monitoring (the default method of continuous indirect integrity monitoring under the LT2ESWTR) is nonetheless required for compliance with the various existing surface water treatment rules. In some cases, utilities may need to purchase additional equipment to comply with the integrity verification requirements of the rule. Detailed guidance on the use of direct integrity testing and continuous indirect integrity monitoring for LT2ESWTR compliance is provided in Chapters 4 and 5, respectively.

Another consideration for existing membrane facilities required to meet the LT2ESWTR criteria is replacement membrane modules. When replacement modules are installed, it is necessary to verify that the specific modules used meet the quality control release value of the non-destructive performance test as a means of indirectly verifying removal efficiency. Additional guidance regarding non-destructive performance testing is provided in section 3.6.

The regulatory framework developed for membrane filtration under the LT2ESWTR addresses many of the specific capabilities and requirements of the technology, and thus may introduce new concepts that might not be included in a given State's current regulatory approach for membrane processes, particularly if the State currently considers membrane filtration as an AFT, as described the *Guidance Manual for Compliance With the Filtration and Disinfection Requirements for Public Water Systems Using Surface Water Sources* (USEPA 1991). Although States may choose to adopt aspects of the LT2ESWTR framework for broader regulation of membrane filtration systems, USEPA only requires that this regulatory framework be applied to systems that utilize membrane filtration to meet the additional *Cryptosporidium* treatment requirements of the LT2ESWTR.

1.5   Membrane Terminology Used in the Guidance Manual

In the development of the regulatory language and associated guidance for the LT2ESWTR, it was necessary to select the most appropriate terminology for various aspects of membrane treatment, with the understanding that use of such terminology can vary widely throughout the industry. The purpose of this section is to clarify the use of membrane treatment terminology associated with the LT2ESWTR and note generally synonymous terms that are also in common use, where applicable. This section also presents some new terms defined under the rule that are critical to the regulatory framework.

The term "membrane filtration" is formally defined under the rule language for the LT2ESWTR, as follows (40 CFR 141.2):

- **Membrane filtration** – a pressure- or vacuum-driven separation process in which particulate matter larger than 1 μm is rejected by an engineered barrier, primarily through a size exclusion mechanism, and which has a measurable removal efficiency of a target organism that can be verified through the application of a direct integrity test. The definition of a membrane filtration process and what it represents are further discussed in section 1.3.1.

In addition, there are a number of terms defined in the context of the rule language that are also used throughout this guidance manual. These terms are as follows:

- **Challenge particulate** – the target organism or acceptable surrogate used to determine the log removal value during a challenge test

- **Challenge test** – a study conducted to determine the removal efficiency (i.e., log removal value) of the membrane filtration media. Challenge testing is discussed in detail in Chapter 3, and the requirements for challenge testing under the LT2ESWTR are summarized briefly in section 1.3.2.

- **Control limit** – an integrity test result that, if exceeded, indicates a potential problem with the system and triggers a response. In the context of this guidance manual the terms *upper control limit* (UCL) and *lower control limit* (LCL) are also used. The term upper control limit is always used in reference to the control limit that is mandated under the LT2ESWTR. The term lower control limit was established to distinguish any more conservative voluntary or additional State-mandated control limits that may trigger increased monitoring or other action.

- **Direct integrity test** – a physical test applied to a membrane unit in order to identify and isolate integrity breaches. Direct integrity testing is discussed in detail in Chapter 4, and the requirements for direct integrity testing under the LT2ESWTR are summarized briefly in section 1.3.3.

- **Flux** – the throughput of a pressure-driven membrane filtration process expressed as flow per unit of membrane area

- **Indirect integrity monitoring** – monitoring some aspect of filtrate water quality that is indicative of the removal of particulate matter. In the context of indirect integrity monitoring, *continuous* is defined as a frequency of no less than once every 15 minutes (40 CFR 141.719(b)(4)(ii)). Continuous indirect integrity monitoring is discussed in detail in Chapter 5, and the requirements for continuous indirect integrity monitoring under the LT2ESWTR are summarized briefly in section 1.3.4.

- **Integrity breach** – one or more leaks in a membrane filtration system that could result in the contamination of the filtrate with unfiltered feed water

- **Membrane unit** – a group of membrane modules that share common valving that allows the unit to be isolated from the rest of the system for the purpose of integrity testing or maintenance. For the purposes of the LT2ESWTR, "membrane unit" is intended to include the commonly-used synonymous terms rack, train, and skid.

- **Module** – the smallest component of a membrane unit in which a specific membrane surface area is housed in a device with a filtrate outlet structure. For the purposes of the LT2ESWTR, this term encompasses hollow-fiber modules and cassettes, spiral-wound elements, cartridge filter elements, plate-and-frame modules, and tubular modules, among other membrane devices of similar scope and purpose.

- **Recovery** – the volumetric percent of feed water that is converted to filtrate over the course of an operating cycle uninterrupted by events such as chemical cleaning or a solids removal process (i.e., backwashing). In the context of the LT2ESWTR, the term recovery does not consider losses that occur due to the use of filtrate in backwashing or cleaning operations.

- **Resolution** – the size of the smallest integrity breach that contributes to a response from a direct integrity test

- **Sensitivity** – the maximum log removal value that can be reliably verified by the direct integrity test associated with a given membrane filtration system

In addition to these terms, it is important to note that the term filtrate, as used in both the rule language and this guidance manual, includes the synonymous term permeate, which is commonly used in the industry in association with the treated water from NF and RO semi-permeable membrane processes. All of the terms clarified in this section, as well as numerous others used in the context of this manual, are defined in the glossary.

1.6   Guidance Manual Objectives

The purpose of this manual is to establish a clear and consistent framework for the application and regulation of membrane filtration for compliance with the requirements of the LT2ESWTR. Specifically, the objective of the manual is to provide utilities, State regulators, membrane module and system manufacturers, and consulting engineers with guidance in the following respective areas:

- **Utilities and consulting engineers**: specific guidance on meeting the criteria specified in the LT2ESWTR in order to receive removal credit for *Cryptosporidium* through the application of membrane filtration, as well as additional guidance regarding industry practices for pilot testing, implementing, and starting up a membrane filtration facility

- **State regulators**: guidance on evaluating membrane filtration systems and determining appropriate *Cryptosporidium* removal credits for these processes under the LT2ESWTR

- **Manufacturers**: guidance on qualifying membrane filtration systems for *Cryptosporidium* removal credits under the LT2ESWTR

Note that the guidance provided in this manual applies to systems that meet the definition of membrane filtration under the LT2ESWTR, including MF, UF, NF, and RO processes, as well as qualifying MCF systems. Electrodialysis (ED) and electrodialysis reversal (EDR) processes are *not* considered in this document, since they do not provide a physical barrier to pathogens and consequently are not considered membrane filtration under the rule.

1.7   Guidance Manual Organization

The *Membrane Filtration Guidance Manual* explains the requirements for utilizing membrane filtration for compliance with the LT2ESWTR and provides guidance to facilitate compliance with these requirements. In addition, this manual describes some recommended industry practices for the application of membrane filtration systems for the removal of pathogens in the drinking water treatment process. These recommended practices are *not* required under the LT2ESWTR and are provided as general guidance only.

Chapters 1, 3, 4, and 5 elaborate on LT2ESWTR requirements and associated guidance, while Chapters 6, 7, and 8 discuss recommended industry practices for membrane filtration that are not specifically related to the rule. Chapter 2 presents an overview of membrane filtration for readers unfamiliar with the technology. It is recommended that even readers with significant membrane process experience review Chapter 2 to better understand how the concepts and terminology are used in the context of the guidance manual.

The guidance manual is organized as follows:

Chapter 1:   Introduction
Chapter 1 presents the objectives of the guidance manual and provides an overview of the regulatory requirements for membrane filtration under the LT2ESWTR.

Chapter 2:   Overview of Membrane Filtration
Chapter 2 provides an overview of the basic theory and concepts of membrane filtration, including: types of membrane processes; types of membrane materials, modules, and systems; fundamental principles; and hydraulic models describing various configurations.

Chapter 3:   Challenge Testing
Chapter 3 provides guidance on designing a challenge study to demonstrate the removal efficiency of a membrane module with respect to *Cryptosporidium*, as required under the LT2ESWTR.

Chapter 4:   Direct Integrity Testing
Chapter 4 provides guidance on meeting the LT2ESWTR performance-based requirements for direct integrity testing of membrane filtration systems and describes the commonly used direct integrity test methods in the context of these requirements.

Chapter 5:   Continuous Indirect Integrity Monitoring
Chapter 5 provides guidance on meeting the LT2ESWTR requirements for continuous indirect integrity monitoring, including both the default approach using turbidity monitoring as well as a discussion of potential alternative methods.

Chapter 6:   Pilot Testing
Chapter 6 discusses aspects of pilot testing membrane filtration systems, including objectives, planning, operation, and data collection and analysis.

Chapter 7:   Implementation Considerations
Chapter 7 discusses a variety of design and operational considerations for implementing membrane filtration processes.

Chapter 8:   Initial Start-Up
Chapter 8 discusses issues associated with the start-up and shakedown of a new membrane filtration system, as well as some recommended practices to facilitate this process.

Appendix A: Development of a Comprehensive Integrity Verification Program
Appendix A presents a framework for developing an integrity testing and monitoring program that utilizes a variety of tools to ensure system performance.

Appendix B: Overview of Bubble Point Theory
Appendix B presents an overview of bubble point theory, which serves as the basis for pressure-based direct integrity tests.

Appendix C: Calculating the Air-Liquid Conversion Ratio
Appendix C provides supplemental guidance for calculating the air-liquid conversion ratio (ALCR) (as described in Chapter 4), a method for converting the results of pressure-based direct integrity tests to the flow of water through an integrity breach during normal system operation for the purpose of determining both test sensitivity and appropriate control limits.

Appendix D: Empirical Method for Determining the Air-Liquid Conversion Ratio for a Hollow-Fiber Membrane Filtration System
Appendix D describes an empirical method called the correlated airflow measurement (CAM) procedure for determining the ALCR, as described in Chapter 4.

Appendix E: Application of Membrane Filtration for Virus Removal
Appendix E provides a general overview of the issue of applying membrane filtration, and in particular UF, for virus removal, including the applicability of the LT2ESWTR regulatory framework for this objective.

2.0 Overview of Membrane Filtration

2.1 Introduction

There are several classes of treatment processes that constitute membrane filtration for the purposes of Long-Term 2 Enhanced Surface Water Treatment Rule (LT2ESWTR) compliance. These processes include: microfiltration (MF), ultrafiltration (UF), nanofiltration (NF), and reverse osmosis (RO). In addition, cartridge filtration devices that meet the criteria for a membrane filtration process as defined under the rule (see sections 1.3.1 and 2.2) would also be eligible for *Cryptosporidium* removal credit as membrane filtration (40 CFR 141.719(b)(1)). For the purposes of this guidance manual, these devices are termed membrane cartridge filtration (MCF).

Each of these technologies utilizes a membrane barrier that allows the passage of water but removes contaminants. The membrane media is generally manufactured as flat sheets or as hollow fibers and then configured into membrane modules. The most common membrane module configurations are hollow-fiber (consisting of hollow-fiber membrane material), spiral-wound (consisting of flat sheet membrane material wrapped around a central collection tube), and cartridges (consisting of flat sheet membrane material that is often pleated to increase the surface area). Although the spiral-wound and cartridge configurations are also termed as "elements" and "cartridges," respectively, under the LT2ESWTR the term "module" – defined as the smallest component of a membrane unit in which a specific membrane surface area is housed in a device with a filtrate outlet structure (see section 1.5) – is used to refer to all of the various membrane module configurations for simplicity of nomenclature.

In addition to the various module configurations, there are a number of different types of membrane materials, hydraulic modes of operation, and operational driving forces (i.e., pressure or vacuum) that can vary among the different classes of membrane filtration (i.e., MF, UF, NF, RO, and MCF). Each of these characteristics of membrane filtration systems may be considered tools that a manufacturer may utilize to meet the particular treatment objectives for a given application.

The purpose of Chapter 2 of the *Membrane Filtration Guidance Manual* is to provide an overview of the various membrane filtration processes, including descriptions of the various classes, membrane materials, geometry, module construction, driving forces, basic principles of design and operation, and hydraulic configurations. Emphasis is given to the manner in which each of these characteristics relates to membrane filtration applied for pathogen removal, as would be the case for compliance with the LT2ESWTR.

This chapter is divided into the following sections:

> Section 2.2: Basic Principles of Membrane Filtration
> This section reviews the basic treatment mechanisms of the various classes of membrane treatment processes, as well as the common principle of

operation that distinguishes membrane filtration systems, particularly in the context of the LT2ESWTR.

Section 2.3: <u>Membrane Materials, Modules, and Systems</u>
This section describes the types of membrane materials, modules, and types of systems associated with the various classes of membrane filtration.

Section 2.4: <u>Basic Principles of Membrane Filtration System Design and Operation</u>
The section presents some the basic concepts and equations governing membrane filtration system design and operation in order to facilitate a more complete general understanding of membrane processes.

Section 2.5: <u>Hydraulic Configurations</u>
This section describes the various hydraulic modes of operation for membrane filtration systems and provides equations for calculating the volumetric concentration factor.

2.2 Basic Principles of Membrane Filtration

For the purposes of the LT2ESWTR, a membrane filtration process is defined by two basic criteria (40 CFR 141.2):

1. The filtration system must be a pressure- or vacuum-driven process and remove particulate matter larger than 1 μm using an, engineered barrier, primarily via a size exclusion mechanism.

2. The process must have a measurable removal efficiency of a target organism that can be verified through the application of a direct integrity test.

The ability of each of type of membrane filtration system to remove various drinking water pathogens of interest on the basis of size is illustrated in Figure 2.1. The figure shows the approximate size range of viruses, bacteria, *Cryptosporidium* oocysts, and *Giardia* cysts, as well as the ability of MF, UF, NF, RO, and MCF, respectively, to remove each of these pathogens on the basis of size exclusion. Overlap between the range covered by a membrane filtration process with a given pathogen size range indicates the ability of that process to remove the pathogen. Note that the molecular weights listed do not correspond precisely to the indicated pathogen size range, but are rough generalizations depicted as a result of the fact that NF, RO, and some UF processes are rated according to a "molecular weight cutoff" on the basis of their ability to remove dissolved phase constituents.

Although each of the classes of membrane filtration functions as a filter for various sizes of particulate matter, the basic principles of operation vary between MF/UF, NF/RO, and MCF systems. Each of these types of systems is described in the following sections.

Figure 2.1 Filtration Application Guide for Pathogen Removal

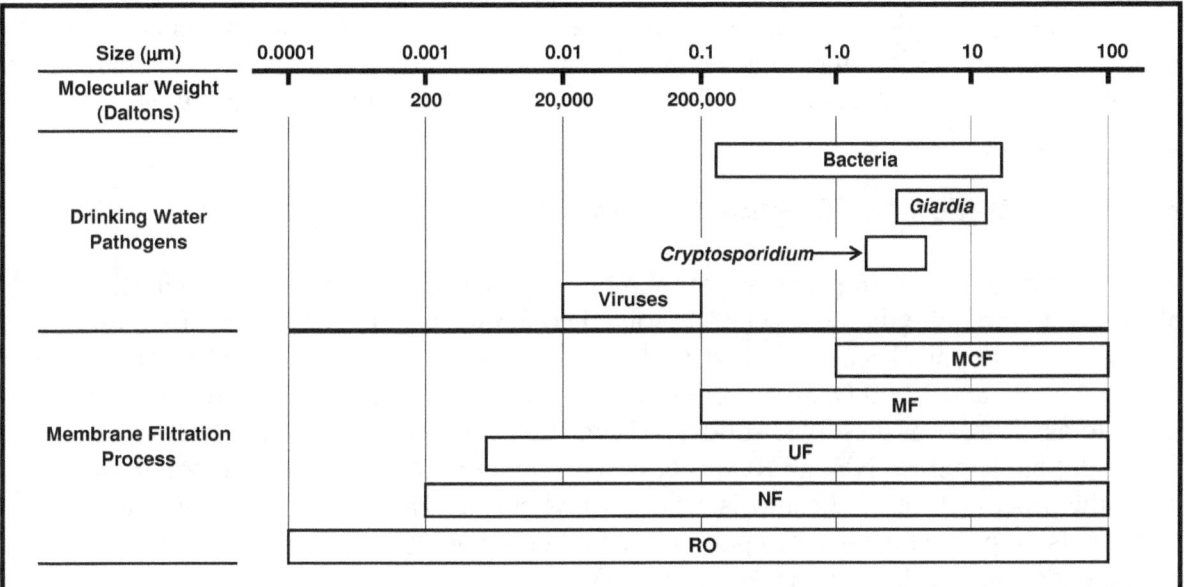

2.2.1 Microfiltration and Ultrafiltration

MF and UF are the two processes that are most often associated with the term "membrane filtration." MF and UF are characterized by their ability to remove suspended or colloidal particles via a sieving mechanism based on the size of the membrane pores relative to that of the particulate matter. However, all membranes have a distribution of pore sizes, and this distribution will vary according to the membrane material and manufacturing process. When a pore size is stated, it can be presented as either nominal (i.e., the average pore size), or absolute (i.e., the maximum pore size) in terms of microns (μm). MF membranes are generally considered to have a pore size range of 0.1 – 0.2 μm (nominally 0.1 μm), although there are exceptions, as MF membranes with pores sizes of up to 10 μm are available. For UF, pore sizes generally range from 0.01 – 0.05 μm (nominally 0.01μm) or less, decreasing to an extent at which the concept of a discernable "pore" becomes inappropriate, a point at which some discrete macromolecules can be retained by the membrane material. In terms of a pore size, the lower cutoff for a UF membrane is approximately 0.005 μm.

Because some UF membranes have the ability to retain larger organic macromolecules, they have been historically characterized by a molecular weight cutoff (MWCO) rather than by a particular pore size. The concept of the MWCO (expressed in Daltons – a unit of mass) is a measure of the removal characteristic of a membrane in terms of atomic weight (or mass) rather than size. Thus, UF membranes with a specified MWCO are presumed to act as a barrier to compounds or molecules with a molecular weight exceeding the MWCO. Because such organic macromolecules are morphologically difficult to define and are typically found in solution rather than as suspended solids, it may be convenient in conceptual terms to use a MWCO rather than a particular pore size to define UF membranes when discussed in reference to these types of compounds. Typical MWCO levels for UF membranes range from 10,000 to 500,000 Daltons, with most membranes used for water treatment at approximately 100,000 MWCO. However, UF membranes remove particulate contaminants via a size exclusion mechanism and not on the basis of weight or mass; thus, UF membranes used for drinking water treatment are also characterized according to pore size with respect to microbial and particulate removal capabilities.

2.2.2 Nanofiltration and Reverse Osmosis

NF and RO constitute the class of membrane processes that is most often used in applications that require the removal of dissolved contaminants, as in the case of softening or desalination. The typical range of MWCO levels is less than 100 Daltons for RO membranes, and between 200 and 1,000 Daltons for NF membranes. While NF and RO are sometimes referred to as "filters" of dissolved solids, NF and RO utilize semi-permeable membranes that do not have definable pores. NF and RO processes achieve removal of dissolved contaminants through the process of reverse osmosis, as described below. However, these membrane processes also represent a barrier to particulate matter and thus are considered membrane filtration under the LT2ESWTR (40 CFR 141.2).

NF/RO membranes are designed to remove dissolved solids through the process of reverse osmosis. Osmosis is the natural flow of a solvent, such as water, through a semi-permeable membrane (acting as a barrier to dissolved solids) from a less concentrated solution to a more concentrated solution. This flow will continue until the chemical potentials (or concentrations, for practical purposes) on both sides of the membrane are equal. The amount of pressure that must be applied to the more concentrated solution to stop this flow of water is called the osmotic pressure. An approximate rule of thumb for the osmotic pressure of fresh or brackish water is approximately 1 psi for every 100 mg/L difference in total dissolved solids (TDS) concentration on opposite sides of the membrane.

Reverse osmosis, as illustrated in Figure 2.2, is the reversal of the natural osmotic process, accomplished by applying pressure in excess of the osmotic pressure to the more concentrated solution. This pressure forces the water through the membrane against the natural osmotic gradient, thereby increasingly concentrating the water on one side (i.e., the feed) of the membrane and increasing the volume of water with a lower concentration of dissolved solids on the opposite side (i.e., the filtrate or permeate). The required operating pressure varies depending on the TDS of the feed water (i.e., osmotic potential), as well as on membrane properties and temperature, and can range from less than 100 psi for some NF applications to more than 1,000 psi for seawater desalting using RO.

Figure 2.2 Conceptual Diagram of Osmotic Pressure

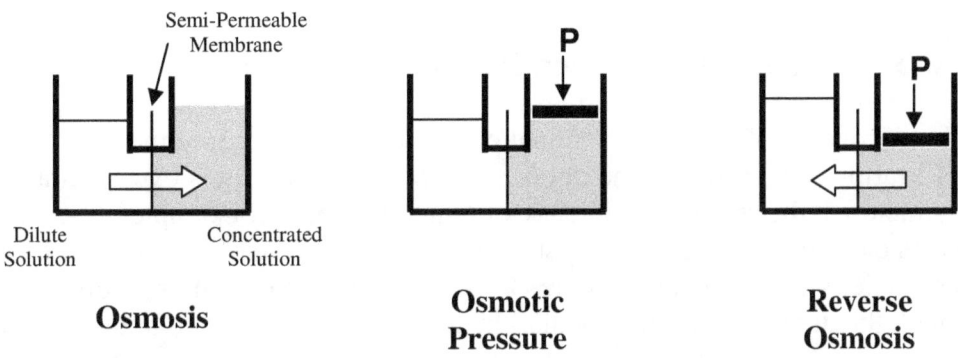

Both NF and RO are pressure-driven separation processes that utilize semi-permeable membrane barriers. NF differs from RO only in terms of its lower removal efficiencies for dissolved substances, particularly for monovalent ions. This results in unique applications of NF, such as the removal of hardness ions at lower pressures than would be possible using RO. Consequently, NF is often called "membrane softening." The differences between NF and RO are irrelevant with respect to the removal of particulate matter, and as a result, these two membrane processes are functionally equivalent for the purposes of the LT2ESWTR.

Because semi-permeable NF and RO membranes are not porous, they have the ability to screen microorganisms and particulate matter in the feed water; however, they are not necessarily absolute barriers. NF and RO membranes are specifically designed for the removal of TDS and not particulate matter, and thus the elimination of all small seal leaks that have only a minor impact on the salt rejection characteristics is not the primary focus of the manufacturing process. Consequently, NF and RO spiral-wound elements are not intended to be sterilizing filters and some passage of particulate matter may occur despite the absence of pores in the membrane, which can be attributed to slight manufacturing imperfections (Meltzer 1997). Nonetheless, NF and RO are eligible for *Cryptosporidium* removal credit under the LT2ESWTR based on the demonstrated ability of these technologies to remove pathogens, as well as on the high probability that these processes can meet the requirements for membrane filtration specified in the rule.

2.2.3 Membrane Cartridge Filtration

The principles of MCF system operation are similar to those for MF/UF systems in that MCF removes particles via a sieving mechanism. While cartridge filtration has not traditionally been considered a membrane treatment process, a cartridge filtration device that utilizes membrane filtration media capable of removing particles 1.0 µm and larger and which can be subjected to direct integrity testing would satisfy the definition of membrane filtration under the LT2ESWTR (40 CFR 141.719(b)(1)). However, many cartridge filters have pore sizes that are larger than 1.0 µm, and the utilization of a direct integrity test with these filters is relatively new in drinking water treatment applications. Thus, it is important to note that only cartridge filters that meet both criteria of the definition are considered membrane filtration under the rule.

2.2.4 Membrane Pore Size and Filtration Removal Efficiency

Although the concept of using the nominal or absolute pore size is sometimes used in reference to the filtration capabilities of membrane material, this concept is overly simplistic and does not fully characterize the removal efficiency of a membrane. For example, the mechanisms of filtering particles close in size to the pore distribution of a membrane becomes more complex than sieving, as particles smaller than most pores may be removed through probabilistic interception through the depth of the filter media. In addition, for some membrane materials, particles may be rejected through electrostatic repulsion and adsorption to the membrane material. The filtration properties of the membrane may also depend on the formation of a cake layer during operation, as the deposition of particles can obscure the pores over the course of a filter run, thus increasing the removal efficiency.

Because there are currently no standard methods for characterizing and reporting the pore sizes for the various membrane filtration processes, the meaning of this information can vary between different membrane manufactures, thus limiting its value. In addition, the concept of pore size has no significance for NF and RO, which utilize semi-permeable membranes that do not have pores. The concept of pore size also does not address the integrity of the manufactured membrane module assembly, which could potentially pass particles larger than the indicated pore size.

Consequently, for the purpose of LT2ESWTR compliance, the rule requires that membrane filtration performance be determined by challenge testing in which the ability of the membrane module to reject *Cryptosporidium* (or a suitable surrogate) is demonstrated. These studies provide a means to empirically determine the actual exclusion characteristic of the membrane module for *Cryptosporidium* (or other contaminant(s) of interest) and thus account for all the factors that contribute to removal efficiency. Thus, unlike a measure of pore size, the use of an empirically determined exclusion characteristic facilitates the direct comparison of different membrane filtration systems for the removal of a target contaminant and provides a direct measure of membrane performance. Challenge testing is described in further detail in Chapter 3.

2.2.5 Electrodialysis and Electrodialysis Reversal

Although both electrodialysis (ED) and electrodialysis reversal (EDR) utilize membranes and are classified as membrane processes, these treatment technologies do not constitute *membrane filtration* as defined by the LT2ESWTR (40 CFR 141.2). Unlike NF and RO, which use pressure to force water through the membranes while rejecting dissolved solids, the driving force for separation in ED and EDR processes is electric potential, and an applied current is utilized to transport ionic species across selectively permeable membranes. Because the water does not physically pass through the membrane in either the ED or EDR process, particulate matter is not removed. Thus, ED and EDR membranes are specifically applied for the removal of dissolved ionic constituents and are not considered filters. Consequently, ED/EDR processes are not addressed further in this guidance manual.

2.3   Membrane Materials, Modules, and Systems

There are a number of different types of membrane materials, modules, and associated systems that are utilized by the various classes of membrane filtration. While several different types of membrane modules may be employed for any single membrane filtration technology, each class of membrane technology is typically associated with only one type of membrane module in water treatment applications. In general, MF and UF use hollow-fiber membranes, and NF and RO use spiral-wound membranes. MCF systems use flat sheet material configured into a cartridge filtration device. The terms hollow-fiber, spiral-wound, and cartridge refer to the module in which the membrane media is manufactured. Section 2.3 describes each of these types of membrane modules, as well as the materials from which the membranes are made and the systems into which they are configured.

2.3.1 Membrane Materials

The membrane material refers to the substance from which the membrane itself is made. Normally, the membrane material is manufactured from a synthetic polymer, although other forms, including ceramic and metallic "membranes," may be available. Currently, almost all membranes manufactured for drinking water production are made of polymeric material, since they are significantly less expensive than membranes constructed of other materials.

The material properties of the membrane may significantly impact the design and operation of the filtration system. For example, membranes constructed of polymers that react with oxidants commonly used in drinking water treatment should not be used with chlorinated feed water. Mechanical strength is another consideration, since a membrane with greater strength can withstand larger transmembrane pressure (TMP) levels allowing for greater operational flexibility and the use of higher pressures with pressure-based direct integrity testing (see section 4.7). Similarly, a membrane with bi-directional strength may allow cleaning operations or integrity testing to be performed from either the feed or the filtrate side of the membrane. Material properties influence the exclusion characteristic of a membrane as well. A membrane with a particular surface charge may achieve enhanced removal of particulate or microbial contaminants of the opposite surface charge due to electrostatic attraction. In addition, a membrane can be characterized as being hydrophilic (i.e., water attracting) or hydrophobic (i.e., water repelling). These terms describe the ease with which membranes can be wetted, as well as the propensity of the material to resist fouling to some degree.

MF and UF membranes may be constructed from a wide variety of materials, including cellulose acetate (CA), polyvinylidene fluoride (PVDF), polyacrylonitrile (PAN), polypropylene (PP), polysulfone (PS), polyethersulfone (PES), or other polymers. Each of these materials has different properties with respect to surface charge, degree of hydrophobicity, pH and oxidant tolerance, strength, and flexibility.

NF and RO membranes are generally manufactured from cellulose acetate or polyamide materials (and their respective derivatives), and there are various advantages and disadvantages associated with each. While cellulose membranes are susceptible to biodegradation and must be operated within a relatively narrow pH range of about 4 to 8, they do have some resistance to continuous low-level oxidant exposure. In general, for example, chlorine doses of 0.5 mg/L or less may control biodegradation as well as biological fouling without damaging the membrane. Polyamide (PA) membranes, by contrast, can be used under a wide range of pH conditions and are not subject to biodegradation. Although PA membranes have very limited tolerance for the presence of strong oxidants, they are compatible with weaker oxidants such as chloramines. PA membranes require significantly less pressure to operate and have become the predominant material used for NF and RO applications.

A characteristic that influences the performance of all membranes is the trans-wall symmetry, a quality that describes the level of uniformity throughout the cross-section of the membrane. There are three types of construction that are commonly used in the production of membranes: symmetric, asymmetric (including both skinned and graded density variations), and composite. Cross-sectional diagrams of membranes with different trans-wall symmetry are shown in Figure 2.3. Symmetric membranes are constructed of a single (i.e., homogeneous) material, while composite membranes use different (i.e., heterogeneous) materials. Asymmetric membranes may be either homogeneous or heterogeneous.

In a symmetric membrane, the membrane is uniform in density or pore structure throughout the cross-section, while in an asymmetric membrane there is a change in the density of the membrane material across the cross sectional area. Some asymmetric membranes have a graded construction, in which the porous structure gradually decreases in density from the feed to the filtrate side of the membrane. In other asymmetric membranes, there may be a distinct transition between the dense filtration layer (i.e., the skin) and the support structure. The denser skinned layer is exposed to the feed water and acts as the primary filtration barrier, while the thicker and more porous understructure serves primarily as mechanical support. Some hollow-fibers may be manufactured as single- or double- skinned membranes, with the double skin providing filtration at both the outer and inner walls of the fibers. Like the asymmetric skinned membranes, composite membranes also have a thin, dense layer that serves as the filtration barrier. However, in composite membranes the skin is a different material than the porous substructure onto which it is cast. This surface layer is designed to be thin so as to limit the resistance of the membrane to the flow of water, which passes more freely through the porous substructure. NF and RO membrane construction is typically either asymmetric or composite, while most MF, UF, and MCF membranes are either symmetric or asymmetric.

Figure 2.3 Membrane Construction and Symmetry

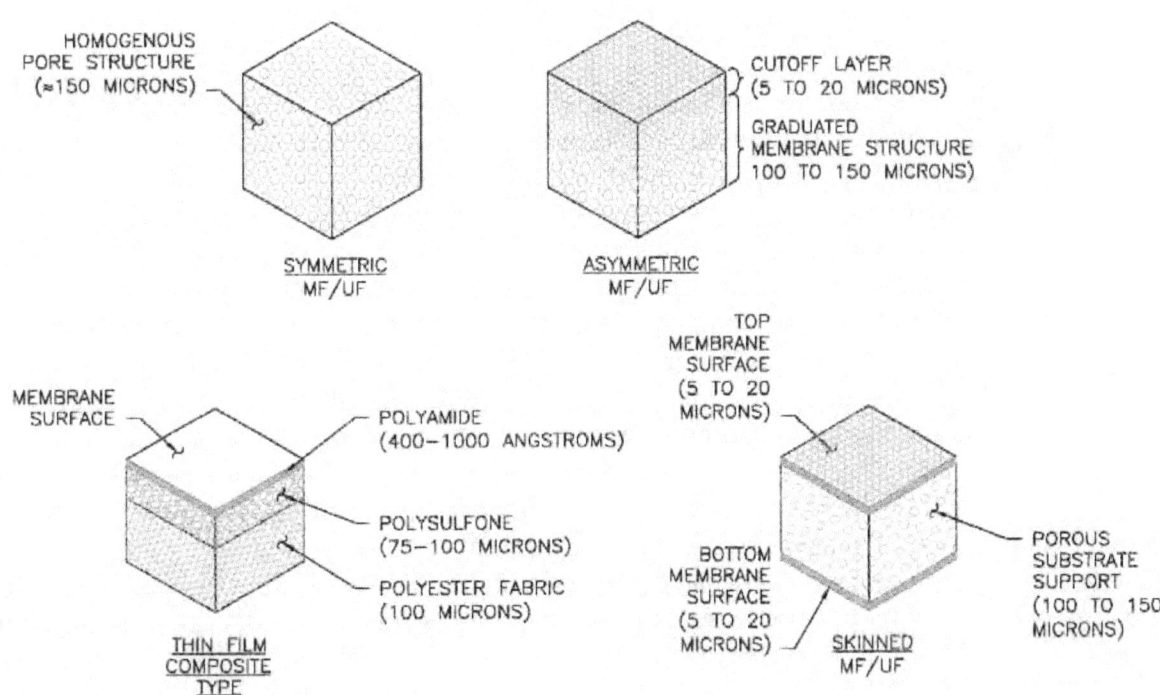

2.3.2 Membrane Modules

Membrane filtration media is usually manufactured as flat sheet stock or as hollow fibers and then configured into one of several different types of membrane modules. As defined for the purposes of the LT2ESWTR, a membrane module represents the smallest discrete filtration unit in a membrane system. (Terminology associated with the LT2ESWTR and used in the *Membrane Filtration Guidance Manual* is described in section 1.5.) Module construction typically involves potting or sealing the membrane material into a corresponding assembly, which may incorporate an integral containment structure, such as with hollow-fiber modules. These types of modules are designed for long-term use over the course of a number of years. Spiral-wound modules are also manufactured for long-term use, although the design of membrane filtration systems that utilize spiral-wound modules requires that the modules be encased in a separate pressure vessel that is independent of the module itself. Alternatively, a module may be configured as a disposable cartridge with a useful life that is typically measured in weeks or months rather than years. Membrane cartridges may either be inserted into pressure vessels that are separate from the module (as with spiral-wound modules) or manufactured within a casing that serves as an integral pressure vessel. Each of these three types of modules, along with some other less common module designs, is discussed in the following subsections. (The membrane filtration systems that utilize these types of modules are subsequently described under section 2.3.3.)

2.3.2.1 Hollow-Fiber Modules

Most hollow-fiber modules used in drinking water treatment applications are manufactured to accommodate porous MF or UF membranes and designed to filter particulate matter. As the name suggests, these modules are comprised of hollow-fiber membranes, which are long and very narrow tubes that may be constructed of any of the various membrane materials described in section 2.3.1. The fibers may be bundled in one of several different arrangements. In one common configuration used by many manufacturers, the fibers are bundled together longitudinally, potted in a resin on both ends, and encased in a pressure vessel that is included as a part of the hollow-fiber module. These modules are typically mounted vertically, although horizontal mounting may also be utilized. One alternate configuration is similar to spiral-wound modules in that both are inserted into pressure vessels that are independent of the module itself. These modules (and the associated pressure vessels) are mounted horizontally. Another configuration in which the bundled hollow fibers are mounted vertically and submerged in a basin does not utilize a pressure vessel. A typical commercially available hollow-fiber module may consist of several hundred to over 10,000 fibers. Although specific dimensions vary by manufacturer, approximate ranges for hollow-fiber construction are as follows:

- Outside diameter: 0.5 – 2.0 mm

- Inside diameter: 0.3 – 1.0 mm

- Fiber wall thickness: 0.1 – 0.6 mm

- Fiber length: 1 – 2 meters

A cross section of a symmetric hollow-fiber is shown in Figure 2.4.

Hollow-fiber membrane modules may operate in either an "inside-out" or "outside-in" mode. In inside-out mode, the feed water enters the fiber lumen (i.e., center or bore of the fiber) and is filtered radially through the fiber wall. The filtrate is then collected from outside of the fiber. During outside-in operation, the feed water passes from outside the fiber through the fiber wall to the inside, where the filtrate is collected in the lumen. Although inside-out mode utilizes a well-defined feed flow path that is advantageous when operating under a crossflow hydraulic configuration (see section 2.5), the membrane is somewhat more subject to plugging as a result of the potential for the lumen to become clogged. The outside-in mode utilizes a less well-defined flow feed flow path, but increases the available membrane surface area for filtration per fiber and avoids potential problems with clogging of the lumen bore.

Figure 2.4 Hollow Fiber Cross-Section Photomicrograph

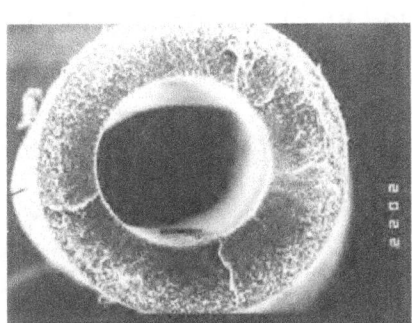

Both the inside-out and outside-in operating modes for hollow-fiber modules utilizing pressure vessels are illustrated in Figure 2.5. When a hollow-fiber module is operated in an inside-out mode, the pressurized feed water may enter the fiber lumen at either end of the module, while the filtrate exits through a filtrate port located at the center or end of the module. In outside-in mode, the feed water typically enters the module through a inlet port located in the center and is filtered into the fiber lumen, where the filtrate collects prior to exiting through a port at one end of the module. Most hollow-fiber systems operate in "dead-end" or direct filtration mode (see section 2.5) and are periodically backwashed to remove the accumulated solids. Note that the submerged hollow-fiber membranes operate in outside-in mode, but do not utilize the pressure vessels (and the associated inlet ports) that are illustrated in Figure 2.5.

2.3.2.2   Spiral-Wound Modules

Spiral-wound modules were developed as an efficient configuration for the use of semi-permeable membranes to remove dissolved solids, and thus are most often associated with NF/RO processes. The basic unit of a spiral-wound module is a sandwich arrangement of flat membrane sheets called a "leaf" wound around a central perforated tube. One leaf consists of two membrane sheets placed back to back and separated by a fabric spacer called a permeate carrier. The layers of the leaf are glued along three edges, while the unglued edge is sealed around the perforated central tube. A single spiral-wound module 8 inches in diameter may contain up to approximately 20 leaves, each separated by a layer of plastic mesh called a spacer that serves as the feed water channel.

Feed water enters the spacer channels at the end of the spiral-wound element in a path parallel to the central tube. As the feed water flows across the membrane surface through the spacers, a portion permeates through either of the two surrounding membrane layers and into the permeate carrier, leaving behind any dissolved and particulate contaminants that are rejected by the semi-permeable membrane. The filtered water in the permeate carrier travels spirally inward around the element toward the central collector tube, while the water in the feed spacer that does not permeate through the membrane layer continues to flow across the membrane surface, becoming increasingly concentrated in rejected contaminants. This concentrate stream exits the element parallel to the central tube through the opposite end from which the feed water entered. A diagram of a spiral-wound element is shown in Figure 2.6.

Figure 2.5 Inside-Out and Outside-In Modes of Operation
(Using Pressure Vessels)

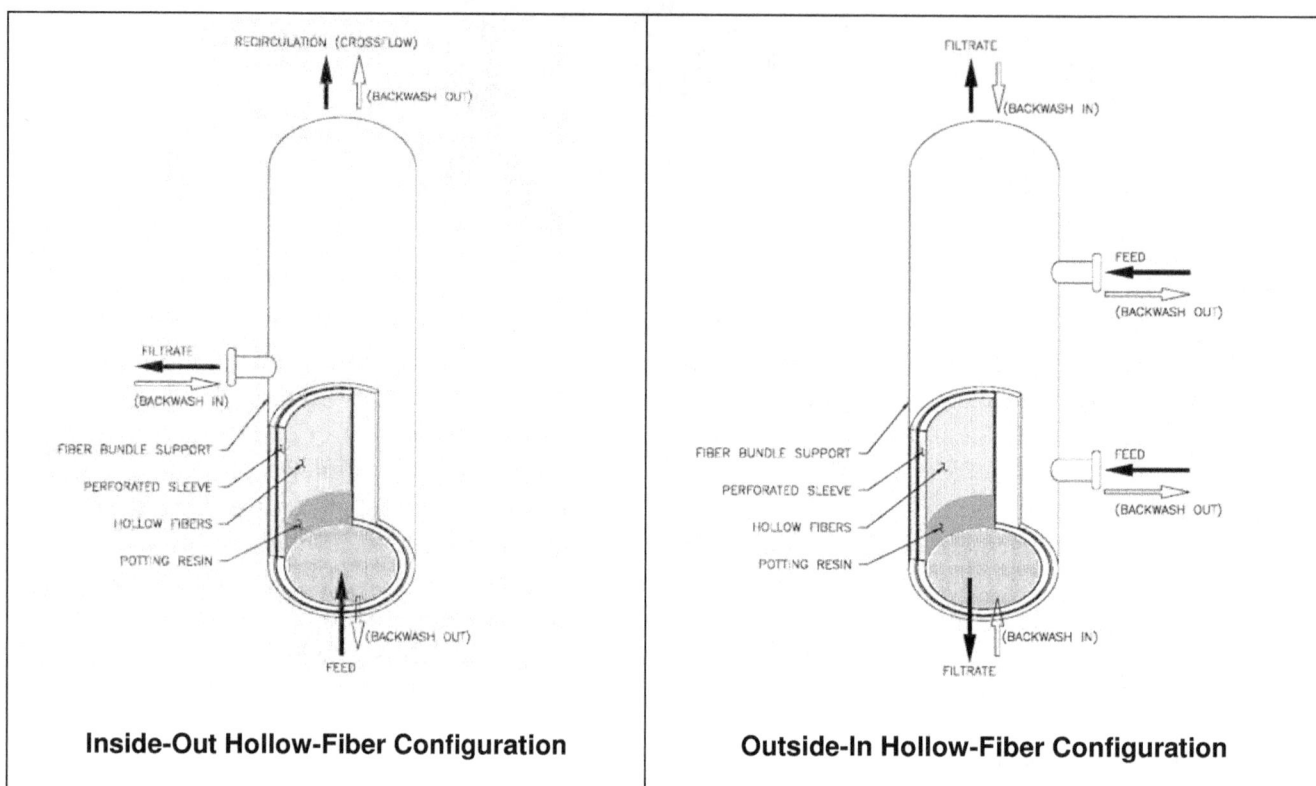

| Inside-Out Hollow-Fiber Configuration | Outside-In Hollow-Fiber Configuration |

Figure 2.6 Spiral-Wound Membrane Module

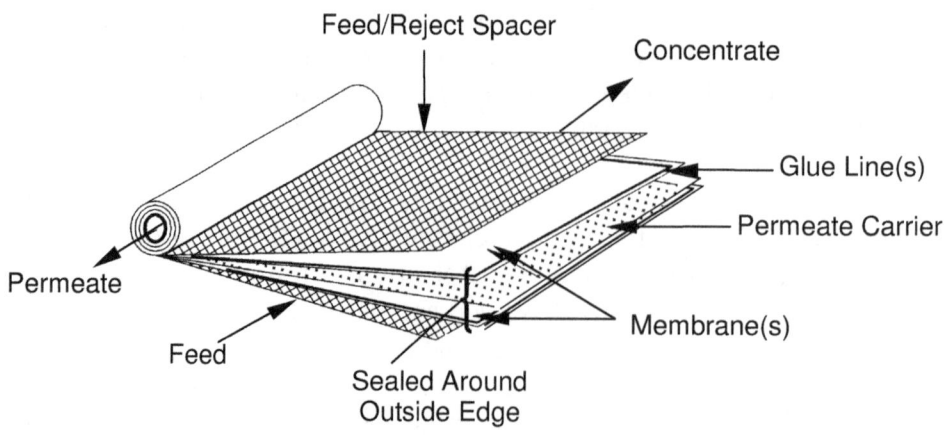

Spiral-wound membranes for drinking water treatment are commercially available in a variety of sizes. Modules that are either 4 or 8 inches in diameter and either 40 or 60 inches long are most common, although other sizes may be used. Some bench- and pilot-scale applications utilize modules that are 2.5 inches in diameter, while modules up to 16 inches in diameter or more may be used in full-scale facilities.

2.3.2.3 Membrane Cartridges

Under the LT2ESWTR, cartridge filters that meet the criteria specified in section 2.2 would be eligible to receive *Cryptosporidium* removal credit as a membrane filtration process. In this case, the cartridge filter element would constitute a membrane module for the purposes of the rule. The ability of these modules to be subjected to direct integrity testing in the field during the course of normal operation, a feature that has not been widely utilized in association with cartridge filters in municipal water treatment applications, is a critical aspect of these systems that distinguishes what is considered to be MCF under the LT2ESWTR.

Membrane cartridge filters are manufactured by placing flat sheet membrane media between a feed and filtrate support layer and pleating the assembly to increase the membrane surface area within the cartridge. The pleat pack assembly is then placed around a center core with a corresponding outer cage and subsequently sealed, via adhesive or thermal means, into its cartridge configuration. End adapters, typically designed with a double o-ring sealing mechanism, are attached to the filter to provide a positive seal with the filter housing. A representative diagram of membrane cartridge filter is shown in Figure 2.7.

Figure 2.7 Membrane Cartridge Filter

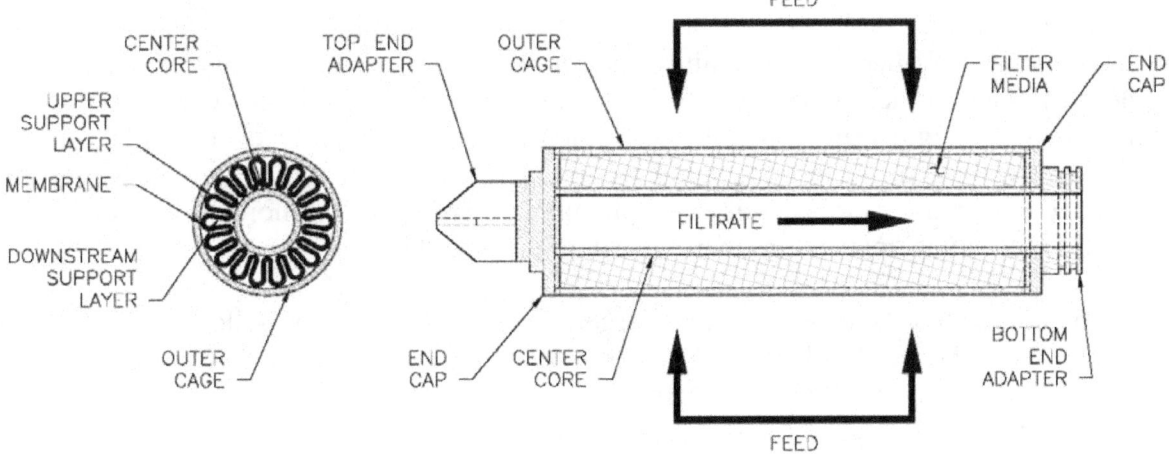

Most membrane cartridge filters are manufactured as disposable components that are inserted into a housing. Once the filter fouls to the point at which the maximum TMP is reached, the cartridge is replaced. Because the cartridges are designed to be disposable, and thus relatively inexpensive to replace, cartridge filtration systems have not historically utilized backwashing or chemical cleaning. However, some systems that feature these processes have recently been introduced. Cartridge filters are available in various sizes and pore sizes, although the device would have to be capable of filtering particulate matter larger than 1 μm to comply with the definition of a membrane filter under the LT2ESWTR (see section 1.3.1).

2.3.2.4 Other Module Configurations

In addition to hollow-fiber modules, spiral-wound modules, and membrane cartridge filters, there are several other types of less common configurations that may be used in membrane filtration systems. These configurations include hollow-fine-fiber (HFF), tubular, and plate-and-frame type modules. While these configurations are seldom employed in membrane filtration systems applied for drinking water treatment, each of these modules could be used for LT2ESWTR compliance, and thus are briefly described in this section.

Semi-permeable HFF membranes were the original hollow-fiber type membranes and were developed for desalting (i.e., RO) applications. With the development of widely used porous hollow-fiber MF and UF membranes for particulate filtration with much larger fiber diameters, the semi-permeable variety gradually became known as hollow-*fine*-fiber membranes. HFF membranes are bundled length-wise and shaped into a "U" arrangement (called a "U-tube"), which is potted in a cylindrical pressure vessel. Feed water enters a HFF module via a perforated tube in the middle of the vessel and flows radially outward through the membrane bundle. The water that permeates the membrane is collected in the fiber lumen and exits the element at the open end of the U-tube. The remaining water that does not permeate into the fiber lumen carries the concentrated salts and suspended solids out of the pressure vessel through the concentrate port. Typical hollow-fine fibers are only about 40 μm in diameter (inside), allowing a very large number of fibers to be contained in a single pressure vessel and maximizing the available membrane surface area per unit volume in the pressure vessel. However, the high packing density also significantly increases the potential for fouling and reduces the flux to levels well below those possible using spiral-wound membranes, typically more than offsetting the advantage in increased surface area. HFF membranes are most commonly used today in some seawater desalting applications, particularly in the Middle East.

Tubular membranes are essentially a larger, more rigid version of hollow-fiber membranes. With diameters as large as 1-2 inches, the tubes are not prone to clogging and the membrane material (i.e., the tube wall) is comparatively easy to clean. However, the large tubes also result in a very inefficient amount of membrane surface area per unit volume in the pressure vessel. Both porous (for MF/UF) and semi-permeable (for NF/RO) membranes have been manufactured in tubular configurations. Ceramics have been considered as non-traditional material for tubular MF/UF membranes, although there are currently no commercially promoted ceramic MF/UF systems for drinking water applications.

A plate-and-frame configuration, one of the earliest membrane modules developed, is simply a series of flat sheet membranes separated by alternating filtrate spacers and feed/concentrate spacers. Because of the very low surface area to volume ratio, the plate-and-frame configuration is considered inefficient and is therefore seldom used in drinking water applications. One notable exception (although not a membrane filtration process under the LT2ESWTR) is in the case of EDR systems, which utilize a design that lends itself well to the use of plate-and-frame membranes.

2.3.3 Types of Membrane Filtration Systems

In drinking water treatment applications, each of the four traditional types of pressure-driven membrane processes (i.e., MF, UF, NF, and RO) is generally associated with a single type of membrane filtration system that is designed around a specific type of module. MF and UF systems typically utilize hollow-fiber modules, while NF and RO systems typically utilize spiral-wound modules. An overview of each of the two types of systems that utilize these respective modules is provided in the following subsections. Because the concept of a MCF system as defined under the LT2ESWTR is a new concept introduced with the rule, a standard type of MCF system has not yet been developed.

2.3.3.1 Hollow-Fiber (MF/UF) Systems

With few exceptions, most MF/UF processes utilize systems designed around hollow-fiber modules. Hollow-fiber membrane filtration systems are designed and constructed in one or more discrete water production units, also called racks, trains, or skids. A unit consists of a number of membrane modules that share feed and filtrate valving, and each respective unit can usually be isolated from the rest of the system for testing, cleaning, or repair. A typical hollow-fiber system is composed of a number of identical units that combine to produce the total filtrate flow.

Most of the currently available hollow-fiber membrane systems are proprietary, such that a single supplier will manufacture the entire filtration system, including the membranes, piping, appurtenances, control system, and other features. The manufacturer also determines the hydraulic configuration and designs the associated operational sub-processes – such as backwashing, chemical cleaning, and integrity testing – that are specific to its particular system. As a result, there are significant differences in the proprietary hollow-fiber membrane systems produced by the various manufacturers, and the membranes and other components are not interchangeable.

Although each manufacturer's system is distinct, all of the hollow-fiber membrane systems fall into one of two categories – pressure-driven or vacuum-driven – according to the driving force for operation. In a pressure-driven system, pressurized feed water is piped directly to the membrane unit, where it enters the module and is filtered through the membrane. Typical operating pressures range from 3 to 40 psi. Most applications require designated feed pumps to generate the required operating pressure, although there are some water treatment plants that take advantage of favorable hydraulic conditions to operate a MF or UF system via gravity flow.

A schematic of a typical pressure-driven hollow-fiber membrane filtration system is shown in Figure 2.8. In the example shown, the system is operated in a "dead-end" hydraulic configuration (see section 2.5) and uses a liquid backwash.

Figure 2.8 Schematic of a Typical Pressure-Driven Hollow-Fiber (MF/UF) System

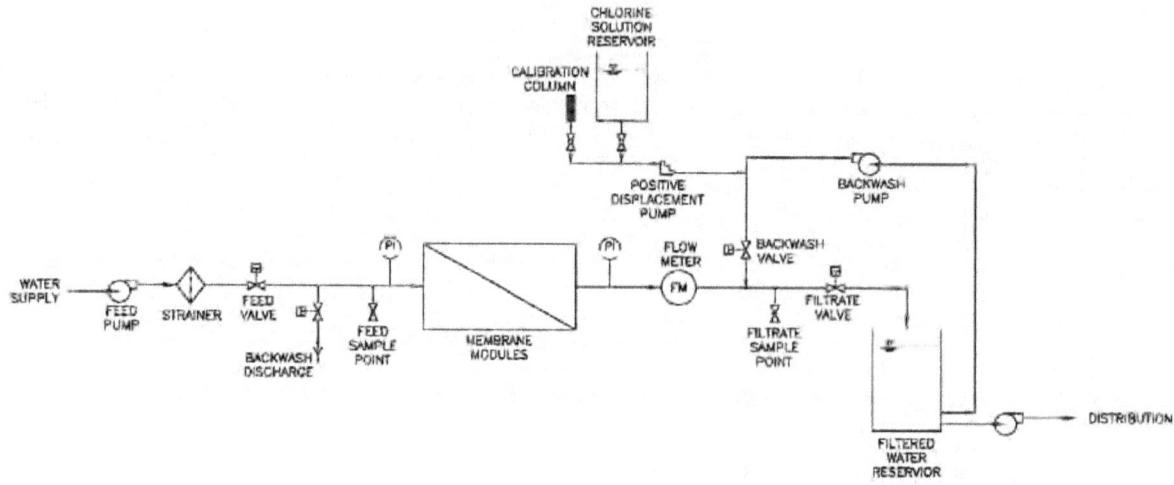

While all hollow-fiber systems employ pressure as a fundamental driving force, a vacuum-driven system is distinguished by its utilization of negative pressure and, consequently, its significantly different design and configuration. Unlike pressure-driven systems, in which each membrane module incorporates a pressure vessel, vacuum-driven systems utilize hollow-fiber modules that are "submerged" or "immersed" in an open tank or basin. While the ends are fixed, the lengths of the hollow-fibers are exposed to the feed water in the basin.

Because the feed water is contained in an open basin, the outside of the fibers cannot be pressurized above the static head in the tank. Therefore, a vacuum of approximately -3 to -12 psi is induced at the inside of the fibers via pump suction. The water in the tank is drawn through the fiber walls, where it is filtered into the lumen. By design, vacuum-driven membrane filtration systems cannot be operated via gravity nor in an inside-out mode. However, a favorable hydraulic gradient might enable the use of a gravity-based siphon to generate the suction required to drive the filtration process in a vacuum-driven system. In some cases with a substantial hydraulic gradient, the large amount of available head could be used to generate the power for suction pumps via on-site turbines.

A representative schematic of a vacuum-driven system is shown in Figure 2.9. In the example shown, the membrane process may be designed with either continuous (Option A) or intermittent discharge (Option B) of concentrated waste.

Figure 2.9 Schematic of a Typical Vacuum-Driven Hollow-Fiber (MF/UF) System

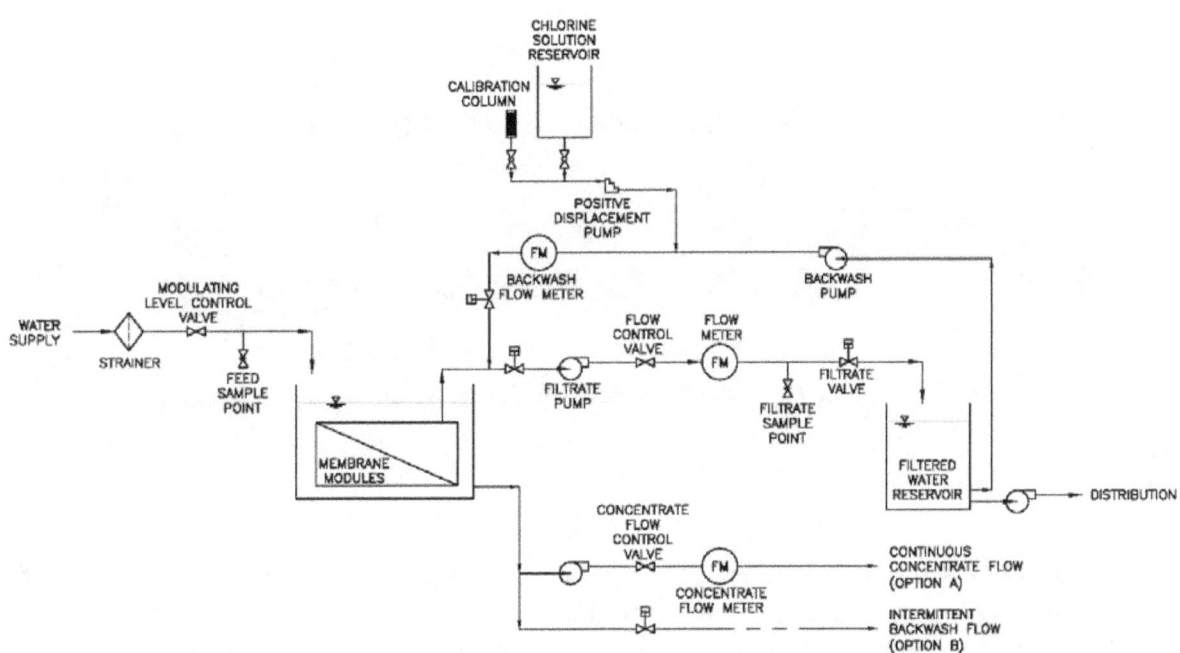

2.3.3.2   Spiral-Wound (NF/RO) Systems

Virtually all NF and RO membrane processes applied for potable water treatment in the United States utilize systems designed for spiral-wound membrane modules. Although some MF and UF membranes may also be manufactured as spiral-wound modules, these are seldom used in municipal drinking water applications. Consequently, the discussion in this section is focused on NF/RO spiral-wound systems.

In a spiral-wound membrane filtration system, the spiral-wound modules are contained in a pressure vessel that is independent of the module itself. Typically, a single pressure vessel houses six or seven modules, although vessels that accommodate other numbers of modules can be custom manufactured. The modules are arranged in series in the pressure vessel such that the concentrate from each preceding element represents the feed water for the next. A brine seal around the outside of the feed end of each element separates the feed water from the concentrate and prevents the feed water from bypassing the element. Although the recovery for a single NF/RO module is typically less than 15 percent, the cumulative recovery associated with a six-module pressure vessel may be 50 percent or more. A diagram of a typical pressure vessel containing spiral-wound modules is shown in Figure 2.10.

Figure 2.10 Typical Spiral-Wound (NF/RO) Module Pressure Vessel

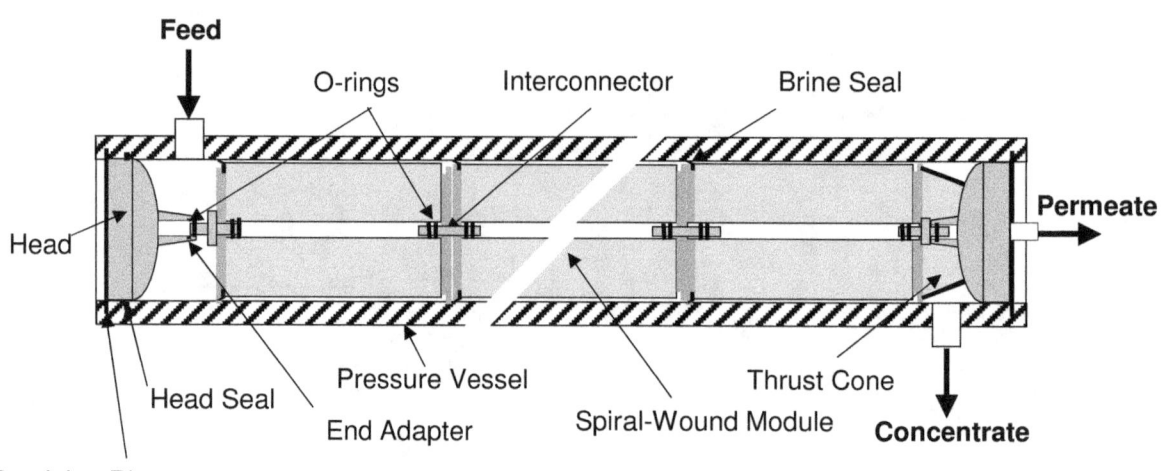

A group of pressure vessels operating in parallel collectively represent a single stage of treatment in a NF/RO spiral-wound system. The total system recovery is increased by incorporating multiple stages of treatment in series, such that the combined concentrate (or reject) from the first stage becomes the feed for the second stage. In some cases in which higher recovery is an objective, a third stage may also be used. This configuration is sometimes referred to as "concentrate staging." Because some fraction of the feed to the first stage has been collected as filtrate (or permeate), the feed flow to the second stage will be reduced by that fraction. As a result, the number of total pressure vessels (and hence the number of modules) in the second stage is also typically reduced by approximately that same fraction. Similar flow, module, and pressure vessel reductions are propagated through all successive stages, as well. Although the potential system recovery is a function of the feed water quality, as a rough approximation, a two-stage design may allow recoveries up to 75 percent, while the addition of a third stage can potentially achieve recoveries up to 90 percent.

Although concentrate staging is most often used in drinking water applications, another arrangement called "permeate staging" may also be employed. In this configuration the filtrate (or permeate) from a stage (rather than the concentrate) becomes the feed water for the subsequent stage. While this arrangement is more commonly employed in ultra-pure water applications (typically in industry), it may also be used for drinking water treatment when the source water salinity is very high, such as with seawater desalination. In these cases, the product water must pass through multiple stages to remove a sufficient amount of salinity to make the water potable quality.

The combination of two or more stages in series is called an array, which is identified by the ratio of pressure vessels in the sequential stages. An array may be defined by the ratio of either the actual number or relative number of pressure vessels in each stage. For example, a 32:16:8 array expressed as the actual number of pressure vessels may be alternatively called a 4:2:1 array in relative terms. Two-stage arrays, such as 2:1 and 3:2 (relative), are most common in drinking water treatment, although the specific array required for a particular application is dictated in part by the feed water quality and targeted overall system recovery. Figure 2.11 illustrates the configuration of a typical 2:1 (relative) array, showing both plan and end-perspective views.

Figure 2.11 Typical 2:1 (Relative) Array of Pressure Vessels

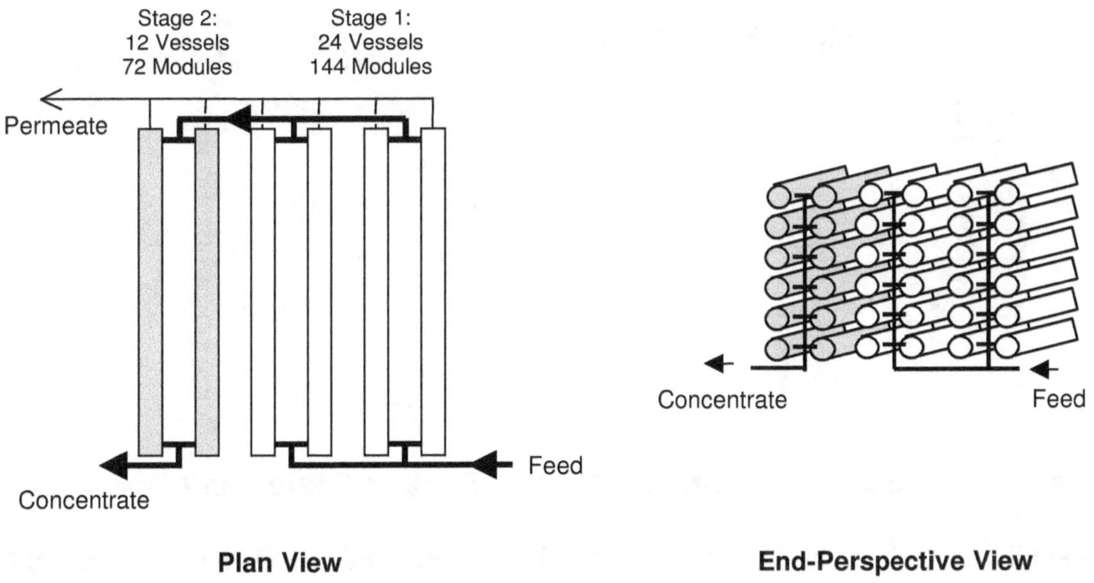

Plan View   End-Perspective View

As with hollow-fiber systems, spiral-wound membrane systems are designed and constructed in discrete units that share common valving and which can be isolated as a group for testing, cleaning, or repair. For spiral-wound systems these uniform units are typically called trains, or alternatively racks or skids. NF and RO treatment processes consist of one or more trains that are typically sized to accommodate a feed flow of up to about 5 MGD per train. A schematic of a typical NF/RO system is shown in Figure 2.12.

Unlike hollow-fiber systems, spiral-wound membrane filtration systems are not manufactured as proprietary equipment. With the exception of the membrane modules, spiral-wound systems are generally custom-designed by an engineer or an original equipment manufacturer (OEM) to suit a particular application. Although the membrane modules are proprietary, standard-sized spiral-wound NF/RO modules share the same basic construction, and thus membranes from one manufacturer are typically interchangeable with those from others.

Figure 2.12 Schematic of a Typical Spiral-Wound (NF/RO) System

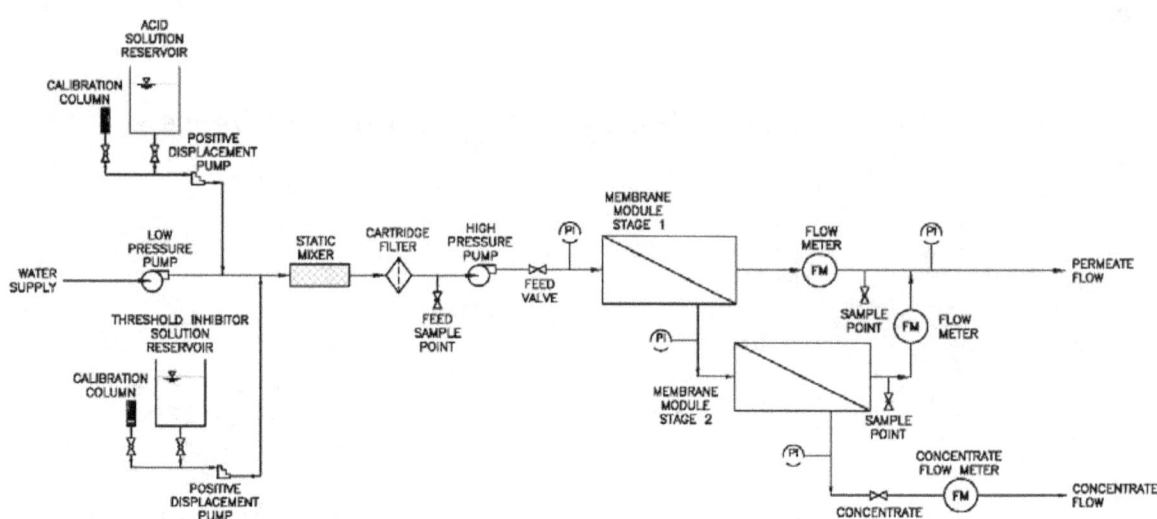

2.4   Basic Principles of Membrane Filtration System Design and Operation

Familiarity with the basic principles underlying membrane filtration system design and operation is important to the comprehension and interpretation of the material presented in this guidance manual. The material presented in this subsection is intended to provide an overview of these basic principles. Although all of the types of membrane filtration addressed in this manual (i.e., MF, UF, NF, RO, and MCF) utilize pressure (or vacuum) as a driving force, there are fundamental differences in the models used to describe systems using porous membranes (MF, UF, and MCF) and semi-permeable membranes (NF and RO). The basic principles of these respective models, along with some general concepts that are applicable to all membrane filtration systems, are discussed in the following subsections. Note that these discussions are intended to be informative for the purposes of fostering an understanding the technology and not a specific guide for system design and operation.

2.4.1 General Concepts

There are a number of general concepts that are applicable to all types of pressure-driven membrane filtration systems and which serve as the underlying basic principles for system design and operation. These concepts include flux, recovery, and flow balance, each of which is discussed in general terms below. Additional concepts that are specific to either MF, UF, and MCF systems or NF and RO systems are discussed in sections 2.4.2 and 2.4.3, respectively.

Membrane filtration system throughput is typically characterized by the system flux, which is defined as the filtrate flow per unit of membrane filtration area, as shown in Equation 2.1:

$$J = \frac{Q_p}{A_m}$$

Equation 2.1

Where: J =   flux (gfd)
         Q_p =   filtrate flow (gpd)
         A_m =   membrane surface area (ft^2)

The recovery of a membrane unit is defined under the LT2ESWTR as the amount of feed flow that is converted to filtrate flow, expressed as a decimal percent, as shown in Equation 2.2:

$$R = \frac{Q_p}{Q_f}$$

Equation 2.2

Where: R =   recovery of the membrane unit (decimal percent)
         Q_p =   filtrate flow produced by the membrane unit (gpd)
         Q_f =   feed flow to the membrane unit (gpd)

The recovery as defined under the rule does not account for the use of filtrate for routine maintenance purposes (such as chemical cleaning or backwashing) or lost production during these maintenance operations. Because the definition of recovery is not necessarily consistent throughout the water treatment industry, it is important to identify how recovery is defined in any particular discussion. However, the use of the term recovery as defined in Equation 2.2 is consistent throughout the LT2ESWTR rule language and the *Membrane Filtration Guidance Manual*. Note that for some types of membrane systems, particularly those that operate in suspension mode that can be modeled as plug flow reactors (see section 2.5), recovery can also be defined as a function of position within a membrane unit. This is simply a variation of Equation 2.2 for systems in which the cumulative volume of filtrate increases in the direction of feed flow through the membrane unit, thus increasing the recovery in direct proportion. The limit of this recovery in the direction of flow (i.e., the recovery at the furthest position in the unit) is equivalent to the overall membrane unit recovery, as defined in Equation 2.2.

A general flow balance that can be applied to all membrane filtration systems is shown in Equation 2.3:

$$Q_f = Q_c + Q_p$$

Equation 2.3

Where:
Q_f = feed flow to the membrane unit (gpd)
Q_c = concentrate flow from the membrane unit (gpd)
Q_p = filtrate flow from the membrane unit (gpd)

Note that the concentrate (i.e., bleed or reject) flow, Q_c, is zero for systems operating without a concentrate waste or "bleed" stream (e.g., systems operated in deposition (or "dead-end) mode or crossflow systems in which 100 percent of the concentrate is recirculated – see section 2.5). For the purpose of sizing a membrane filtration system, it may be desirable to account for the additional filtered water used for both backwashing and chemical cleaning in the determination of the filtrate flow, Q_p. Similarly, an estimate of the total required feed flow Q_f to the system should incorporate any raw water that may be used in these routine maintenance processes.

2.4.2 MF, UF, and MCF Processes

The driving force for the transport of water across a microporous membrane – utilized by MF, UF, and MCF processes – is a pressure gradient across the membrane, or the TMP. The TMP is defined by the pressure on the feed side of the membrane minus the filtrate pressure, commonly called the backpressure, as shown in Equation 2.4:

$$TMP = P_f - P_p$$

Equation 2.4

Where:
TMP = transmembrane pressure (psi)
P_f = feed pressure (psi)
P_p = filtrate pressure (i.e., backpressure) (psi)

For systems that operate in suspension mode and thus utilize a concentrate stream that is wasted or recirculated (as described in section 2.5), the pressure on the feed side of the membrane is not constant, but can instead be approximated by a linear pressure gradient from the feed inlet to the concentrate outlet. In this case, the pressure on the feed side of the membrane may be represented by the average of the feed and concentrate pressures, as shown in Equation 2.5:

$$TMP = \frac{P_f + P_c}{2} - P_p$$
Equation 2.5

Where: TMP = transmembrane pressure (psi)
P_f = feed pressure (psi)
P_c = concentrate pressure (psi)
P_p = filtrate pressure (i.e., backpressure) (psi)

The resistance to flow acting in opposition to the driving force and inhibiting the transport of water across the membrane can also be quantified. This resistance has two components: the intrinsic resistance of the membrane and the resistance attributable to the accumulated foulant layer at the membrane surface at any point during operation. The total resistance is represented by the sum of these two components, as shown in Equation 2.6:

$$R_t = R_m + R_f$$
Equation 2.6

Where: R_t = total membrane resistance (psi/gfd-cp)
R_m = intrinsic membrane resistance (psi/gfd-cp)
R_f = resistance of the foulant layer (psi/gfd-cp)

While the intrinsic resistance of the membrane should remain constant for all practical purposes and can generally be obtained from the membrane manufacturer (if necessary), the increase in fouling during normal operation and the decrease in fouling as a result of backwashing and chemical cleaning causes the fouling resistance to fluctuate.

If the total membrane resistance is known, the flux can be calculated as a function of the TMP and water viscosity, as shown in Equation 2.7:

$$J_T = \frac{TMP}{R_t \bullet \mu_T}$$
Equation 2.7

Where: J_T = flux at temperature T (gfd)
TMP = transmembrane pressure (psi)
R_t = total membrane resistance (psi/gfd-cp)
μ_T = viscosity of water at temperature T (cp)

Because the viscosity of water increases with decreasing temperature, larger TMPs (by application of increased pressure or vacuum) are required to maintain constant flux. Values for water viscosity can be found in the literature or approximated using the empirical relationship expressed in Equation 2.8:

$$\mu_T = 1.784 - (0.0575 \bullet T) + (0.0011 \bullet T^2) - (10^{-5} \bullet T^3) \qquad \text{Equation 2.8}$$

Where: μ_T = viscosity of water at temperature T (cp)

T = water temperature ($^\circ$C)

Since water temperature can have a significant impact on flux, it is common practice to "normalize" the flux to a reference temperature during operation for the purpose of monitoring system productivity independent of changes in water temperature. For MF/UF and MCF processes, a reference temperature of 20 $^\circ$C is typically used for convenience, since the viscosity of water is approximately 1 cp at 20 $^\circ$C. For constant TMP and total membrane resistance, a form of Equation 2.7 can be used to illustrate the relationship between the normalized flux and viscosity at 20 $^\circ$C and the actual flux and viscosity at a given temperature of interest T, as shown in Equation 2.9:

$$J_{20} \bullet \mu_{20} = J_T \bullet \mu_T \qquad \text{Equation 2.9}$$

Where: J_{20} = normalized flux at 20 $^\circ$C (gfd)

μ_{20} = viscosity of water at 20 $^\circ$C (cp)

J_T = actual flux at temperature T (gfd)

μ_T = viscosity of water at temperature T (cp)

Substituting the value of 1 cp for the viscosity at 20 $^\circ$C (μ_{20}) and Equation 2.5 for the viscosity of water at temperature T (μ_T) yields and expression for normalized flux at 20 $^\circ$C as a function of the actual flux and the temperature, as shown in Equation 2.10:

$$J_{20} = J_T \bullet [1.784 - (0.0575 \bullet T) + (0.0011 \bullet T^2) - (10^{-5} \bullet T^3)] \qquad \text{Equation 2.10}$$

Where: J_{20} = normalized flux at 20 $^\circ$C (gfd)

J_T = actual flux at temperature T (gfd)

T = water temperature ($^\circ$C)

It is important to note that the normalized flux (J_{20}) does not represent an actual operating condition. This term simply represents what the flux would be at 20 $^\circ$C for a the same TMP and total membrane resistance, such that changes in the value of J_{20} during the course of normal operation are indicative of changes in pressure and/or membrane resistance due to fouling. Thus, this term is only intended to normalize for temperature effects in order to illustrate the influence of fouling or changes in TMP on system operation.

If values for viscosity are known, the polynomial expression for viscosity as a function of temperature in Equation 2.10 may be simplified to a temperature correction factor (TCF). For a MF, UF, or MCF process, the TCF is defined as the ratio of the viscosity at temperature T to the viscosity at 20 °C, as shown in Equation 2.11:

$$TCF = \frac{\mu_T}{\mu_{20}}$$

Equation 2.11

Where:
- TCF = temperature correction factor (dimensionless)
- μ_T = viscosity of water at temperature T (cp)
- μ_{20} = viscosity of water at 20 °C (cp)

Note that the term TCF is often used generically to refer to any type of correction factor used to adjust a parameter for temperature. Thus, the specific equation for the TCF may vary depending on the parameter to which it is applied. For example, in the context of membrane filtration, the TCF applied to reference MF, UF, and MCF flux to a standard temperature, as defined in Equation 2.11, is different than that applied to NF and RO flux to a standard temperature, as shown in Equation 2.19. Thus, it is important to always consider the context in which the term TCF is used.

Because the TCF is a dimensionless ratio, the values for viscosity can be expressed in any convenient and consistent units. Thus, the temperature-normalized flux can be expressed in simplified as terms, as shown in Equation 2.12:

$$J_{20} = J_T \bullet TCF$$

Equation 2.12

Where:
- J_{20} = normalized flux at 20 °C (gfd)
- J_T = actual flux at temperature T (gfd)
- TCF = temperature correction factor (dimensionless)

Generally, in order to identify changes in productivity (as measured by flux) that are specifically attributable to membrane fouling, it is desirable to normalize the flux for pressure as well as temperature, as shown in Equation 2.13. Note that the temperature- and pressure-normalized flux is often referred to as the specific flux.

$$M_{20} = \frac{J_{20}}{TMP}$$

Equation 2.13

Where:
M_{20} = temperature- and pressure-normalized flux (gfd/psi)
J_{20} = normalized flux at 20 °C (gfd)
TMP = transmembrane pressure (psi)

The effect of temperature on a MF, UF, or MCF system can also be expressed in terms of a temperature-corrected TMP value. An expression for such a term can be derived by starting with a modified form of Equation 2.7, as shown below in Equation 2.14:

$$J \bullet R_t = \frac{TMP}{\mu}$$

Equation 2.14

Where:
J = flux (gfd)
R_t = total membrane resistance (psi/gfd-cp)
TMP = transmembrane pressure (psi)
μ = viscosity of water (cp)

For constant flux and membrane resistance, equating two expressions of Equation 2.14 illustrate the relationship between two pairs of TMP and viscosity data at different temperatures, as shown in Equation 2.15:

$$\left(\frac{TMP}{\mu} \right)_1 = J \bullet R_t = \left(\frac{TMP}{\mu} \right)_2$$

Equation 2.15

Where:
TMP = transmembrane pressure (psi)
μ = viscosity of water (cp)
J = flux (gfd)
R_t = total membrane resistance (psi/gfd-cp)

Rearranging Equation 2.15 and assigning a reference temperature of 20 °C yields an expression for the TMP at 20 °C as a function of any given TMP data point and the ratio of the water viscosity at 20 °C to that at the given data point, as shown in Equation 2.16:

$$TMP_{20} = TMP_T \bullet \left(\frac{\mu_{20}}{\mu_T} \right)$$

Equation 2.16

Where:		
TMP_{20}	=	transmembrane pressure at 20 °C (psi)
TMP_T	=	transmembrane pressure at temperature T (psi)
μ_{20}	=	viscosity of water at 20 °C (cp)
μ_T	=	viscosity of water at temperature T (cp)

Equation 2.16 allows for the normalization of TMP data to 20 °C, or alternatively stated, the value of the TMP that would have been observed at 20 °C for the same flux and degree of fouling. By adjusting all the TMP data to the same reference temperature, any observed increase in TMP is known to be attributable to fouling or other phenomena that impact membrane resistance (such as compaction), assuming operation at constant flux.

2.4.3 NF and RO Processes

As with the microporous MF, UF, and MCF membranes, the driving force for the transport of water across a semi-permeable membrane – such as that utilized by NF and RO processes – is a pressure gradient across the membrane. However, because NF and RO processes reject dissolved salts, the resulting osmotic pressure gradient, which acts against the transport of water from the feed to the filtrate side of the membrane, must also be taken into account. Typically, the osmotic pressure gradient is approximated from the concentration of TDS on the feed and filtrate sides of the membrane. The corrected driving force across semi-permeable membrane is termed the net driving pressure (NDP) and can be calculated using Equation 2.17 (AWWA 1999):

$$NDP = \left[\left(\frac{P_f + P_c}{2} \right) - \left(P_p \right) \right] - \left[\left\{ \left(\frac{TDS_f + TDS_c}{2} \right) - TDS_p \right\} \bullet 0.01 \frac{psi}{mg/L} \right]$$

Equation 2.17

Where:		
NDP	=	net driving pressure (psi)
P_f	=	feed pressure (psi)
P_c	=	concentrate pressure (psi)
P_p	=	filtrate pressure (i.e., backpressure) (psi)
TDS_f	=	feed TDS concentration (mg/L)
TDS_c	=	concentrate TDS concentration (mg/L)
TDS_p	=	filtrate TDS concentration (mg/L)

Equation 2.17 can be considered as two distinct components, each shown above in square brackets. The first term represents difference between the average pressure on the feed side of the membrane and the filtrate backpressure; the second term represents the average osmotic backpressure. The conversion factor of 0.01 in the osmotic pressure term comes from a widely used rule of thumb for fresh and brackish waters indicating that there is approximately 1 psi of osmotic pressure for every 100 mg/L of TDS, as discussed in section 2.2.2. In many cases the filtrate TDS concentration (TDS_p) is small and can be neglected.

Equation 2.17 is often given in simplified form by combining the respective parameters associated with the two components into two consolidated terms representing differential pressure (ΔP) and the transmembrane osmotic pressure differential ($\Delta\pi$). This simplified form is shown as Equation 2.18.

$$NDP = \Delta P - \Delta \pi$$

Equation 2.18

Where:

NDP	=	net driving pressure (psi)
ΔP	=	transmembrane differential pressure (psi)
$\Delta\pi$	=	transmembrane osmotic pressure differential (psi)

Note that there are a number of equations throughout this document equally applicable to MCF, MF/UF, and NF/RO systems which include a term for net pressure differential across the membrane. Although in each of these equations the nomenclature "TMP" is used for simplicity, this term should be considered the functional equivalent of "NDP" when applied to NF/RO systems.

While the flux associated with MF, UF, and MCF systems is typically referenced to a temperature of 20 °C for the purposes of assessing operational performance, it is common to reference the flux associated with NF and RO systems to 25 °C (298 K). Accordingly, the appropriate TCF for NF and RO systems is shown in Equation 2.19:

$$TCF = \exp\left[U \bullet \left(\frac{1}{T + 273} - \frac{1}{298} \right) \right]$$

Equation 2.19

Where:

TCF	=	temperature correction factor (dimensionless)
T	=	water temperature (°C)
U	=	membrane-specific manufacturer-supplied constant (1/K)

Once the TCF has been determined, the flux normalized to 25 °C can be calculated according to Equation 2.20:

$$J_{25} = J_T \bullet (TCF)$$ Equation 2.20

Where: J_{25} =   normalized flux at 25 $^{\circ}$C (gfd)
                J_T =   actual flux at temperature T (gfd)
                TCF   =   temperature correction factor (dimensionless)

As with MF, UF, and MCF systems, it is important to note that the normalized flux (J_{25}) for NF and RO systems does not represent an actual operating condition. This term simply represents what the flux would be at 25 $^{\circ}$C for the purposes of comparing membrane performance independent of temperature-related affects. Similarly, it is also common to normalize the flux for pressure in order to identify changes in productivity that are attributable to fouling, as shown in Equation 2.21:

$$M_{25} = \frac{J_{25}}{NDP}$$ Equation 2.21

Where: M_{25} =   temperature- and pressure-normalized flux (gfd/psi)
                J_{25} =   normalized flux at 25 $^{\circ}$C (gfd)
                NDP   =   net driving pressure (psi)

2.5   Hydraulic Configurations

The term hydraulic configuration is used to describe the manner in which the feed water and associated suspended solids are processed by a membrane filtration system. Although there are a number of different hydraulic configurations in which the various membrane filtration systems can operate, each of these configurations can be categorized into one of two basic modes of operation: deposition mode and suspension mode. The hydraulic configuration of a system is determined from operational conditions such as backwash, concentrate flow, and recycle flow, where applicable.

For the purposes of the LT2ESWTR, one of the most important implications of a system's hydraulic configuration is its impact on the degree to which suspended solids are concentrated on the feed side of the membrane. This concentration effect is characterized by the volumetric concentration factor (VCF), a dimensionless parameter representing the ratio of the concentration of suspended solids on the feed side of the membrane relative to that of the influent feed to the membrane filtration process, as shown in Equation 2.22:

$$VCF = \frac{C_m}{C_f}$$

Equation 2.22

Where: VCF   =   volumetric concentration factor (dimensionless)

C_m =   concentration of suspended solids maintained on the feed side of the membrane (number or mass / volume)

C_f =   concentration of suspended solids in the influent feed water to the membrane system (number or mass / volume)

By definition, the VCF is equal to 1 for a system that does not concentrate suspended solids on the feed side of the membrane (i.e., $C_m = C_f$); these are defined as deposition mode systems. However, some hydraulic configurations concentrate suspended solids on the feed side of the membrane to degrees much greater than the influent feed concentration, with a corresponding VCF greater than one; these are the suspension mode systems. Consequently, the VCF can significantly affect the quantity of particulate matter that can flow through an integrity breach, and thus must be considered in the determination of direct integrity test method sensitivity, as discussed in section 4.3 (40 CFR 141.719(b)(3)(iii)(A)). For example, for the same size integrity breach, systems with higher VCFs will allow increased passage of pathogens to the filtrate. This effect is taken into account in the determination of sensitivity for pressure-based tests in that systems with higher VCFs have lower test sensitivity. Thus, even though systems with higher VCFs allow increased filtrate contamination for a given integrity breach, the lower sensitivity associated with a pressure-based direct integrity test means that the maximum permissible breach size is smaller. Note that the VCF does not factor directly into the methodology associated with marker-based tests, since the measurement of filtrate marker concentration inherently accounts for increases in pathogen passage resulting from any feed-side concentration effects.

The primary purposes of this section are to describe the various hydraulic configurations in which membrane filtration systems can operate and present the associated equations used to determine the respective VCFs. The discussion of the various hydraulic configurations in this section is predicated on three basic assumptions:

1. In the absence of a hydraulic force tangential to the membrane surface, particulate matter in the feed stream is deposited on the membrane and held in place by the TMP.

2. In the presence of a hydraulic force tangential to the membrane surface, significant particulate matter remains in suspension, resulting in elevated concentrations of suspended particulate matter on the feed side of the membrane. This increase in concentration is characterized by the VCF, which can vary as a function of position and/or time for various hydraulic configurations.

3. The membrane is a complete barrier to the passage of the particulate contaminants (assuming fully integral conditions).

Deposition mode systems and the various types of suspension mode systems are discussed in sections 2.5.1 and 2.5.2, respectively. Although many systems utilize one of the hydraulic configurations described in this section, there may be cases in which a system-specific analysis is necessary to characterize the VCF. For example, some systems that are designed to operate in deposition mode may still exhibit some degree of particle suspension, and conversely, some degree of particle deposition may occur in a system operating primarily in suspension mode. In such cases in which characteristics of both types of hydraulic configurations may be observed, the VCF should be calculated using the most conservative applicable assumptions that result in the highest anticipated VCF values (if mathematical models are used to determine the VCF). Note that the most conservative condition for a particular system is that which results in the highest estimated concentration of suspended particulate matter.

Some membrane filtration systems may not necessarily conform to the theoretical descriptions of any of the basic hydraulic configurations discussed in sections 2.5.1 and 2.5.2, but which may be operated in such a way as to largely emulate the particulate matter concentration profile of one of these configurations. Such alternative systems are addressed in section 2.5.3.

As an alternative to the theoretical calculations described in sections 2.5.1 and 2.5.2, the VCF may be measured experimentally under typical, full scale operating conditions. This approach may be advantageous for systems that may not necessarily be well described by mathematical modeling. The experimental evaluation of hydraulic configurations to determine the VCF is discussed in section 2.5.4.

In addition to both the theoretical modeling and experimental approaches, there may be some cases in which the manufacturer of a proprietary membrane filtration system has developed a system-specific method for determining the VCF. For these cases the manufacturer's methodology may be used, subject to State approval.

2.5.1 Deposition Mode

Membrane filtration systems operating in deposition mode utilize no concentrate stream such that there is only one influent (i.e., the feed) and one effluent (i.e., the filtrate) stream, as shown in the schematic in Figure 2.13. These systems are also commonly called "dead-end" or "direct" filtration systems and are analogous to conventional granular media filters in terms of hydraulic configuration. In the deposition mode of operation, contaminants suspended in the feed stream accumulate on the membrane surface and are held in place by hydraulic forces acting perpendicular to the membrane, forming a cake layer, as illustrated in Figure 2.14.

Figure 2.13: Schematic of a System Operating in Deposition Mode

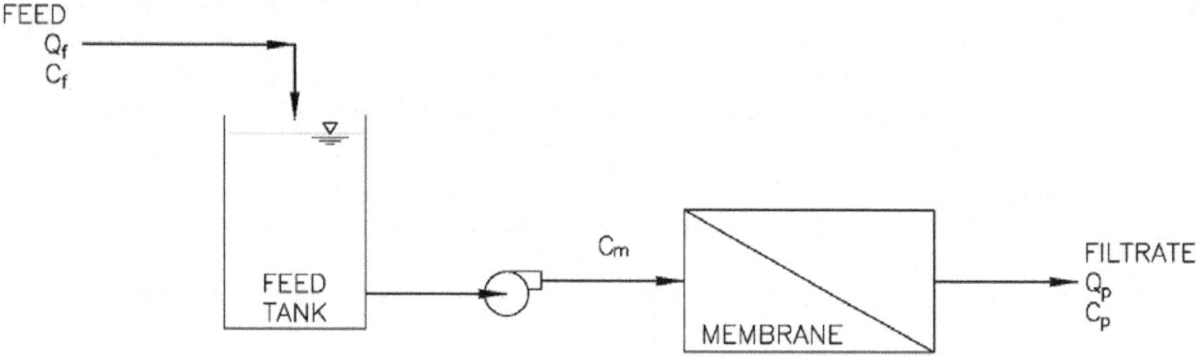

Figure 2.14 Conceptual Illustration of Deposition Mode Operation

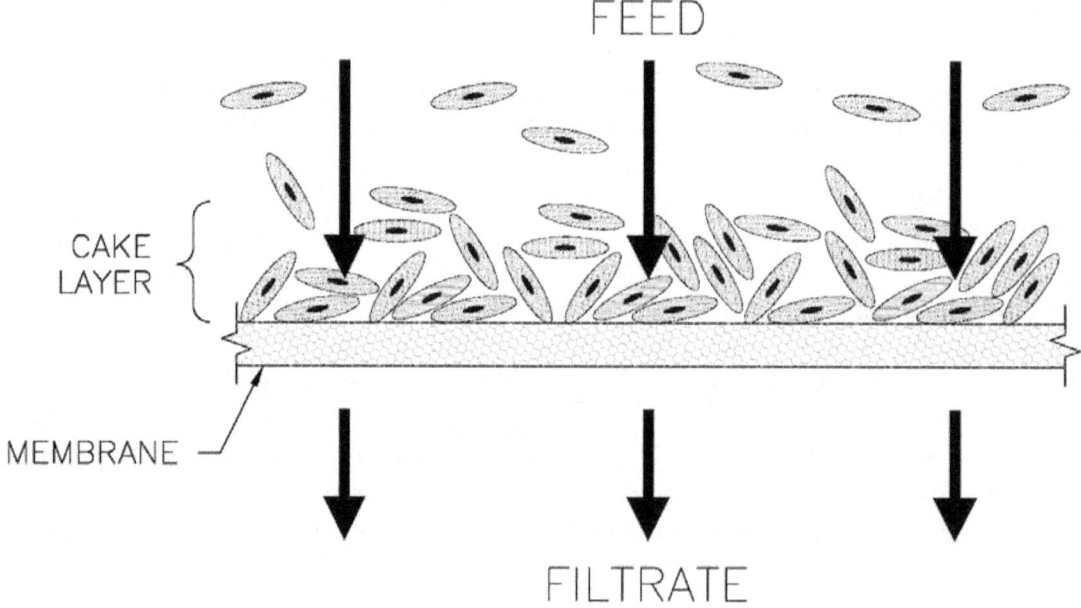

In a deposition mode hydraulic configuration, the concentration of suspended material on the feed side of the membrane (C_m) is assumed to be equivalent to the concentration of suspended material in the feed stream (C_f), independent of time or position in the membrane system, as the suspended contaminants are removed from the process stream and deposited in the accumulated cake layer. Therefore, all systems operating in deposition mode have a VCF equal to one. MCF and most hollow-fiber MF and UF systems operate in deposition mode. Typically, the accumulated solids are removed from MF/UF systems by backwashing, while most MCF systems simply operate until the accumulated solids reduce the flow and/or TMP to an unacceptable level, at which point the membrane cartridge is replaced.

Some MF/UF systems utilize a periodic "backpulse" – a short interval of reverse flow (which may include air and/or the addition of small doses of oxidants) designed to dislodge particles from the membrane surface without removing these solids from the system. This process re-suspends particles, effectively concentrating the suspended solids in the feed near the membrane surface and increasing the potential for pathogens or other particulate to pass through an integrity breach and contaminate the filtrate. Consequently, systems that do not utilize a concentrate stream but still practice backpulsing may be more appropriately and conservatively modeled as operating in suspension mode.

2.5.2 Suspension Mode

In membrane filtration systems that operate in suspension mode, a scouring force using water and/or air is applied parallel (i.e., tangential) to the membrane surface during the production of filtrate in a continuous or intermittent manner, as illustrated in Figure 2.15. The objective of operating in this mode is to minimize the accumulation of contaminants at the membrane surface or boundary layer, thus reducing fouling. As shown in Equation 2.22, the VCF quantifies the increase in the feed side concentration of suspended solids relative to that of the influent feed stream that occurs in a suspension mode of operation.

Figure 2.15 Conceptual Illustration of Suspension Mode Operation

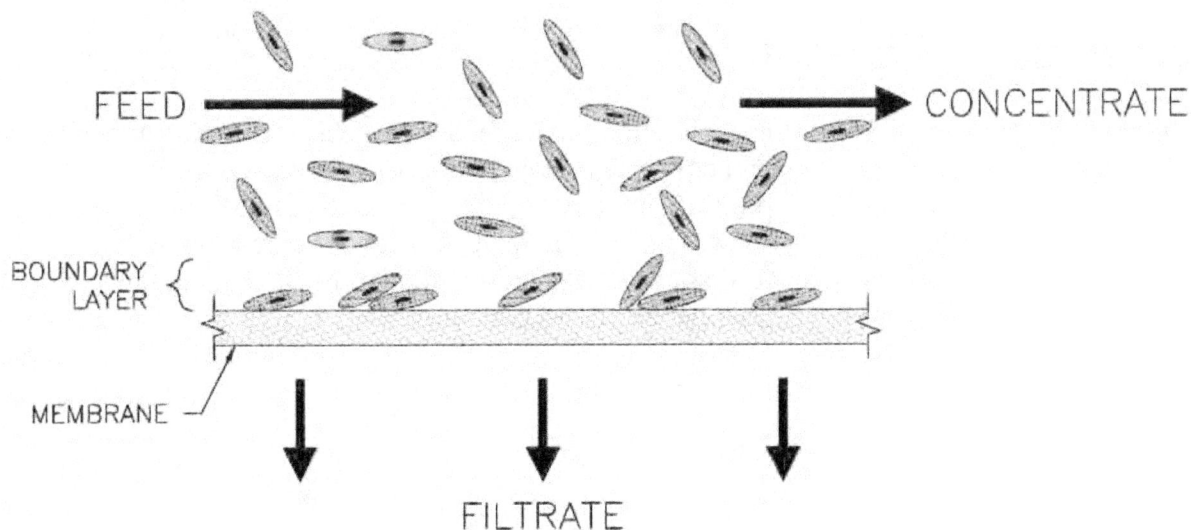

The three most common suspension mode hydraulic configurations are the plug flow reactor (PFR) model, the crossflow model, and the continuous stirred tank reactor (CSTR) model. Systems operating under a crossflow hydraulic configuration may be further categorized as either small volume or large volume systems, since the volume affects the manner in which suspended solids are concentrated in such a system. Depending on the particular hydraulic configuration, the VCF may vary temporally or spatially. In a PFR, the VCF increases in the direction of feed flow as a function of position in the system; however, in a CSTR or crossflow reactor, the VCF increases with time over the course of a filtration cycle. Theoretical methods for calculating both the average and maximum VCF are presented for PFR, crossflow, and CSTR systems as discussed in the following subsections. Note that the mechanism for achieving suspension of particulate matter (e.g., aeration, velocity of water tangential to the membrane surface, etc.) may not be sufficient to produce the suspended solids concentrations predicted by the theoretical models presented in this section. Therefore, experimental determination of the VCF may be appropriate, as described in section 2.5.4.

2.5.2.1 Plug Flow Reactor Model

Membrane filtration systems that operate as PFRs have a VCF that varies as a function of position in the system. Examples of such systems include spiral-wound NF/RO systems and vacuum-driven MF/UF systems submerged in tanks with large length-to-width ratios. The concentration profile of these two types of PFR systems is illustrated in Figure 2.16.

The concentration of suspended solids on the feed side of the membrane in a PFR can be expressed as a function of the recovery at any position within the system, as shown in Equation 2.23:

$$C_m(x) = \frac{C_f}{1 - R(x)}$$

Equation 2.23

Where:

$C_m(x)$ = concentration on the feed side of the membrane at position x in the membrane unit (number or mass / volume)

x = position in the membrane unit in the direction of tangential flow (i.e., $x = 0$ at the entrance to the first module)

C_f = feed concentration at the inlet to the membrane unit (number or mass / volume)

$R(x)$ = recovery as a function of position within the membrane unit (decimal percent)

Figure 2.16 Flow Diagram for a Plug Flow Reactor

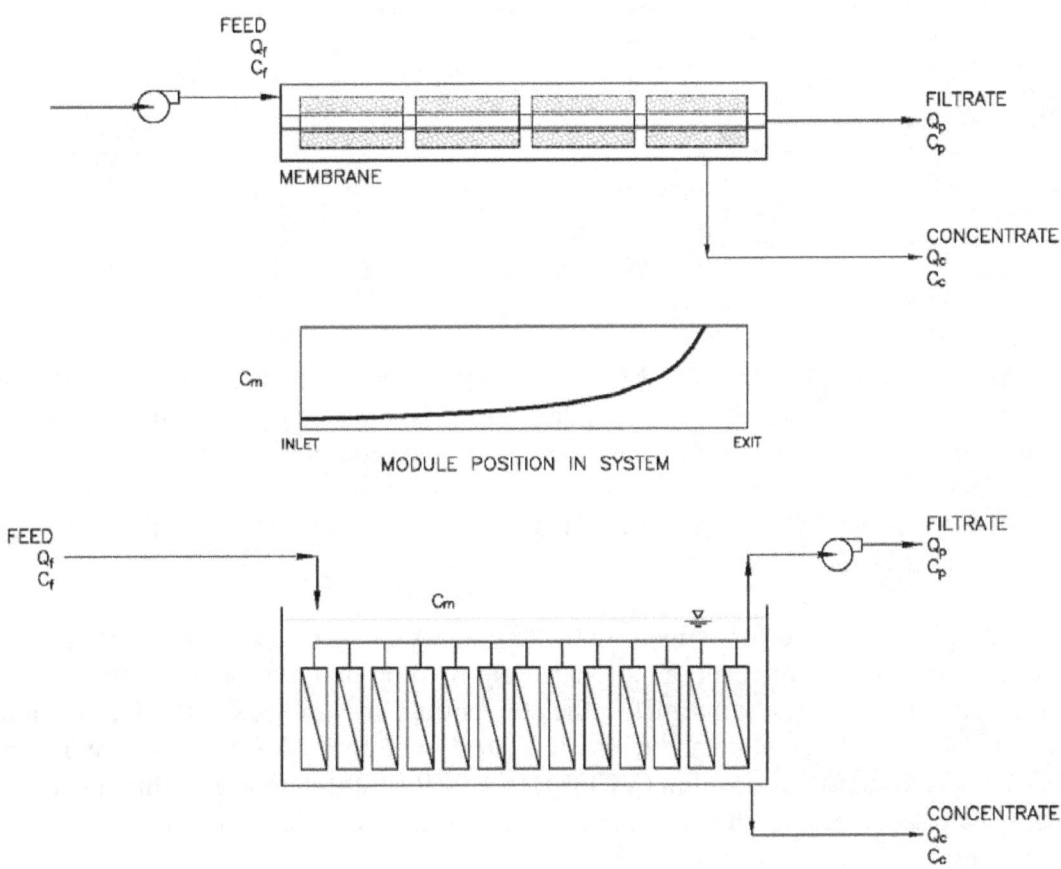

Equations 2.22 and 2.23 can be combined to create an expression for the VCF as a function of position in a PFR, as shown in Equation 2.24:

$$VCF(x) = \frac{1}{1 - R(x)}$$

Equation 2.24

Where: VCF(x) = VCF as a function of position in the membrane unit (dimensionless)

    x = position in the membrane unit in the direction of tangential flow (i.e., $x = 0$ at the entrance to the first module)

    R(x) = recovery as a function of position within the membrane unit (decimal percent)

The recovery in a PFR varies from zero at the inlet to the membrane unit to the value of the overall unit recovery at the outlet of the tangential flow stream in the membrane unit. Accordingly, between these inlet and outlet boundary conditions, the VCF increases from 1.0 at the system inlet (where the recovery is equal to zero), to a maximum value (VCF_{max}) at the concentrate outlet of the membrane unit, as shown in Equation 2.25:

$$VCF_{max} = \frac{1}{1-R}$$ Equation 2.25

Where $\quad$ VCF_{max} $\quad$ = $\quad$ maximum VCF (dimensionless)
$\qquad\qquad$ R $\qquad$ = $\quad$ membrane unit recovery (decimal percent)

The value of the VCF yielded by Equation 2.25 represents the largest (and thus most conservative) such value in a PFR. Suspended particulate matter is concentrated to this degree only at the far end of the unit (in the direction of flow) at the point of the tangential flow stream outlet; the rest of the membrane unit may be characterized by a lower VCF. Thus, it may be appropriate to consider an average value of the VCF that is more representative of concentrations over the membrane unit as a whole.

A simplistic method for estimating the average VCF (VCF_{avg}) is to simply divide the maximum VCF by two. However, a more accurate method involves calculating a position-averaged VCF(x) from the system inlet to the concentrate stream outlet. If a mathematical expression for R(x) is known, it can be inserted into Equation 2.24 to yield a known expression for the VCF as a function of position (VCF(x)) in a PFR. Subsequently, if this resulting expression can be integrated, then the average value function can be applied to determine an accurate value for VCF_{avg}, as shown in Equation 2.26:

$$VCF_{avg} = \frac{1}{x_{max}} \bullet \int_{0}^{x_{max}} \frac{1}{1-R(x)}(dx)$$ Equation 2.26

Where: $\quad$ VCF_{avg} $\quad$ = $\quad$ average VCF (dimensionless)
$\qquad\qquad$ x_{max} $\qquad$ = $\quad$ end position of the membrane unit in the direction of tangential flow (i.e., at the outlet of the tangential flow stream)
$\qquad\qquad$ R(x) $\qquad$ = $\quad$ recovery as a function of position within the membrane unit (decimal percent)
$\qquad\qquad$ x $\qquad\qquad$ = $\quad$ position in the membrane unit in the direction of tangential flow (i.e., $x = 0$ at the entrance to the first module)

The parameter R(x) in Equation 2.24 is typically a function of pressure, membrane area, and flow, all of which may vary as a function of position within the membrane unit. For example, NF and RO systems experience a decrease in NDP due to pressure losses through the system and increasing osmotic pressure gradients. For MF and UF membrane systems operating as PFRs, there may also be a flow imbalance across the membrane surface in the direction of flow resulting from the increasing concentration of suspended solids, which may increase resistance to the flow of water through the membrane (i.e., the filtrate flow) toward the outlet the tangential flow stream. In addition, the back-mixing of suspended solids can also create non-ideal flow and concentration conditions (Côté et al. 2003). However, in general, the VCF increases exponentially in the direction of the tangential flow stream in a PFR.

If an expression for the VCF as a function of position (VCF(x)) cannot be developed or if the resulting equation cannot be integrated, a flow-weighted average can be computed using numerical techniques, as shown in Equation 2.27:

$$VCF_{avg} = \frac{\sum[VCF(x) \bullet Q_p(x)]}{Q_p} \qquad \text{Equation 2.27}$$

Where:

VCF_{avg}	=	(flow-weighted) average VCF (dimensionless)
$VCF(x)$	=	VCF at discrete position "x" within the membrane unit (dimensionless)
x	=	position in the membrane unit in the direction of tangential flow (i.e., $x = 0$ at the entrance to the first module)
$Q_p(x)$	=	filtrate flow at position "x" within the membrane unit (gpd)
Q_p	=	total filtrate flow from the membrane unit (gpd)

The use of Equation 2.27 first requires determination of the filtrate flow over a differential membrane area in the unit at a specific location. The resulting filtrate flow at position "x" ($Q_p(x)$) can be used to calculate the recovery at position "x" ($R(x)$) within the membrane unit using Equation 2.2 (see section 2.4). The VCF at position "x" can then be calculated using Equation 2.24. These calculation steps are repeated for numerous positions throughout the membrane unit, and a flow-weighted average is computed from this data using Equation 2.27. Typically, values for the filtrate flow at discrete intervals within a membrane unit may be obtained from modeling software available from the membrane module manufacturer, and the various calculations can be facilitated by the use of a spreadsheet.

Both the maximum and flow-weighted average VCFs (as calculated using Equations 2.25 and 2.27, respectively) for a typical NF/RO system over a range of overall system recoveries from 70 to 99 percent are summarized in Table 2.1.

Table 2.1 shows that the maximum VCF is substantially higher than the average VCF, particularly for higher recoveries. For example, a NF/RO system operating at an overall recovery of 85 percent has a maximum VCF approximately three times larger than its average VCF. Thus, the concentration of suspended particulate matter at the outlet of the tangential flow stream in the membrane unit is similarly three times higher than the average concentration. As a result, a membrane breach located close to the outlet of the tangential flow stream in membrane unit operating as a PFR has the potential to allow a significantly larger concentration of contaminants to pass into the filtrate than a breach occurring at a location further upstream.

Table 2.1 Summary of VCF Values for a NF/RO System Modeled as a PFR

Recovery (percent)	VCF		Recovery (percent)	VCF		Recovery (percent)	VCF	
	Avg.[1]	Max.		Avg.[1]	Max.		Avg.[1]	Max.
70	1.73	3.33	80	2.02	5.0	90	2.56	10.0
71	1.75	3.45	81	2.06	5.26	91	2.65	11.1
72	1.78	3.57	82	2.11	5.56	92	2.75	12.5
73	1.80	3.70	83	2.15	5.88	93	2.87	14.3
74	1.83	3.84	84	2.20	6.25	94	3.00	16.7
75	1.86	4.0	85	2.25	6.67	95	3.16	20.0
76	1.89	4.17	86	2.31	7.14	96	3.37	25.0
77	1.92	4.35	87	2.37	7.69	97	3.63	33.3
78	1.95	4.55	88	2.43	8.33	98	4.01	50
79	1.99	4.76	89	2.49	9.09	99	4.68	100

1 Ideal, theoretical average

If information regarding the filtrate flow as a function of membrane position cannot be obtained or approximated, the VCF data presented in Table 2.1 may be used as an estimate for NF/RO systems. However, it is strongly recommended that Equation 2.24, 2.26, or 2.27 be used to obtain a more accurate, system-specific estimate for the average VCF. Alternatively, the maximum VCF (i.e., the most conservative value) can be calculated from the overall recovery using Equation 2.25.

2.5.2.2 Crossflow Model

The objective of crossflow filtration is to maintain a high scour velocity across the membrane surface to minimize particle deposition and membrane fouling. Crossflow membrane processes operate in an unsteady-state manner in which suspended solids accumulate on the feed side of the membrane over the course of a filtration cycle. Thus, in crossflow systems, the VCF varies as a function of time. At the end of each filtration cycle the membrane unit is backwashed to remove the accumulated solids. Crossflow filtration has traditionally been used in conjunction with inside-out hollow-fiber membrane processes to increase the scouring velocity in the fiber lumen in order to minimize fouling.

For crossflow systems, a portion of the concentrate stream (i.e., the tangential flow) is recirculated or recycled to the system inlet and mixed with the incoming feed stream. Because the concentrate stream has a higher concentration of suspended solids than the influent feed stream to the membrane process, the VCF for crossflow systems is greater than one. Although the recycled concentrate stream in a crossflow system may be less than 10 percent of the combined feed flow in some cases, it is more typically 5 to 20 times higher than the influent feed flow. The manner in which crossflow systems concentrate suspended particulate matter depends on the volume of the feed side of a membrane unit between the point at which the recirculated stream joins the influent flow and the inlet to the membrane module(s). For simplicity, two variations of the crossflow configuration are considered in this discussion: small volume and large volume systems.

Small Volume Crossflow Systems

In a small volume crossflow system, the unfiltered concentrate stream is returned to the inlet of the membrane system and blended with the incoming feed after any prefiltration that may be incorporated into the treatment process. The line that connects the concentrate outlet to the feed inlet is termed the recirculation loop. A schematic of a typical small volume crossflow system configuration is shown in Figure 2.17. The small volume crossflow configuration, in which the contaminants accumulated during a filtration cycle are removed via backwashing, is relatively common among hollow-fiber MF/UF membrane systems.

Figure 2.17 Schematic of a Typical Small Volume Crossflow System

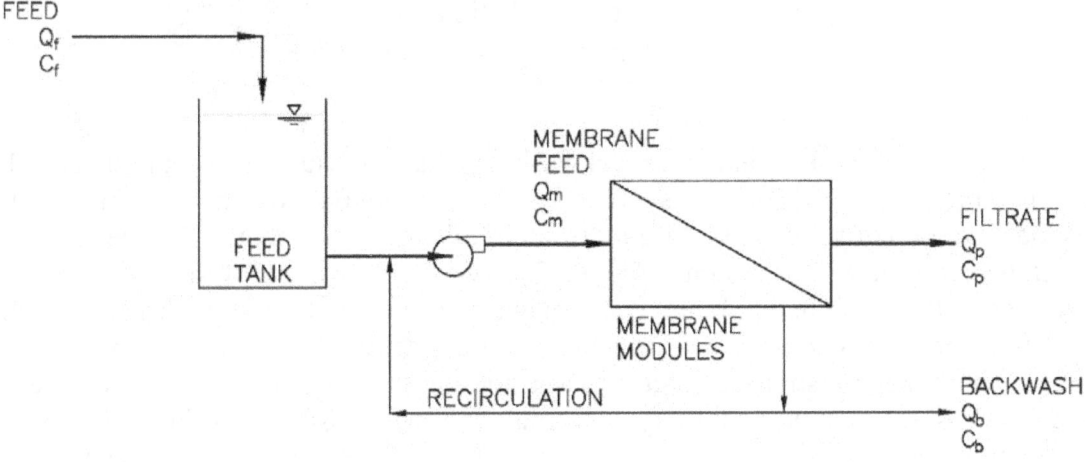

In crossflow systems, the concentration of suspended solids on the feed side of the membrane (C_m in Figure 2.17) increases linearly over the filtration cycle. The rate at which the concentration increases is a function of the feed flow (Q_f in Figure 2.17) and volume of the recirculation loop (V_r). As shown in Figure 2.17, this volume would include the volume on the feed side of the membrane modules back to the recirculation blending point, as well as that of the recirculation loop piping.

In analyzing such systems, it is useful to define the volumetric turnover time (τ), which represents the time required for a one-fold increase in the concentration of suspended particulate matter on the feed side of the membrane. The volumetric turnover time is calculated as the ratio of the total volume of the recirculation loop to the feed flow, as shown in Equation 2.28:

$$\tau = \frac{V_r}{Q_f}$$
Equation 2.28

Where:
τ = volumetric turnover time (min)
V_r = total volume of the recirculation loop (gallons)
Q_f = feed flow (gpm)

The VCF is a function of the volumetric turnover time and the operational time within a filtration cycle, as shown in Equation 2.29:

$$VCF(t) = \frac{t}{\tau}$$
Equation 2.29

Where:
$VCF(t)$ = VCF as a function of filtration cycle time (dimensionless)
t = filtration cycle time (min)
τ = volumetric turnover time (min)

At the end of the filtration cycle, backwashing is used to flush the accumulated solids from the membrane system. Since most crossflow systems are backwashed with filtrate water, and the concentration of solids in the filtrate is negligible relative to that of the feed and recirculation streams, it can be assumed that the backwash removes all of the solids from the entire volume of the recirculation loop. Consequently, the concentration of suspended solids at the membrane surface is equal to zero immediately after backwash (i.e., at $t = 0$, both $C_m = 0$ and $VCF = 0$). Figure 2.18 illustrates the periodic variation in the concentration of suspended solids on the feed side of the membrane for a typical small volume crossflow system with a 20-minute filtration cycle. (Note that the filtration cycle time depicted in Figure 2.18 represents an arbitrarily selected example; actual times could vary widely from system to system.)

As shown in Figure 2.18, for a small volume crossflow system, the maximum VCF occurs at the end of the filtration cycle (i.e., $t = t_f$) and is calculated according to Equation 2.30.

$$VCF_{max} = \frac{t_f}{\tau}$$

Equation 2.30

Where:

VCF$_{max}$ = maximum VCF (dimensionless)
t$_f$ = filtration cycle duration (min)
τ = volumetric turnover time (min)

Figure 2.18 Concentration Profile in a Small Volume Crossflow System

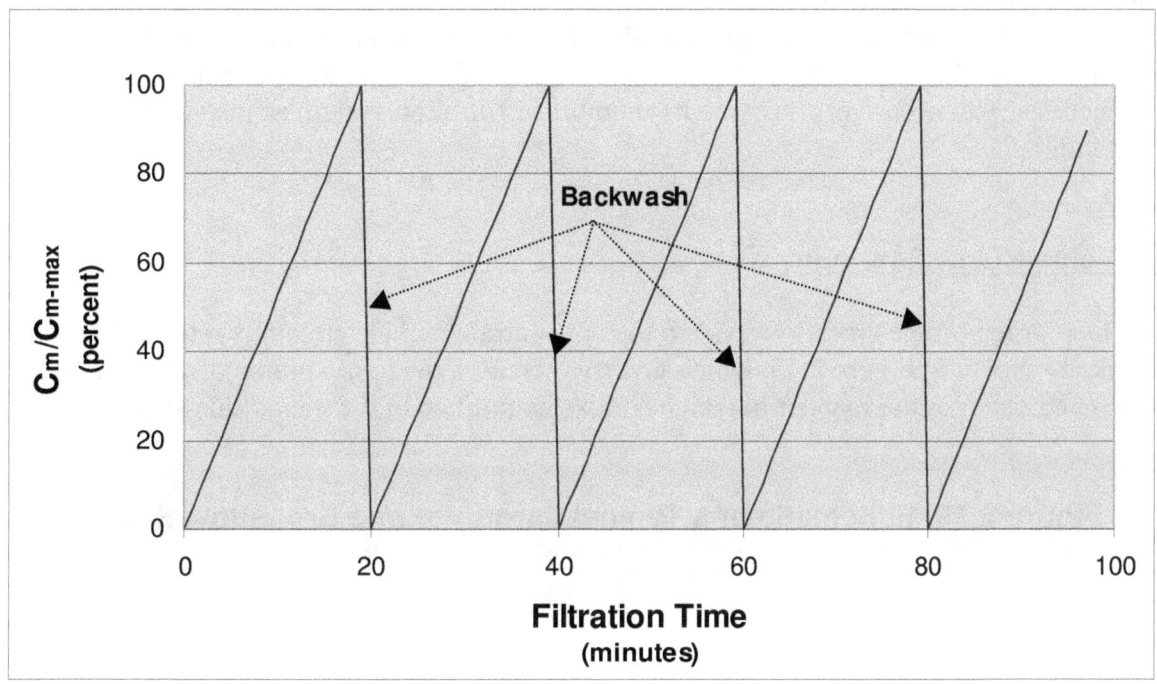

Since the VCF increases linearly from zero to VCF$_{max}$ over the course of a filtration cycle in a crossflow system, the average VCF is simply one half of the maximum value, as shown in Equation 2.31:

$$VCF_{avg} = 0.5 \bullet \left(\frac{t_f}{\tau} \right)$$

Equation 2.31

Where:

VCF$_{avg}$ = average VCF (dimensionless)
t$_f$ = filtration cycle duration (min)
τ = volumetric turnover time (min)

Note that some small volume crossflow systems may be operated at lower crossflow velocities, resulting in incomplete particle suspension. In these cases the mathematical model may overestimate the VCF. While this calculated VCF would represent a conservative estimate, the VCF could also be measured experimentally (as discussed in section 2.5.4) under realistic operating conditions in order to determine a more accurate value. Also, this simple small volume crossflow model assumes that the concentration of suspended solids maintained on the feed side of the membrane (C_m) reverts to zero following a backwash just prior to beginning a new filtration cycle; however, this assumption may not be applicable to all such membrane filtration systems. For example, one potentially common variant is the case in which the concentration of suspended solids maintained on the feed side of the membrane reverts to that of the influent feed water following a backwash (i.e., $C_m = C_f$), such that the VCF is equal to 1 at the beginning of the subsequent filtration cycle. The manufacturer should be able to provide information about the applicability of this model and whether some adjustments to the model may be necessary to accurately describe its membrane filtration system as installed and operated at a particular site.

Large Volume Crossflow Systems

In a large volume crossflow system the concentrate recycle stream is returned to a large feed tank, as shown in Figure 2.19, which greatly increases the total volume of the recirculation loop (V_r) and changes the manner in which solids accumulate in the membrane system.

Figure 2.19 Schematic of a Typical Large Volume Crossflow System

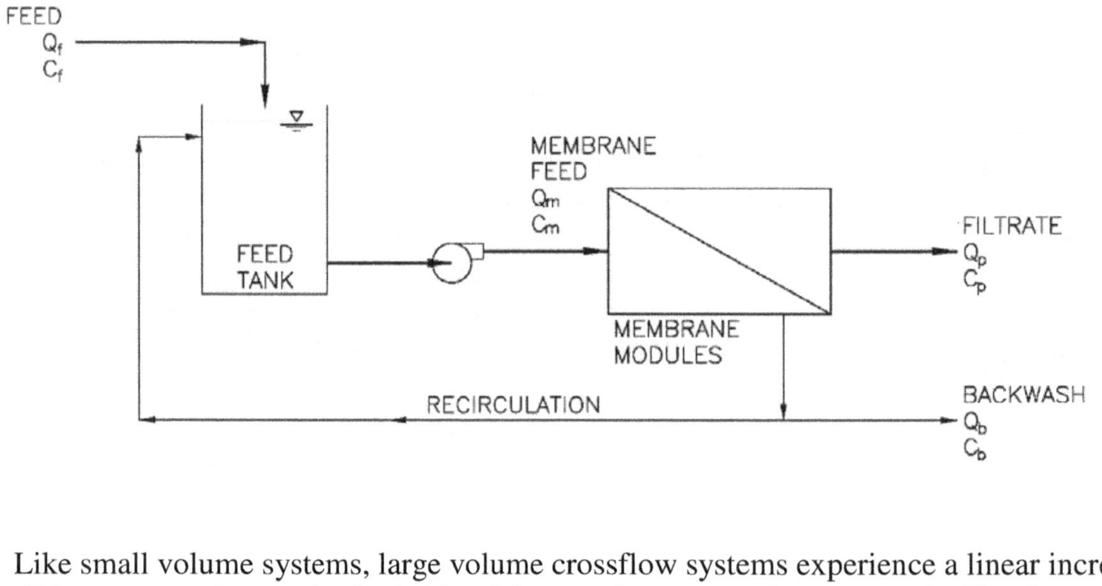

Like small volume systems, large volume crossflow systems experience a linear increase in the solids concentration on the feed side of the membrane over the course of a filtration cycle. However, in a large volume system, backwashing does not remove all of the contaminants that have accumulated in the recirculation loop and on the feed side of the membrane; thus concentration on the feed side of the membrane (C_m) and the VCF do not return to zero after a backwash operation. As a result, the solids concentration gradually increases over multiple filtration cycles, a phenomenon which must be considered when calculating the VCF. A periodic decrease in concentration occurs during a backwash operation when a portion of the suspended solids is removed. The suspended particulate matter continues to accumulate over successive filtration cycles until a stable process condition is reached, in which the amount of solids delivered to the membrane system during a filtration cycle is equal to the amount of solids that are discharged during the backwash process. The "saw-tooth pattern" concentration profile and the gradual establishment of a stable process condition for a large volume crossflow system are illustrated in Figure 2.20.

Figure 2.20 Concentration Profile in a Large Volume Crossflow System

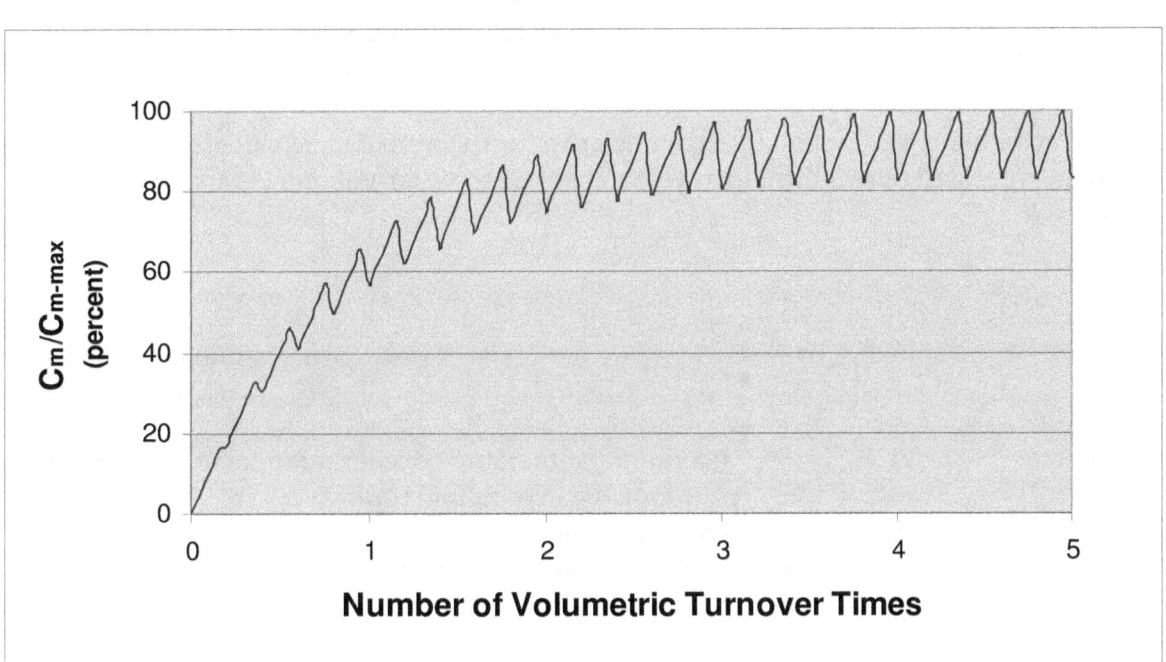

Equation 2.32 describes VCF as a function of time for a large volume crossflow system, taking into account both the buildup of solids over a filtration cycle and the partial removal of solids during backwash:

$$VCF(t) = \frac{t}{\tau} + \frac{C_m(t-1)}{C_f} \bullet \left(1 - \frac{Q_b \bullet t_b}{V_r}\right)$$

Equation 2.32

Where:

$VCF(t)$	=	VCF as a function of filtration cycle time (dimensionless)
t	=	filtration cycle time (min)
τ	=	volumetric turnover time (min)
$C_m(t-1)$	=	concentration on the feed side of the membrane immediately after the previous backwash operation (number or mass / volume)
C_f	=	feed concentration at the inlet to the membrane unit (number or mass / volume)
Q_b	=	backwash flow (gpm)
t_b	=	backwash duration (min)
V_r	=	total volume of the recirculation loop (gallons)

The maximum VCF for a large volume crossflow system can be calculated as the ratio of feed volume processed over a filtration cycle to backwash water volume, as shown in Equation 2.33:

$$VCF_{max} = \frac{t_f \bullet Q_f}{t_b \bullet Q_b}$$

Equation 2.33

Where:

VCF_{max}	=	maximum volumetric concentration factor (dimensionless)
t_f	=	filtration cycle duration (min)
Q_f	=	feed flow (gpm)
t_b	=	backwash duration (min)
Q_b	=	backwash flow (gpm)

The average VCF for a large volume crossflow system can be calculated using Equation 2.34:

$$VCF_{avg} = \frac{t_f \bullet Q_f}{t_b \bullet Q_b} \bullet \left[1 - \frac{t_f}{2 \bullet \tau \bullet \dfrac{t_f \bullet Q_f}{t_b \bullet Q_b}} \right]$$

Equation 2.34

Where:

VCF_{avg}	=	average VCF (dimensionless)
t_f	=	filtration cycle duration (min)
Q_f	=	feed flow (gpm)
t_b	=	backwash duration (min)
Q_b	=	backwash flow (gpm)
τ	=	volumetric turnover time (min)

2.5.2.3 Continuous Stirred Tank Reactor Model

The CSTR (also known as "feed-and-bleed" in some applications) hydraulic configuration is similar to that of a crossflow system in that the particulate matter is held in suspension and increases in concentration on the feed side of the membrane as a function of time. However, the CSTR incorporates a continuous concentrate waste stream (also referred to as the reject or bleed stream) that removes suspended solids from the system. Since the solids are continuously removed from the system, steady-state operation is achieved when the rate at which solids are removed with the concentrate stream is equal to the rate at which solids enter the system with the feed water. Although a PFR also utilizes a continuous concentrate stream and operates at steady-state, the CSTR model assumes complete mixing and thus a uniform concentration of suspended solids on the feed side of the membrane. A PFR, by contrast, has a feed side concentration profile that increases in the direction of feed flow through the membrane unit. Thus, the CSTR model is more appropriate than the PFR model for systems that are expected to have a relatively uniform suspended solids concentration on the feed side of the membrane.

CSTR theory generally describes systems that operate continuously without backwashing. For example, some submerged MF and UF systems may utilize a periodic short duration reverse flow operation (i.e., a backpulse) to remove solids from the membrane surface but collect the dislodged solids in the feed tank. Since this operation does not constitute backwashing (i.e., removal of solids from the system), these MF and UF systems can be modeled as CSTRs without backwashing. In addition, a small-scale NF or RO membrane unit that is too small to be considered a PFR (e.g., a single stage system using a high recirculation flow or a single module system) can also be modeled as CSTR that does not utilize backwashing. However, there may be types of membrane filtration systems that can be effectively modeled as a CSTR with backwashing, such as a small, submerged vacuum-driven membrane system in which the backwash water is removed from the tank after the backwash operation. Thus, CSTR models both with and without backwashing, respectively, are described as follows.

CSTR Without Backwashing

Figures 2.21 and 2.22 provide representative schematics illustrating the types of pressure- and vacuum-driven membrane filtration systems, respectively, that can be modeled as CSTRs without backwashing. Note that a submerged, vacuum-driven system with mechanical agitation (e.g., due to aeration) can be modeled as either a CSTR without backwashing or a PFR, depending on system specific factors such as the number of modules and/or the length to width ratio of the basin housing the membrane modules (as applicable). Systems with a very small number of modules and/or a relatively small length to width ratio may be modeled as a CSTR without backwashing, as shown in Figure 2.22. However, vacuum-driven systems with a larger number of modules and/or length to width ratio may be more accurately modeled as a PFR, as shown in Figure 2.16. Alternatively, the VCF may be measured experimentally (as described in section 2.5.4) if neither model is sufficient to describe the system within an acceptable degree of accuracy.

Figure 2.21 Schematic of a Typical Pressure-Driven CSTR Without Backwashing

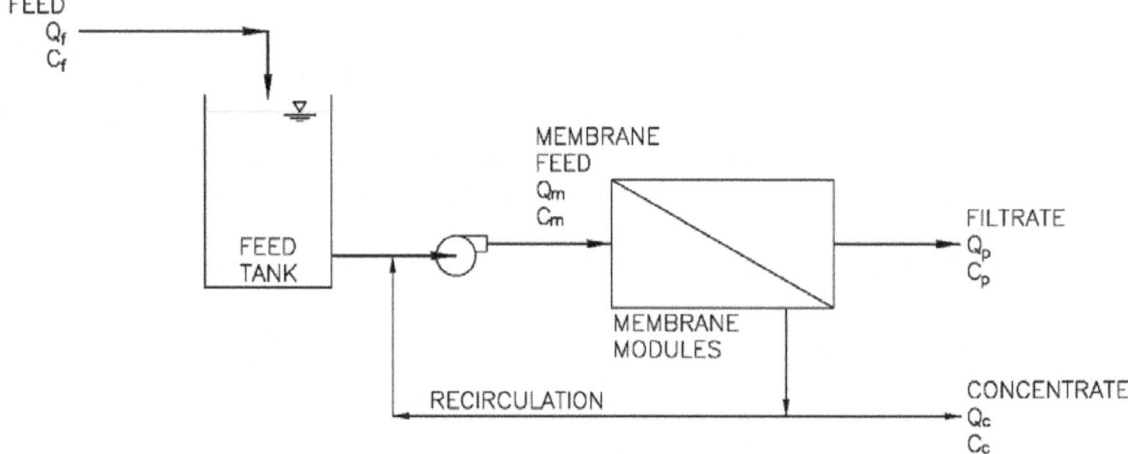

Figure 2.22 Schematic of a Typical Vacuum-Driven CSTR Without Backwashing

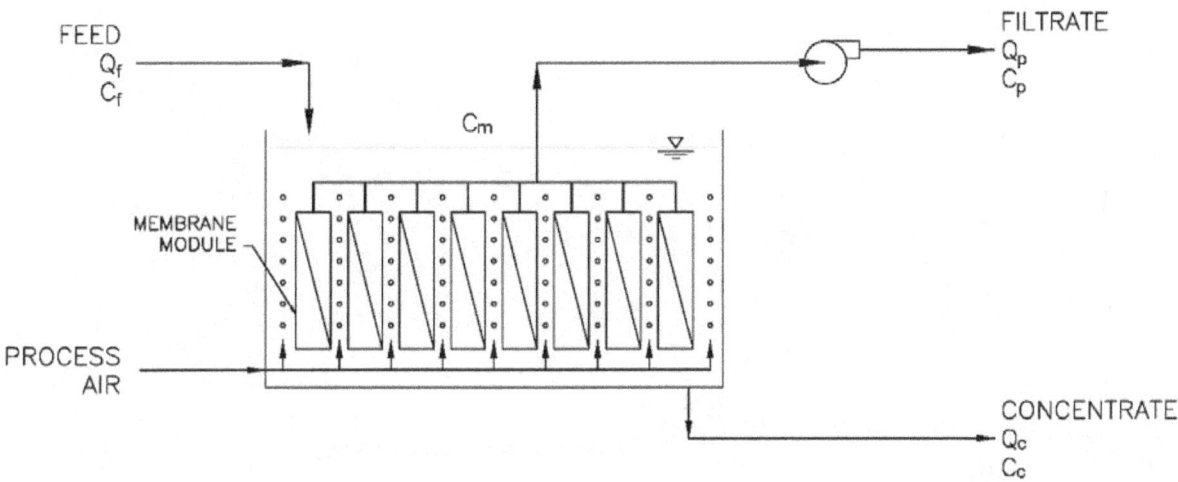

The VCF for a CSTR without backwashing increases as a function of filtration cycle time and can be defined in terms of the number of volumetric turnover times (i.e., t/τ) and the recovery, as shown in Equation 2.35:

$$VCF(t) = \left[\frac{1}{1-R}\right] \bullet \left[1 - \exp\left(\frac{-t}{\tau}\right)\right]$$

Equation 2.35

Where:

VCF(t)	=	VCF as a function of filtration cycle time (dimensionless)
R	=	recovery (dimensionless decimal)
t	=	filtration cycle time (min)
τ	=	volumetric turnover time (min)

The concentration profile for a CSTR without backwash exhibits an exponential rise in the VCF over time to an equilibrium value of VCF_{max}, as illustrated in Figure 2.23. Additional detail is provided in Table 2.2, which shows the percent of equilibrium value (i.e., VCF_{max}) that is achieved in a CSTR without backwashing as a function of the number of volumetric turnover times. In general, a CSTR without backwashing approaches equilibrium rapidly – Table 2.2 shows that the concentration of suspended solids within a CSTR without backwashing reaches 95 percent of the equilibrium value after only three volumetric turnover times have elapsed.

Figure 2.23 Concentration Profile in a CSTR Without Backwashing

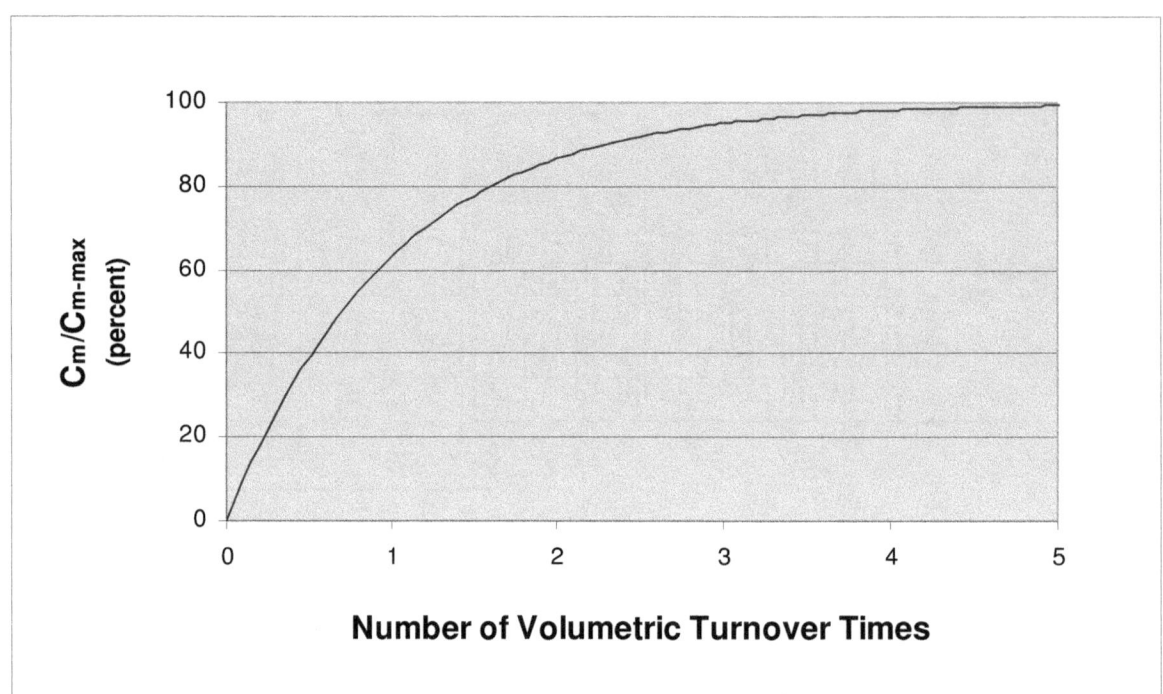

Table 2.2 Increase in VCF as a Function of Number of Turnover Times

No. of Turnover Times (t / τ)	Fraction of VCF$_{max}$
0	0
1	.632
2	.865
3	.950
4	.982
5	.993

Since most systems are operated significantly longer than three volumetric turnover times before being purged and cleaned, a CSTR without backwashing generally operates under steady-state conditions at the maximum VCF. Consequently, the average and maximum VCF are equivalent, and both can be calculated as shown in Equation 2.36:

$$VCF_{max} = VCF_{avg} = \frac{1}{1-R}$$

Equation 2.36

Where: VCF_{max} =   maximum VCF (dimensionless)
         VCF_{avg} =   average VCF (dimensionless)
         R   =   recovery (decimal percent)

CSTR With Backwashing

Figure 2.24 shows a schematic of a representative pressure-driven membrane system that can be modeled as a CSTR with periodic backwashing. Note that the concentration profile for a CSTR with backwashing is similar to that for a small volume crossflow system, although with the latter the maximum VCF is approached linearly rather than exponentially. Another important difference between these two types of hydraulic configurations is that the CSTR with backwashing utilizes a concentrate waste (i.e., bleed) stream, whereas a crossflow system does not use a concentrate bleed.

Figure 2.24 Schematic of a Typical Pressure-Driven CSTR With Backwashing

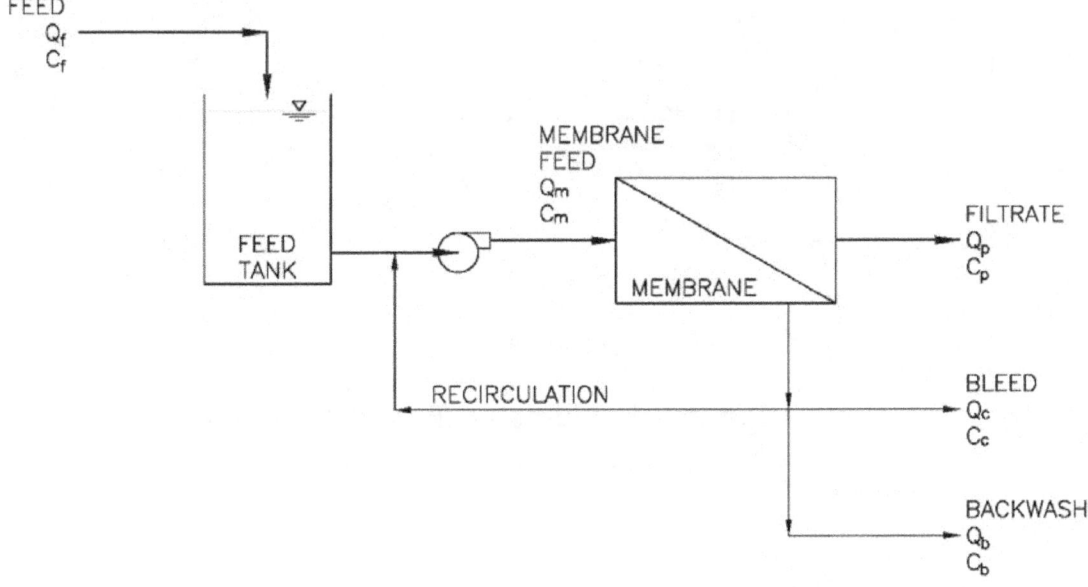

As with a CSTR without backwashing, the VCF for a CSTR with backwashing increases over time according to the first order exponential function given in Equation 2.35. However, in a CSTR with backwashing, the concentration of suspended solids on the feed side of the membrane is periodically reduced. Because filtrate is typically used in the backwash process, the backwash water introduces a negligible amount of particulate matter into the system; thus, it is reasonable to assume that the backwash process removes all suspended solids from the feed side of the membrane. As a result, the VCF is reduced to zero at beginning of each successive filtration cycle (i.e., at t = 0). However, as discussed for the small volume crossflow model, this assumption may not be applicable in all cases. As previously noted, one potentially common variant is the case in which the concentration of suspended solids maintained on the feed side of the membrane reverts to that of the influent feed water following a backwash (i.e., $C_m = C_f$), such that the VCF is equal to one at the beginning of the subsequent filtration cycle. The manufacturer should be consulted in regard to the applicability of this model and whether some adjustments to the model may be necessary to accurately describe its membrane filtration system as installed and operated at a particular site. Note that the VCF can also be measured experimentally, as discussed in section 2.5.4.

Figure 2.25 illustrates the concentration profile of a CSTR with backwashing over time for a system that is backwash once every four volumetric turnover times.

Figure 2.25 Concentration Profile in a CSTR With Backwashing

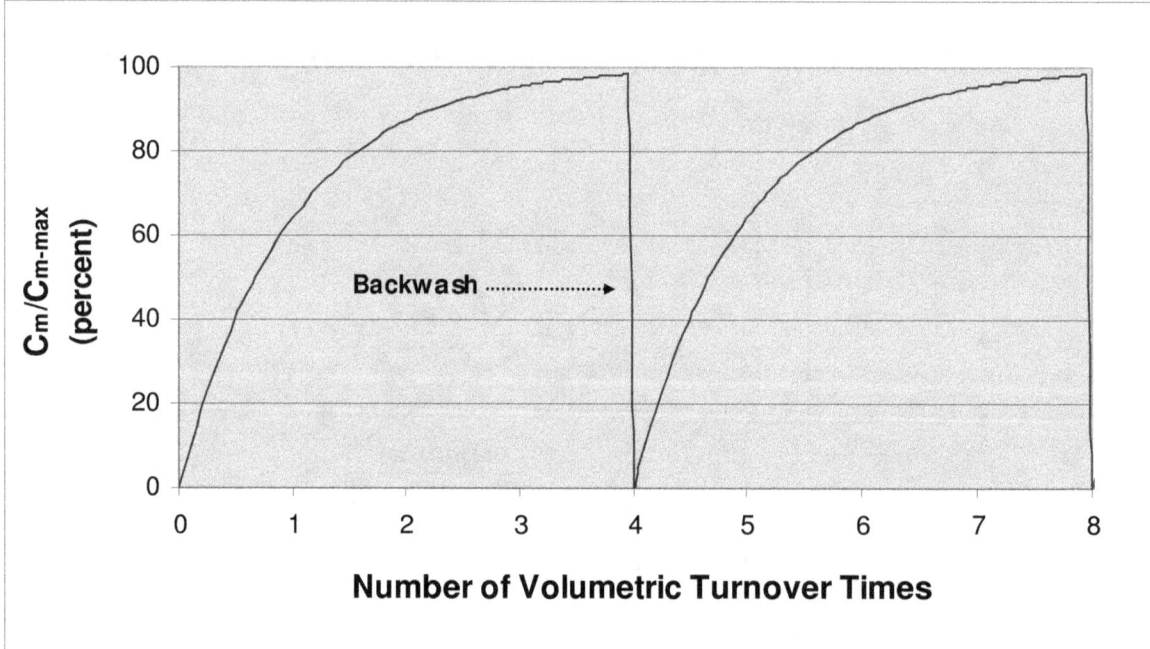

The maximum VCF for a CSTR with backwashing is determined by evaluating Equation 2.35 at the end of the filtration cycle (i.e., $t = t_f$), just prior to backwashing, as shown in Equation 2.37:

$$VCF_{max} = \left[\frac{1}{1-R}\right] \bullet \left[1 - \exp\left(\frac{-t_f}{\tau}\right)\right]$$

Equation 2.37

Where:

VCF_{max}	=	maximum VCF (dimensionless)
R	=	recovery (decimal percent)
t_f	=	filtration cycle duration (min)
τ	=	volumetric turnover time (min)

The average VCF for a CSTR with backwashing is determined by applying the mathematical average value function to Equation 2.37 over a filtration cycle. The resulting expression for VCF_{avg} is shown in Equation 2.38:

$$VCF_{avg} = \frac{1}{1-R} \bullet \left[1 - \left(\frac{\tau}{t_f} \bullet \left[1 - \exp\left(\frac{-t_f}{\tau}\right)\right]\right)\right]$$

Equation 2.38

Where:

VCF_{avg}	=	average VCF (dimensionless)
R	=	water recovery (decimal percent)
t_f	=	duration of the filtration cycle (min)
τ	=	volumetric turnover time (min)

2.5.3 Alternative Configurations

Sections 2.5.1 and 2.5.2 discuss the basic theoretical hydraulic configurations for deposition and suspension mode operation, respectively, including mathematical modeling to determine the VCF and examples of the most common types of currently available membrane filtration systems associated with each respective configuration. However, these examples are not intended to exclude other types of membrane filtration systems that may also be operated in a particular hydraulic configuration.

In addition, there may be cases in which a membrane filtration system does not necessarily conform to the theoretical description of a particular hydraulic configuration (i.e., the mathematical modeling does not apply), but which could be operated in such a way as to emulate the concentration profile of particulate matter associated with that configuration. In these cases, the manufacturer of the membrane filtration system should be able to provide guidance for determining the VCF.

One case of particular interest is represented by those systems in which a second stage of membrane filtration is used to treat the concentrate (NF/RO systems) or residuals (MCF or MF/UF systems) from a primary stage for filtrate blending as a means to augment the system recovery and minimize the waste stream (i.e., concentrate staging). For these systems each stage of membrane filtration must be considered independently such that the membrane units in the second stage would have a distinct and higher VCF than those in the first as a result of the increased particulate concentration in the water feeding the second stage. In this case, for the purposes of calculating the VCF, the applicable concentration effect is represented by the concentration of suspended solids maintained on the feed side of the membranes associated with the second (or subsequent) stage relative to that of the influent feed water to the initial stage (i.e., the feed water to the overall membrane treatment process). This calculation is illustrated in Equation 2.39.

$$VCF = \frac{(C_m)_i}{(C_f)_1}$$
Equation 2.39

Where:

	VCF	=	volumetric concentration factor (dimensionless)
	$(C_m)_i$	=	concentration of suspended solids maintained on the feed side of the membrane associated with the second or subsequent stage (i.e., stage "i") of a multi-stage membrane filtration process (number or mass / volume)
	$(C_f)_1$	=	concentration of suspended solids in the influent feed water to the first stage of a multi-stage membrane filtration process (number or mass / volume)

Note that the concentration of suspended solids in the feed water to the second (or subsequent) stage does not factor into the VCF determination. This same methodology for determining the VCF would be applicable to filtrate staging applications (i.e., cases in which the filtrate from one stage becomes the feed for the subsequent stage(s)), as well. The various equations for calculating the VCF associated with each model described in this section may need to be adapted for these cases of subsequent stages, as appropriate. As with both applicable concentrate and filtrate staging applications, the VCF must also be determined independently for dissimilar membrane units operating in parallel within the same stage (e.g., different manufacturers, capacities, operating modes, etc.).

It is also possible that none of the hydraulic configurations discussed in sections 2.5.1 and 2.5.2 accurately describe the operation and VCF of a particular membrane filtration system. In these cases the membrane manufacturer should provide experimental data demonstrating the appropriate VCF. However, site-specific experimental validation is also recommended, as discussed in section 2.5.4.

2.5.4 Experimental Evaluation

In addition to use of the theoretical models developed in sections 2.5.1 and 2.5.2, the VCF may also be determined experimentally on a site-specific basis. Experimental evaluation of the VCF may be necessary for any membrane filtration systems that do not conform to any of these theoretical models, as a result of either the system design or the manner in which a particular system is operated. It may also be necessary to measure the VCF experimentally for systems that are conceptually similar to one of the models in sections 2.5.1 and 2.5.2, but which do not exhibit ideal operation to an extent sufficient to enable the system to be accurately described by that model. For example, some systems that operate in a mode conceptually similar to a crossflow hydraulic configuration may not generate a scouring force (e.g., using air or water) sufficient to generate suspension mode conditions. In this case, the actual behavior of the system may not fully represent either deposition or suspension mode as a result of the incomplete suspension of particulate matter. Even if a membrane filtration system is expected to generally correspond well to one the theoretical models, it may still be advantageous to measure the VCF experimentally to verify the modeling results. If there is a discrepancy between the modeling and the measured results, the experimentally determined VCF should be used. Note that any experimental method utilized must be conservative and technically defensible to the satisfaction of the State.

Determining the VCF experimentally involves measuring the concentration of suspended solids both in the influent feed water to the membrane filtration system (i.e., C_f in Equation 2.22) and that maintained on the feed side of the membrane (i.e., C_m in Equation 2.22) under conditions that are representative of normal operation, including (as applicable) typical values for: flux; recovery; crossflow velocity; backwash frequency, flow, and duration; recirculation and concentrate flows; outside-in vs. inside-out operation; use of mechanically-induced turbulence; etc. Operating parameters that result in a more conservative (i.e., higher) VCF could also be used. Because the site-specific design of a membrane filtration system and associated appurtenances can affect the concentration of suspended solids maintained on the feed side of the membrane (i.e., C_m), the VCF for any system using a hydraulic configuration that is not accurately described by one of the theoretical models presented in sections 2.5.1 and 2.5.2 should be measured on the full-scale system as installed.

In some hydraulic configurations, the concentration of suspended solids on the feed side of the membrane (i.e., C_m) can vary as a function of filtration cycle time and/or position within the membrane filtration system. In these cases it is important to measure the concentration of suspended solids as a function of time or position within the practical limitations of the test. At a minimum, the concentration of suspended solids should be measured at a point (in terms of position or time, as applicable) that results in the maximum (i.e., most conservative) VCF when evaluating the systems experimentally, thus accounting for conditions with the potential to allow the greatest passage of pathogens to the filtrate in the event of an integrity breach. For example, concentration varying with time is typically characteristic of systems that periodically waste solids (i.e., removing solids from the system, not simply pulsing them from the membrane surface) using a backwash process. For membrane filtration systems utilizing such hydraulic configurations, the concentration of suspended solids should be measured just prior to the backwash cycle, the point at which the concentration should be maximized. Alternatively, if the system does not backwash and continually wastes solids (e.g., NF/RO systems and some submerged MF/UF systems (depending on the operating scheme)) such that the concentration of suspended solids varies by position, the concentration should be measured at the point within the system at which it is greatest. It may also be useful to characterize the average VCF during the experimental evaluation, as this value could be utilized in the determination of direct integrity test sensitivity (see section 4.3).

2.5.5 Summary

The hydraulic configuration of a membrane filtration system governs the concentration of suspended solids on the feed side of membrane and whether it increases as a function of time or position in a membrane unit. Accordingly, the VCF has a significant impact on the amount of particulate matter that could pass through an integrity breach and thus on the sensitivity of a pressure-based direct integrity test (as discussed in section 4.3). Systems that operate in deposition mode have a VCF equal to one, while the various suspension mode hydraulic configurations have VCFs greater than one. Table 2.3 summarizes the general range of VCFs that might be expected for the hydraulic configurations discussed in sections 2.5.1 and 2.5.2, based on typical values for system operating parameters. Note that the ranges given in Table 2.3 are intended to be illustrative only and not a substitute for more rigorous VCF determination using the theoretical models or experimental evaluation.

Table 2.3 Typical Range of VCF Values for Various Hydraulic Configurations

Hydraulic Configuration		VCF
Deposition mode	Dead-end	1
Suspension mode	PFR	3 - 20
Suspension mode	Crossflow[1]	4 - 20
Suspension mode	CSTR[2]	4 - 20

1 Encompasses both large and small volume crossflow systems
2 Encompasses CSTR systems both with and without backwash

A summary of the equations used to calculate the average and maximum VCFs for each of the various hydraulic configurations is presented in Table 2.4. The LT2ESWTR does not require that a specific value for the VCF (i.e., average, maximum, or other statistical parameter) be utilized in determining the sensitivity of a pressure-based direct integrity test; any value that is representative of the membrane filtration system under anticipated operating conditions may be used at the discretion of the State. However, the VCF used should be appropriately conservative.

The table also indicates the types of membrane filtration systems that are most commonly associated with the respective hydraulic configurations. It is also possible to determine the VCF experimentally for a particular system (either full-scale or representative pilot-scale) by measuring the influent feed concentration (C_f) and the concentration on the feed side of the membrane (C_m). Experimental evaluation should be used to determine the VCF if a membrane filtration system's hydraulic configuration does not conform to any of the models developed in sections 2.5.1 and 2.5.2; it may also be used to verify a VCF calculated using theoretical modeling. Note that because the hydraulic configuration depends on site-specific design features, the VCF is not characteristic of a particular proprietary product, but rather a site-specific parameter.

Table 2.4 Summary of VCF Equations for Various Hydraulic Configurations

Hydraulic Configuration	Typical Membrane Process(es)	VCF Equation	
		Average[1]	Maximum
Deposition Mode	• MCF • MF/UF	1	1
PFR	• NF/RO	$$\dfrac{\sum VCF(x) \bullet Q_p(x)}{Q_p}$$	$$\dfrac{1}{1-R}$$
Small Volume Crossflow $[(t_b \bullet Q_b) < V_f]$	• MF/UF (w/o feed tank)	$$0.5 \bullet \left(\dfrac{t_f}{\tau}\right)$$	$$\dfrac{t_f}{\tau}$$
Large Volume Crossflow $[(t_b \bullet Q_b) > V_f]$	• MF/UF (w/ feed tank)	$$\dfrac{t_f \bullet Q_f}{t_b \bullet Q_b} \bullet \left[1 - \dfrac{t_f}{2 \bullet \tau \bullet \dfrac{t_f \bullet Q_f}{t_b \bullet Q_b}}\right]$$	$$\dfrac{t_f \bullet Q_f}{t_b \bullet Q_b}$$
CSTR w/o Backwashing	• NF/RO (small-scale) • MF/UF (w/ bleed)	$$\dfrac{1}{1-R}$$	$$\dfrac{1}{1-R}$$
CSTR w/ Backwashing	• MF/UF (w/ bleed)	$$\dfrac{1}{1-R} \bullet \left[1 - \left(\dfrac{\tau}{t_f} \bullet \left[1 - \exp\left(\dfrac{-t_f}{\tau}\right)\right]\right)\right]$$	$$\left[\dfrac{1}{1-R}\right] \bullet \left[1 - \exp\left(\dfrac{-t_f}{\tau}\right)\right]$$

1 Ideal, theoretical average

3.0 Challenge Testing

3.1 Introduction

The Long-Term 2 Enhanced Surface Water Treatment Rule (LT2ESWTR) requires that any membrane filtration system used to comply with the *Cryptosporidium* treatment requirements of the rule undergo challenge testing (40 CFR 141.719(b)(2)). The primary purpose of this challenge testing is to establish the log removal value (LRV) that an integral membrane can achieve. Under the LT2ESWTR, the maximum removal credit that a membrane filtration system is eligible to receive is the lower of the two values established as follows (40 CFR 141.719(b)(1)):

- The removal efficiency demonstrated during challenge testing; *or*

- The maximum LRV that can be verified by the particular direct integrity test used during the course of normal operation

The requirement for challenge testing under the LT2ESWTR is intended to be product-specific such that site-specific demonstration of *Cryptosporidium* removal efficiency is not necessary. Once the LRV of a membrane has been established through a challenge test that meets the requirements of the LT2ESWTR, additional challenge testing is not required unless significant modifications are made to the membrane process (as discussed in section 3.14). The rule specifies criteria for the following aspects of challenge testing:

- Full-scale vs. small-scale module testing

- Appropriate challenge particulates

- Challenge particulate concentrations

- Test operating conditions

- Calculation of removal efficiency

- Verifying characteristic removal efficiency for untested modules

- Module modifications

Specific requirements of the rule are summarized in section 3.2 and further discussed in the appropriate context in subsequent sections of Chapter 3. The discussion of challenge testing in this chapter applies similarly to microfiltration (MF), ultrafiltration (UF), nanofiltration (NF), reverse osmosis (RO), and membrane cartridge filtration (MCF), except as otherwise noted. Although the primary focus of challenge testing as required under the LT2ESWTR is demonstration of *Cryptosporidium* removal, the general framework for challenge testing developed in this guidance manual may be adapted for use in establishing removal efficiencies for other microbial pathogens of concern, including bacteria, viruses, and other protozoa such as *Giardia*.

Chapter 3 is organized into sections that describe the various issues to be considered in the design and implementation of a challenge test. Furthermore, it presents the various aspects of challenge testing in order from the planning and design phase, through test implementation, and finally to the analysis and interpretation of results.

This chapter is divided into the following sections:

Section 3.2: <u>Summary of Challenge Testing Requirements</u>
This section summarizes the requirements for challenge testing under the LT2ESWTR.

Section 3.3: <u>Test Organization Qualification</u>
This section provides an overview of factors to consider when selecting an organization for conducting a challenge test.

Section 3.4: <u>General Procedure for Developing a Challenge Test Protocol</u>
This section describes the general procedures for developing a challenge testing protocol that meets the requirements of the LT2ESWTR.

Section 3.5: <u>Module Specifications</u>
This section outlines the particular module specifications that are important considerations in the development of a challenge test.

Section 3.6: <u>Non-Destructive Performance Testing</u>
This section describes the important role of non-destructive performance testing – a common quality control procedure used in the production of membrane modules – in the challenge testing process.

Section 3.7: <u>Selection of Modules for Challenge Testing</u>
This section discusses some considerations and two potential approaches for both selecting particular modules for challenge testing and identifying an appropriate number of modules to test.

Section 3.8: <u>Small-Scale Module Testing</u>
This section describes some of the issues associated with testing small-scale rather than full-scale modules.

Section 3.9: <u>Target Organisms and Challenge Particulates</u>
This section discusses factors to consider in selecting a challenge particulate for evaluating the removal efficiency of a membrane filtration process, including attributes of potential surrogates for *Cryptosporidium*.

Section 3.10: <u>Challenge Test Solutions</u>
This section describes methods and procedures for preparing a challenge test solution and for seeding challenge particulates into the solution.

Section 3.11: <u>Challenge Test Systems</u>
This section provides some examples of typical challenge test apparatuses used under different conditions, as well as the appropriate operational parameters to use during the testing.

Section 3.12: <u>Sampling</u>
This section describes aspects of sampling during a challenge test, including sampling methods, sample port design and location, process monitoring, and the development of a thorough sampling plan.

Section 3.13: <u>Analysis and Reporting of Challenge Test Results</u>
This section discusses the calculation of removal efficiency based on the results of challenge testing, as well as suggestions for statistical analyses and the preparation of a summary report for State review.

Section 3.14: <u>Re-Testing of Modified Membrane Modules</u>
This section provides guidance with respect to product modifications that would warrant re-testing of a membrane filtration device.

Section 3.15: <u>Grandfathering Challenge Test Data From Previous Studies</u>
This section discusses factors that should be considered when evaluating data from previously conducted removal studies that may not meet all of the specific requirements for challenge testing under the LT2ESWTR for potential acceptance for satisfying the rule requirements.

3.2   Summary of Challenge Testing Requirements

The LT2ESWTR specifies the core requirements that a challenge test must meet in order to demonstrate the removal efficiency of a membrane filtration system with respect to *Cryptosporidium*. These requirements are summarized as follows:

- **<u>Full-Scale vs. Small-Scale Module Testing</u>**:
  Challenge testing must be conducted on a full-scale membrane module identical in material and construction to the membrane modules proposed for use in full-scale treatment facilities. Alternatively, challenge testing may be conducted on a smaller scale module that is identical in material and similar in construction to the full-scale modules (40 CFR 141.719(b)(2)(i)).

- **Appropriate Challenge Particulates**:
Challenge testing must be conducted using *Cryptosporidium* oocysts or a surrogate that has been determined to be removed no more efficiently than *Cryptosporidium* oocysts. The organism or surrogate used during challenge testing is referred to as the ***challenge particulate***. The concentration of the challenge particulate in both the feed and the filtrate must be determined using a method capable of discretely quantifying the specific challenge particulate used in the test; gross water quality measurements such as turbidity or conductivity cannot be used for this purpose (40 CFR 141.719(b)(2)(ii)).

- **Challenge Particulate Concentrations**:
The maximum allowable feed water concentration used during a challenge test is based on the detection limit of the challenge particulate in the filtrate and must be determined according to the following equation:

$$\text{Maximum Feed Concentration} = (3.16 \times 10^6) \times (\text{Filtrate Detection Limit})$$

This expression allows for the demonstration of up to 6.5 log removal during challenge testing if the challenge particulate is removed to the detection limit (40 CFR 141.719(b)(2)(iii)).

- **Test Operating Conditions**:
Challenge testing must be conducted under representative hydraulic conditions at the maximum design flux and maximum design system recovery specified by the membrane module manufacturer (40 CFR 141.719(b)(2)(iv)).

- **Calculation of Removal Efficiency**:
The removal efficiency of a membrane filtration process as determined from the results of the challenge test is expressed in terms of a LRV according to the following equation:

$$LRV = \log(C_f) - \log(C_p)$$

Where:
LRV	=	log removal value demonstrated during challenge testing
C_f	=	feed concentration measured during challenge testing (number or mass / volume)
C_p	=	filtrate concentration measured during challenge testing (number or mass / volume)

In order for this equation to be valid, equivalent units must be used for both the feed and filtrate concentrations. If the challenge particulate is not detected in the filtrate, then the term C_p is set equal to the detection limit. A single LRV is calculated for each module tested. The overall removal efficiency demonstrated during challenge testing is referred to as $LRV_{C\text{-}Test}$. If fewer than 20 modules are tested, then $LRV_{C\text{-}Test}$ is assigned a value equal to the lowest of the representative LRVs among the various modules tested. If 20 or more modules are tested, then $LRV_{C\text{-}Test}$ is assigned a value equal to the 10th percentile of the representative LRVs among the various modules, as defined by [i/(n+1)], where "i" is the rank of "n" individual data points ordered from lowest to highest. It may be necessary to calculate the 10th percentile using linear interpolation (40 CFR 141.719(b)(2)(v), 40 CFR 141.719(b)(2)(vi)).

- **Verifying Characteristic Removal Efficiency for Untested Modules**:
  Because the LT2ESWTR does not require that every membrane module be subject to challenge testing, a non-destructive performance test (NDPT) (e.g., bubble point test, diffusive airflow test, pressure or vacuum decay test, etc.) must be applied to each production membrane module that did not undergo challenge testing in order to verify *Cryptosporidium* removal efficiency. A quality control release value (QCRV) must be established for the NDPT that is directly related to the removal efficiency of the membrane filtration process as demonstrated during challenge testing. Membrane modules that do not meet the established QCRV are not eligible for the removal credit demonstrated during challenge testing (40 CFR 141.719(b)(2)(vii)).

- **Module Modifications**:
  Any significant modification to the membrane media (e.g., a change in the polymer chemistry), hydraulic configuration (e.g., changing from suspension to deposition mode), or any other modification that could potentially affect removal efficiency or NDPT parameters would require additional challenge testing to both demonstrate the removal efficiency of the modified module and define a new QCRV for the NDPT (40 CFR 141.719(b)(2)(viii)).

Beyond these core requirements, the rule provides substantial flexibility in the design of a challenge test. Guidance for planning, designing, and implementing a challenge test, including elaboration on the core requirements in the appropriate context, is provided in the subsequent sections of this chapter.

3.3 Test Organization Qualification

The LT2ESWTR does not specify any requirements with respect to the qualifications of an organization conducting a challenge test, as long as the test is performed according to the criteria mandated under the rule (40 CFR 141.719(b)(2)). Each State has discretion in approving the results from a challenge test conducted by any organization. However, since challenge testing is intended to be product-specific, it is important that there be some consensus regarding what constitutes an acceptable test. Thus, the guidance provided in this section is intended to outline the skills and capabilities that a test organization should possess in order to produce quality data.

In general, to conduct a successful challenge test, the testing organization should demonstrate effective knowledge of the following:

- The various membrane processes commonly used in drinking water treatment

- Operation of membrane filtration equipment and related system processes, including pretreatment, post-treatment, backwashing, chemical cleaning, and integrity testing

- Proper challenge particulate seeding and sampling techniques

- Analytical techniques for the enumeration of the challenge particulate used in the challenge test, including analyses of microorganisms, inert particulate markers, or molecular markers (as applicable)

- Adequate quality assurance (QA)/quality control (QC) procedures to ensure that data quality objectives are achieved

- Basic statistical procedures that may be used in data analysis

- Preparation of reports for regulatory agencies

Historically, many utilities and States have used independent, third party organizations to conduct verification testing (i.e., challenge testing) in order to ensure an unbiased evaluation of the process. While there are advantages to this approach, a membrane manufacturer may have the ability to conduct an acceptable challenge test if it can demonstrate that appropriate QA/QC procedures are used.

3.4   General Procedure for Designing a Challenge Test Protocol

The core challenge test requirements of the LT2ESWTR should be incorporated into a detailed protocol for implementing the test that documents the necessary equipment, procedures, and analyses. Due to the variety of membrane systems available from numerous suppliers, it is not possible to develop a single comprehensive protocol. However, the following general list of procedures describes the basic steps in the development of such a protocol.

1. Document basic membrane module specifications (as indicated in section 3.5), including:

   - Maximum design flux

   - Mode of operation
     (i.e., inside-out or outside-in)

   - Hydraulic configuration
     (i.e., deposition or suspension)

- Module dimensions and filtration area

- Operating constraints
  (e.g., maximum feed pressure, temperature, pH range, oxidant tolerance, etc.)

- Backwash and chemical cleaning procedures

2. Document the manufacturer's procedure for conducting non-destructive performance testing and ensure that the associated QCRV (i.e., the minimal result from the NDPT that constitutes an acceptable product) is indicative of a NDPT resolution of 3 μm in order to demonstrate *Cryptosporidium* removal capability, as discussed in section 3.6. If available, the statistical distribution of the NDPT results for the product line may also be useful.

3. Determine the number of modules that will be evaluated during the challenge test and the method or criteria that will be used to select specific modules for testing (see section 3.7).

4. Determine whether or not small-scale module testing is an option (see section 3.8).

5. Identify the target organism or contaminant for the test. For the purposes of compliance with the LT2ESWTR, the target organism is *Cryptosporidium* (see section 3.9).

6. Establish the target LRV for the challenge test. Because challenge testing is intended to be product-specific under the LT2ESWTR, it is generally advantageous for a manufacturer to set this target at the maximum LRV for which it is anticipated that the system will qualify.

7. Select the challenge particulate to be used for testing. If it is not feasible or desirable to use the target organism as the challenge particulate, it is necessary to identify an acceptable surrogate that is removed on an equivalent or more conservative basis (see section 3.9).

8. Select an analytical method that will be used to discretely quantify (i.e., enumerate) the challenge particulate and collect information relevant to the methodology for use in developing a sampling plan (see section 3.9). Determine the detection limit for the challenge particulate in the filtrate based on the method capabilities and filtrate sample volume.

9. Design the challenge test solution and establish the method for seeding the challenge particulates into the solution (e.g., continuous or batch seeding) (see section 3.10).

10. Design and construct the testing apparatus, and select appropriate operational parameters (see section 3.11).

11. Develop a sampling and monitoring plan that specifies (as described in section 3.12):

- The number of feed and filtrate samples to be collected and analyzed

- The frequency of feed and filtrate sample collection

- The feed and filtrate sample volumes

- Procedures for sample collection

- Additional operating and water quality parameters to be monitored and associated monitoring frequency

After completing the steps outlined above, the specific protocol for conducting challenge testing should be documented and submitted for State approval, if required. Note that the LT2ESWTR does not require that the challenge test protocol be reviewed or approved by USEPA; however, each State may exercise its discretion regarding whether approval of the protocol is required before results of the challenge testing are accepted in that State.

3.5 Module Specifications

Because there are significant differences in the numerous types of membrane modules that are commercially available from various suppliers, it is important to document the specifications of the module of interest prior to developing a product-specific challenge test. Membrane equipment suppliers typically provide product specification sheets that contain general information about a particular product line. These sheets are generally applicable to all the modules in a particular product line and typically contain the following information:

- Membrane module (i.e., product) designation

- Type of membrane material

- Membrane pore size or molecular weight cutoff (MWCO) – both mean and maximum values

- Feed side membrane filtration area within a module

- Module configuration
  (e.g., hollow-fiber, spiral-wound, etc.)

- Module dimensions

- Membrane media dimensions

  - <u>Hollow-fiber modules</u>: inside and outside fiber diameters, membrane thickness, fiber length, etc.

  - <u>Spiral-wound modules</u>: sheet dimensions, sheet thickness, etc.

- Membrane media symmetry
  (e.g., symmetric, asymmetric, composite, etc.)

- Maximum design flux

- Maximum feed and transmembrane pressure (TMP)

- Mode of operation
  (i.e., inside-out or outside-in)

- Hydraulic configuration
  (i.e., deposition or suspension)

- Operating constraints
  (e.g., temperature limit, pH range, oxidant tolerance, etc.)

Although not all of the information listed above or provided with any particular product specification sheet may be necessary for developing a challenge test, it is nevertheless prudent to compile as much available information about the product as possible not only for its potential use in challenge testing, but for long-term use during operation of the full-scale facility.

It is also common for manufacturers to supply data for each specific module produced. Membrane module data sheets are applicable only to the particular module with the listed serial number and typically contain the results from a manufacturer's QC process. If available, it is important that these data sheets be obtained for each module that is to undergo challenge testing, since it is critical to document as much information about these particular modules as possible. For reference, it may also be useful to document the QC procedures associated with the production of modules to be used for challenge testing.

3.6   Non-Destructive Performance Testing

While challenge testing is used to establish the LRV of an integral module of a particular product type, it does not necessarily guarantee that all such modules produced will achieve the same level of performance due to variability in the manufacturing process. In order to address this issue, a NDPT is applied to all subsequently manufactured modules that are not subject to challenge testing to ensure that these modules comply with the minimum standards for *Cryptosporidium* removal under the LT2ESWTR. A NDPT is a physical test applied to the membrane module with the objective of characterizing some aspect of process performance and which does not alter or damage the membrane.

In order to be utilized in a membrane filtration system that is applied for the purpose of receiving *Cryptosporidium* removal credit under the LT2ESWTR, each module must pass a NDPT that is consistent with the 3 μm resolution requirement of the rule. For example, one commonly used type of NDPT is the bubble point test, which characterizes the largest pore (or defect) in a membrane module. (Note that the "bubble point test" – in this context a type of NDPT – is not the same as the "bubble test" – a diagnostic direct integrity test – described in Chapter 4, although these two procedures are both based on the same principles of bubble point theory, as described in Appendix B.) In a bubble point test, a pressure is applied to a fully-wetted membrane module and gradually increased. The pressure at which water is first evacuated from the pores represents the bubble point of the membrane associated with a particular module. If the established bubble point of the membrane is sufficient to demonstrate that there are no pores (or defects) larger than 3 μm (as described in section 4.2.1.), then the NDPT is consistent with resolution specified in the rule for the removal of *Cryptosporidium*.

The minimum passing test result for a NDPT is known as the quality control release value (QCRV). In the context of the LT2ESWTR, a test result that surpasses the QCRV indicates both that a module can adequately remove *Cryptosporidium* and is sufficiently similar in quality to the modules subjected to challenge testing to demonstrate the ability of the module of interest (which would not have been subject to challenge testing) to achieve the same LRV. After a group of modules has been subjected to challenge testing, the NDPT is applied to those modules to determine an appropriate QCRV associated with the removal efficiency observed during the test. Subsequently, all modules that are not subjected to challenge testing must pass the same NDPT by exceeding the established QCRV applicable to *Cryptosporidium* removal under the LT2ESWTR. Modules that do not pass the NDPT at the QCRV would not be eligible for *Cryptosporidium* removal credit under the rule and could not be used in any membrane filtration systems applied for this purpose (40 CFR 141.719(b)(2)(vii)).

The LT2ESWTR does not specify a particular procedure for determining the QCRV from the various modules that are subjected to challenge testing. Thus, the manufacturer or independent testing organization may exercise its discretion in selecting an appropriate methodology. The QCRV may be selected as the average result among the various modules tested, the most conservative result (to establish the most stringent QA/QC standards), or the least conservative result (to maximize the number of modules eligible for removal credit under the LT2ESWTR). Alternatively, a methodology similar to that required by the LT2ESWTR for determining the overall removal efficiency based on the number of LRV observations could be applied to the various NDPT results to yield an appropriate QCRV, as described in section 3.13.1. The method of module selection for challenge testing, as discussed in section 3.7, should also be considered when determining the QCRV from the test data.

Note that the rule does not specify the manner in which the QCRV is determined from the challenge test data; however, the methodology must be acceptable to each State in which the product line is applied for the purpose of receiving *Cryptosporidium* removal credit under the LT2ESWTR. It is recommended that each module subjected to challenge testing also undergo subsequent non-destructive performance testing for the purpose of establishing a QCRV. The manner in which the NDPT results are used to determine the QCRV should reflect the manner in which modules are selected, as well as the range of LRVs observed during the test. If modules are selected in a conservative manner (see section 3.7), and if the range of LRVs observed during challenge testing is small, the average of the respective NDPT results from the modules subjected to challenge testing might be selected to represent the QCRV. However, if a statistical distribution of modules is selected for challenge testing, or if the range of LRVs observed during challenge testing is significant, a more conservative value is recommended for the QCRV, such as the minimum of the NDPT results observed among the tested modules.

Because it is common for manufacturers to conduct some type of NDPT on every module as a routine component of their respective QA/QC programs, the NDPT requirements of the LT2ESWTR simply ensure that the QCRV used by the manufacturer is sufficient to justify the LRV for *Cryptosporidium* demonstrated via challenge testing. Note that because different NDPTs may be used by the various membrane module manufacturers, the rule does not specify a particular type of NDPT. However, the NDPT used must be consistent with the resolution requirements of the LT2ESWTR in order for a module to be eligible for *Cryptosporidium* removal credit.

3.7   Selection of Modules for Challenge Testing

The intent of challenge testing under the LT2ESWTR is to characterize the removal efficiency of a specific membrane product without requiring challenge testing for all production modules. In addition, the rule does not specify a particular number of modules that are required to undergo challenge testing in order to demonstrate *Cryptosporidium* removal efficiency. However, because it is important that manufacturing variability in the product line be considered in the development of an appropriate challenge test, the number of modules subject to challenge testing, as well as the particular modules chosen, should be carefully selected on a scientifically defensible basis. Although manufacturers or independent testing organizations may develop any number of different procedures for module selection, two common approaches are discussed in this guidance as illustrative examples:

1. Selection of modules based on previously collected QC data for the product line

2. Random sampling of membrane modules from several manufactured lots according to a statistical sample design

Use of the first approach listed above is predicated on the existence of significant QC data for the product line accumulated over time by the manufacturer. Since manufacturers typically conduct some kind of NDPT on all modules produced to ensure quality and characterize the variability of a product line independent of the link established between non-destructive performance testing and challenge testing under the LT2ESWTR, such data should generally be available. Because the modules subject to challenge testing will be subsequently re-characterized with the NDPT to establish an acceptable QCRV required for all modules to be eligible for *Cryptosporidium* removal credit under the LT2ESWTR, it may be most advantageous for a manufacturer to select modules for challenge testing that are near the lower end of the statistical distribution of acceptable (i.e., under the manufacturer's in-house QC procedures) NDPT results based on historical data. If these tested modules yield a QCRV that is consistent with the resolution requirement of the rule, then it is likely that the majority of production modules will also meet the established QCRV and thus be eligible for *Cryptosporidium* removal credit. Using this approach, the number of modules selected for challenge testing is generally at the discretion of the manufacturer or independent testing organization.

If historical QC data for the product line is not available from the manufacturer, the second approach listed above may represent an appropriate option. This method involves the evaluation of a statistically significant random sample of modules from a number of production lots. Because the modules at the lower end of the QC data are not artificially selected (as with the first approach described above), it is likely that this method will result in a higher QCRV, resulting in a somewhat higher rejection rate of modules eligible for *Cryptosporidium* removal credit under the LT2ESWTR. The number of modules selected for challenge testing using this approach will likely be dictated by the particular statistical sampling technique used.

Either of these two approaches or other rational approach developed by the manufacturer or independent testing organization could be utilized to select modules for challenge testing. Regardless of the method used, it is suggested that at least five (5) membrane modules from different manufactured lots be evaluated during a challenge test.

3.8   Small-Scale Module Testing

The evaluation of small-scale (as opposed to full-scale) modules during a challenge test is permitted under the LT2ESWTR to allow for cases in which it may not be feasible or practical to test a full-scale module (40 CFR 141.719(b)(2)(i)). For example, if it is desirable to conduct challenge testing using the target organism (i.e., *Cryptosporidium* for the purposes of the LT2ESWTR) rather than a surrogate, the use of a small-scale module may be the only economically viable alternative.

All challenge testing requirements under the LT2ESWTR (as well as the associated guidance) are equally applicable to both full-scale and small-scale modules. Furthermore, any small-scale module tested must be similar in design to the full-scale modules of the product of interest such that it can be operated (and thus tested) under a hydraulic configuration and at a maximum design flux and recovery that are representative of the full-scale modules. Simulating the full-scale recovery and hydraulic configuration are important considerations for small-scale challenge testing, since both of these parameters affect the concentration of suspended solids on the feed side of the membrane. In addition, it is essential that any small-scale module used for the purposes of challenge testing use membrane material that is identical to that utilized by the full-scale modules.

Although the decision to allow the use of small-scale module testing is left to the discretion of the State, the option is permitted under the LT2ESWTR since it is considered a valid approach for characterizing removal efficiencies. For the purposes of consistency, it is recommended that manufacturers or independent testing agencies that opt to subject a product line to challenge testing using small-scale modules utilize a protocol that has been accepted by a wide range of stakeholders. Such a protocol has been developed for use under the National Sanitation Foundation (NSF) Environmental Technology Verification (ETV) program. Information about this protocol may be obtained by contacting the NSF at (800) 673-6275 (NSF 2005).

3.9 Target Organisms and Challenge Particulates

The purpose of a challenge test is to determine the removal efficiency of a membrane module for one or more target organisms or pathogens. Challenge testing can be conducted using either the target organism itself or an appropriate surrogate; the organism or surrogate used in the test is referred to as the *challenge particulate*. The selection of a suitable challenge particulate is critical to the design of a challenge test.

This section provides guidance for selecting an appropriate challenge particulate, including the selection of a target organism for the test and characteristics of suitable surrogates for the target organism. A more detailed discussion of particular surrogates for *Cryptosporidium* is also provided.

3.9.1 Selecting a Target Organism

The target organism or pathogen of interest for the purposes of challenge testing is selected based on the treatment objectives for the membrane filtration system. For example, *Cryptosporidium* would be the target organism in a challenge test conducted to demonstrate the ability of a membrane filtration system to comply with the treatment requirements of the LT2ESWTR. However, in some cases it may be desirable to determine the removal efficiency of a system for multiple target organisms. In such cases, the most conservative target organism should be selected for the purpose of designing a challenge test. For example, if the challenge test is designed to evaluate the removal efficiency of a system for both *Cryptosporidium* and *Giardia*, then the smaller of these two pathogens should be used as the target organism. Although the approximate size ranges for these two organisms overlap to some degree, as shown in Table 3.1, *Cryptosporidium* has the smaller lower bound. Since membrane filtration is a barrier technology based primarily on the principle of size exclusion, the removal efficiency for the smallest organism of interest should be conservative for larger pathogens.

Table 3.1 Potential Target Organisms for Challenge Testing

Target Organism	Size Range (μm)
Enteric viruses	0.03 - 0.1
Fecal coliform	1 - 4
Cryptosporidium	3 - 7
Giardia	7 - 15

3.9.2 Surrogate Characteristics

Although use of the target organism as the challenge particulate offers the advantages of directly measuring removal efficiency for the pathogen of interest and eliminates issues regarding the appropriateness of a surrogate, it may not be practical or feasible as a result of economic considerations or concerns about working directly with the pathogen. Thus, the use of surrogates may be the most viable option for challenge testing. An ***ideal surrogate*** should have characteristics that are likely to affect removal efficiency which are similar to those of the target organism, while a ***conservative surrogate*** would have characteristics that may result in a lower removal efficiency relative to the target organism. In general, it is necessary to use a conservative surrogate unless there are data to support the use of an ideal surrogate.

As a result of the cost and potential health concerns associated with conducting a challenge test using *Cryptosporidium* oocysts, the LT2ESWTR allows challenge testing to be administered with a surrogate that has been verified to be removed no more efficiently than *Cryptosporidium* oocysts (i.e., an ideal or conservative surrogate) (40 CFR 141.719(b)(2)(ii)). The most direct means of demonstrating that a surrogate is ideal or conservative is through a comparative test in which removal of the surrogate and the target organism(s) are evaluated side-by-side. However, if the characteristics of a surrogate are sufficiently conservative, direct verification may not be necessary. Key physical characteristics to consider when evaluating the suitability of a surrogate for *Cryptosporidium* removal using membrane filtration include size, shape, and surface charge. Other considerations include ease of use and measurement, as well as cost. Each of these factors is discussed as follows.

Particle size and shape

The effective size of an appropriate surrogate should be equivalent to or smaller than the lower bound of the size range of the target organism. Furthermore, the effective size of the surrogate should be characterized using an upper bound of its size distribution such the 99th or 99.9[th] percentile rather than the median. Ideally, a surrogate would have a relatively narrow size distribution and a high uniformity coefficient. For example, the lower size range of *Cryptosporidium* is approximately 3 μm, and thus a conservative surrogate might be one in which 99 percent of particles have a diameter of 1 μm or less.

Generally, it is desirable to use a surrogate that is the same shape as the target organism. In the case of *Cryptosporidium*, an appropriate surrogate would have a spherical shape, although in some cases a non-spherical surrogate might be considered. If a non-spherical surrogate is used, it is recommended that the smallest dimension be considered as the effective size since particles can interact with a membrane barrier at any orientation.

Another consideration is the surface structure of the proposed surrogate. A particle that has a highly irregular surface structure may be removed more efficiently than a similarly sized particle that has a smooth surface. While it may be difficult to completely characterize the surface of a potential surrogate, those with rough surfaces that are known to exhibit a high degree of adherence may be removed through mechanisms other than size exclusion, and thus may not provide a conservative estimate of removal efficiency.

The manner in which the surrogate disperses in the challenge test solution has a significant impact on the effective size and shape of the challenge particulate. Some may agglomerate or become attached to other particles while in solution, which would yield larger effective particle sizes. For example, organisms such as *Staphylococci* exist as clumps and *Streptococci* exist as chains. In its aggregate form each of these organisms is too large to be considered conservative surrogates for *Cryptosporidium*. Surface structure also impacts the tendency of particles to agglomerate, and in general particles with a smooth surface are more likely to be mono-dispersed in solution.

Particle surface charge

A conservative challenge particulate should have a neutral surface charge, since charged particles may interact with other particles and surfaces, thus enhancing removal. The solution pH can also affect the charge of some surrogates and thus should be considered in the preparation of a test solution. If there is a concern regarding the charge of the surrogate such that mechanisms of particle retention other than size exclusion may be responsible for surrogate removal in a MF or UF system, a nonionic surfactant could be used in the challenge test solution to significantly reduce the impact of charge-related removal mechanisms.

Ease of Use and Measurement

Although factors such as ease of handling and measurement are not critical to determining the appropriateness of a surrogate, these nevertheless may be important factors to consider. Handling the surrogate could expose personnel to the challenge particulate, and thus the surrogate should be selected to minimize unacceptable risk to the technicians conducting the test. The material should also be easy to work with and dose accurately since repeated tests may be conducted in which reproducibility is desirable. Surrogates that could degrade during the test, resulting in an inconsistent challenge concentration, should be avoided.

It is also desirable to use a surrogate that is easy to enumerate through established analytical techniques. The LT2ESWTR requires that the concentration of challenge particulate be determined using a discrete measuring technique; gross measurements such as turbidity are not acceptable (40 CFR 141.719(b)(2)(ii)). Particle counters may be appropriate for enumerating challenge particulates in some circumstances. For example, because particle counters are susceptible to coincidence errors and clogging at higher solution concentrations (as described in section 5.3.3), these devices may not be as suitable for accurately enumerating challenge particulates in the feed, which would exhibit a concentration several orders of magnitude (or more) greater than that observed in the filtrate. It is important to note that particle counters used to enumerate challenge particulates in a challenge test should be precisely calibrated for detecting particles smaller than 3 µm in size.

Cost

The cost of seeding and analysis may preclude the use of some surrogates. Both the cost of the surrogate itself and cost of the required analytical techniques should be considered, as well as any other miscellaneous costs associated with the surrogate.

3.9.3 Surrogates for *Cryptosporidium*

In the absence of an acceptable surrogate, formalin- or heat-fixed *Cryptosporidium parvum* could be used as the challenge particulate for compliance with the requirements of the LT2ESWTR. However, the rule does permit surrogates for the purpose of challenge testing, and several different surrogates have been successfully used in studies evaluating physical removal of *Cryptosporidium*. There are three general classifications of surrogates: alternate microorganisms, inert particles, and molecular markers. It is important to note that not all of these classes of surrogates are appropriate for each type of membrane filtration system. Generally, particulate surrogates such as alternate microorganisms and inert particles are appropriate for MF, UF, and MCF systems, while molecular markers would not be removed by these types of membranes. It may be necessary to use molecular makers with NF and RO membrane systems that can remove dissolved substances and which are not designed to accommodate large particulate concentrations. Each of these surrogate classes is discussed in further detail in the following subsections. Some of the potential advantages and disadvantages associated with each class are summarized in Table 3.2.

Table 3.2 Comparative Summary of *Cryptosporidium* and Potential Surrogates

Challenge Particulate	Size Range	Advantages	Disadvantages
Cryptosporidium parvum	3 - 7 µm	• No verification of surrogate required	• High cost • Difficult to measure
Alternate microorganisms	0.01 - 1 µm	• Low cost • Easy to measure • Accepted use	• Difficult to handle • Potential clumping
Inert particles	≤ 1 µm	• Moderate cost • High uniformity • Easy to use	• Difficult to measure accurately
Molecular markers	< 100,000 Daltons	• Low cost • Easy to measure	• Inappropriate for some applications

3.9.3.1   *Alternate Microorganisms*

Microorganisms other than the target can be used as surrogates for the purposes of challenge testing for MF, UF, and MCF systems. Numerous organisms that have a history of use in filter evaluation studies are smaller than 1 µm (when mono-dispersed in solution), and these could be considered conservative surrogates for *Cryptosporidium*. A number of these organisms, including both bacteria and viruses, along with appropriate enumeration methods, are listed in Table 3.3. Table 3.3 also includes common surrogates for *Giardia* and enteric viruses. *S. marcessans* and *P. dimunita* have been widely used as surrogates within the membrane filtration industry, and the use of MS2 bacteriophage has generally been accepted as a surrogate for enteric viruses, since it is similar in size and shape to the poliovirus and hepatitis virus.

Table 3.3 Potential Microbiological Surrogates for *Cryptosporidium*

Microorganism	Size Range (µm)	Target Organism	Enumeration Method
Micrococcus l.	7 - 12	*Giardia*	Standard Methods 9222
Bacillus subtilis	~ 1	*Cryptosporidium*	Barbeau et al. (1997)
E. coli	1 - 4	*Cryptosporidium*	Standard Methods 9222
P. dimunita	0.3	*Cryptosporidium*	Standard Methods 9222
S. marcessans	0.5	*Cryptosporidium*	Standard Methods 9222
MS2 bacteriophage	0.01	Enteric virus	Adams (1959)

Although *Bacillus subtilis* has been used as a surrogate for *Cryptosporidium* for testing the removal efficiencies of conventional treatment processes, it has not been used extensively as a *Cryptosporidium* surrogate for challenge testing membrane filtration devices. Because there is limited data currently available regarding the use of *Bacillus subtilis* in membrane challenge studies, a characterization of this organism would be necessary in order to determine whether it could be used as a *Cryptosporidium* surrogate for the purposes of challenge testing under the LT2ESWTR. Based on the size cited in Table 3.3, *Bacillus subtilis* could potentially be considered a conservative surrogate (see section 3.9.2) for *Cryptosporidium,* pending a comparison of other characteristics (e.g., shape, surface charge, etc.) between these two organisms.

The primary advantage of many microbial surrogates is that enumeration is fairly simple and inexpensive, typically involving culturing the test organisms present in the feed and filtrate samples. The ease with which these organisms can be cultured allows many to be grown in a laboratory to produce a stock for use in challenge testing. Bacteria can be cultured to yield stock concentrations in the range of 10^5 to 10^9 organisms per 100 mL, while MS2 bacteriophage can be grown at concentrations in the range of 10^7 to 10^{12} organisms per 100 mL. Any microbial stock used for the purpose of seeding during a challenge test should be enumerated prior to conducting the challenge test in order to facilitate seeding at the target level.

3.9.3.2 *Inert Particles*

Inert particles may also be used as a surrogate for *Cryptosporidium* under the LT2ESWTR. For example, polystyrene latex microspheres (i.e., latex beads) have been used as a surrogate for *Cryptosporidium* in a number of studies. Historically, microspheres have been used in the calibration of particle counters and similar optical equipment in which a challenge particulate of a known size and geometry is required by the investigator. Microspheres can be manufactured with very high particle uniformity and a smooth surface, both of which are important considerations when selecting a conservative surrogate. Microspheres are chemically inert, easy to handle, and relatively inexpensive. Furthermore, microspheres without a significant surface charge can be produced to minimize the potential for adsorption and interaction with either other particles or the membrane surface. Microspheres are also readily available with particle concentrations ranging from 10^7 to 10^9 particles per mL.

The primary difficulty associated with the use of microspheres is particulate enumeration. Although particle counting is a simple means of enumeration, this technique may not meet the rule requirement that the challenge particulate be discretely quantified as a result of the potential for background particles other than the microspheres to affect the results. Furthermore, other problems such as coincidence error and the dynamic range of most particle counting instruments may also skew the results. Any clumping of microspheres may also complicate particulate enumeration.   A more reliable, albeit more expensive, means of enumerating microspheres is through capture (normally on a laboratory grade membrane filter) and direct examination. The use of fluorescent microspheres is recommended to facilitate particulate identification. Methods for microscopic analysis of fluorescent microspheres are reported in the literature (Abbaszadegan et al. 1997; Li et al. 1997).

The appropriateness of microspheres as a surrogate for *Cryptosporidium* could be directly verified through a comparative study; however, microspheres that meet certain criteria might be deemed conservative surrogates that would not require direct verification. For example, neutral, spherical shaped microspheres with a maximum diameter of 1 µm and which are completely mono-dispersed in solution might constitute a conservative surrogate for *Cryptosporidium* that would not require direct verification.

3.9.3.3   Molecular Markers

The suitability of molecular markers as surrogates for *Cryptosporidium* should be considered on a case-by-case basis. While the justification for using microorganisms and inert particles as surrogates for *Cryptosporidium* is more straightforward given that all are particulates, molecular markers are dissolved substances that are fundamentally different from particulate contaminants. As such, the removal mechanisms for molecular markers may be different than for those associated with discrete particles in many cases. However, semi-permeable membranes that are capable of achieving very high removal efficiencies for dissolved substances may be capable of achieving similar removal of particulates such as *Cryptosporidium*. In addition, porous membranes with very fine pore sizes may be able to remove large macromolecules via mechanisms similar to those that filter discrete particles. Thus, the use of molecular challenge particulates is permitted for the purposes of challenge testing under the LT2ESWTR if the molecular marker used is determined to be conservative for *Cryptosporidium* and is discretely quantifiable.

A variety of molecular markers have been historically used to characterize the pore size or removal capabilities of membrane processes. For example, macromolecular protein compounds are used to determine the MWCO for many UF membranes. In addition, fluorescent dyes such as Rhodamine WT and FDC Red #40 are used to characterize NF and RO membranes. These substances have high spectrophotometric absorbance characteristics that allow measurement and detection at the ng/L level (Lozier et al. 2003). However, these low molecular weight (~500 Daltons) solutions could only be used with RO and low MWCO NF membranes. If molecular markers are considered for challenge testing, it is desirable to use compounds that are more similar to discrete particles such as macromolecular proteins. It is also recommended that a mass balance be conducted on the feed, filtrate, and concentrate streams prior to challenge testing to assess the potential for adsorption or other loss of the molecular marker (Lozier et al. 2003).

With some molecular markers, it may difficult to demonstrate removal in excess of 3 log unless sufficiently sensitive instrumentation is used. For challenge tests conducted with molecular markers, the feed and filtrate concentrations are typically quantified in terms of mass per unit volume. If the analytical method is specific for the molecular marker used in the test, the use of a mass-based concentration is acceptable since the mass of a known substance can be related to moles, which is a discrete quantification. As is the case with any challenge particulate, gross measurements cannot be used for the purpose of quantification. This requirement would preclude the use of analytical techniques such as total organic carbon (TOC) monitoring and conductivity monitoring in most cases.

3.10 Challenge Test Solutions

Generation of an appropriate challenge test solution is an essential component of an effective test program. The purpose of the challenge test solution is to deliver the challenge particulate to the module of interest under the established test conditions. The design of the challenge test solution includes establishing acceptable solution water quality, determining volume requirements and the challenge particulate concentration, and selecting a seeding method. These considerations are discussed in the following subsections.

3.10.1 Test Solution Water Quality

While the LT2ESWTR does not stipulate any constraints on the design of the challenge test solution, it is desirable to conduct the test in a manner that would be considered valid under any anticipated conditions to which the product undergoing testing might be applied in the field. Thus, the water quality of the test solution should be taken into consideration. In designing the test solution, it is important to consider that the primary objective of challenge testing is to evaluate the removal efficiency of the challenge particulate during filtration. While water quality may not have a significant impact on the removal efficiency of *Cryptosporidium* in most cases, particulate matter in the feed water can enhance removal of smaller contaminants. Thus, it is generally accepted that a test solution matrix comprised of high-quality water provides the most conservative estimate of removal efficiency. Note that challenge testing is not intended to yield meaningful information regarding membrane productivity and fouling potential, and thus the use of a high quality water matrix should not be considered inhibitory to testing these measures of membrane performance that are unrelated to the primary objective of challenge testing – to evaluate the removal efficiency of the challenge particulate. Productivity and fouling are addressed using site-specific pilot testing (see Chapter 6) and not on a product-specific basis.

Some particular considerations regarding water quality characteristics and their implications for challenge testing are as follows:

- High quality water or "particle–free water" with a low concentration of suspended solids (e.g., membrane filtrate) should be used as the matrix for the challenge solution, minimizing the potential for formation of a fouling layer during the challenge test that would enhance removal of the challenge particulate.

- No oxidants, disinfectants, or other pretreatment chemicals should be added to the test solution unless necessitated by process requirements (e.g., acid addition and/or scale inhibitor which may be necessary with NF/RO processes).

- If the challenge particulate is a molecular marker, the water quality of the matrix for the test solution should not interfere with the introduction, dispersion, or measurement of the marker. Thus, the impact of water quality parameters, such as pH and ionic strength, on the chemical characteristics and speciation of molecular markers should be considered in the design of the test solution. This is particularly critical for NF or RO membrane modules, which can concentrate organic and inorganic solutes, potentially interfering with some molecular markers.

- If a microbial challenge particulate is used, it may be necessary to add buffers or other materials to maintain the viability of the organisms. Any additives used must not interfere with any aspect of the test or result in a change in the concentration of the challenge particulate over the duration of the test. In addition, because water quality parameters such as pH and ionic strength can affect microbial aggregation, these solution characteristics should also be considered.

- It is recommended that the challenge test solution be characterized with respect to basic water quality parameters, such as pH, turbidity, temperature, total dissolved solids (TDS), TOC, and any others that are critical to the test or interpretation of the results.

3.10.2 Test Solution Volume

The solution volume necessary to conduct a challenge test depends on several factors determined during the test design, including:

- Filtrate flow

- Recovery

- Test duration

- Hold-up volume of the test system

- Equilibration time for the test solution

Taking these factors into account, the volume of solution required for challenge testing may be calculated using Equation 3.1:

$$V_{test} = \left(\frac{Q_p \bullet T_{min}}{R} + V_{hold} + V_{eq} \right) \bullet SF \qquad \text{Equation 3.1}$$

Where:

V_{test}	=	minimum challenge test solution volume (gallons)
Q_p	=	filtrate flow (gpm)
T_{min}	=	challenge test duration (min)
R	=	system recovery during test (decimal percent)
V_{hold}	=	hold-up volume of the test system (gallons)
V_{eq}	=	system volume required to attain equilibrium feed concentration (gallons)
SF	=	safety factor (dimensionless)

The safety factor in Equation 3.1 accounts for unanticipated circumstances that might require additional solution volume. The value of the safety factor should be greater than 1.0 and may be as high as 2.0 under conservative conditions.

The parameters Q_p and R are dictated by the system operating conditions (see section 3.11.2). The LT2ESWTR requires that the flux and recovery used during the challenge test be set at the maximum design values for each parameter, as per manufacturer specifications (40 CFR 141.719(b)(2)(iv)).

The hold-up volume of the test system, V_{hold}, is the unfiltered test solution volume that would remain in the system on the feed side of the membrane barrier at the end of the test (i.e., after the system is shut down). At a minimum, this volume would include the feed side volume of the membrane module and associated piping. In general, it is desirable to design the system with a small hold-up volume, which could potentially allow V_{hold} to be ignored in Equation 3.1. However, the hold-up volume can be significant in some systems, and in such cases V_{hold} should taken into account. Another approach for dealing with the system hold-up volume is to finish the test by pumping a "chaser" of clean water (i.e., without the challenge particulate) through the system allowing the entire test solution to be filtered. However, because this approach can dilute the feed concentration of the challenge particulate, the use of a clean water chaser is only recommended when the entire filtrate stream is sampled as a single composite (see Figure 3.1).

The equilibrium volume (V_{eq}) is the quantity of the test solution that must pass through the membrane module(s) at the beginning of the test before the system stabilizes (i.e., the feed side concentration reaches an equilibrium value). In general, filtrate sampling cannot begin until at least this equilibrium volume has passed through the system. For most test apparatuses, a reasonable assumption is that a system achieves 90 percent of its equilibrium condition after three hold-up volumes (i.e., $3 \bullet V_{hold}$) have passed through the system.

The duration of the challenge test, T_{min}, as given in Equation 3.1, does not include the time required for the test solution to come to equilibrium, as this is taken into account by the parameter V_{eq}. Thus, T_{min} represents the time necessary to implement the sampling program associated with the challenge test (see section 3.12), which typically requires less than one hour.

Table 3.4 provides examples of the challenge test solution volumes (i.e., V_{test}) required as calculated using Equation 3.1 for various membrane configurations under the listed conditions, in these cases assuming the system hold-up volume (V_{hold}) and volume required to achieve equilibrium (V_{eq}) are negligible and a safety factor of 1.1. The filtrate flow (Q_p) is not shown specifically in the table, but is calculated simply by multiplying the membrane area and maximum flux. Table 3.4 is intended to be illustrative only. Thus, it is recommended that the solution volume requirements for a specific challenge test be determined according to the procedure described above. Also, note that the values listed in Table 3.4 are examples only, and that particular product specifications will vary by module manufacturer.

Table 3.4 Example Challenge Test Solution Volume for Various Types of Modules

Module		Example Membrane Area $\{ft^2 (m^2)\}$	Example Maximum Flux $\{gfd (Lmh)\}$	Test Duration (min)	Recovery (percent)	Volume (V_{test}) $\{gal (L)\}$
Cartridge filter[1]		5 (0.46)	1,364 (2,318)	30	100	156 (586)
Spiral-wound	4" diameter	75 (7.0)	17.8 (30.3)	30	85	36.0 (137)
	8" diameter	350 (32.5)	17.8 (30.3)	30	85	168 (637)
Hollow-fiber	Outside-in	350 (32.5)	53.5 (90.8)	30	100	429 (1,623)
	Inside-out	1,400 (130)	107 (182)	30	100	3,433 (13,013)

1 Note that because MCF is a new concept introduced with the LT2ESWTR, the example specifications cited are for cartridge filters, in general, not necessarily for MCF devices

3.10.3 Test Solution Concentration

The concentration of the test solution is based on the target LRV to be demonstrated during the challenge test (LRV_t) and the detection limit for the challenge particulate in the filtrate samples. Since challenge testing is intended to be a one-time, product-specific requirement, it is generally advantageous to select a LRV_t at or near the maximum of 6.5 log removal that can be demonstrated under the LT2ESWTR. The detection limit is a function of the analytical technique used to enumerate the challenge particulate and the filtrate sample volume. For example, if the method can detect 1 particle in a sample, and the filtrate sample volume is one liter, the detection limit is 1 particle/L. The detection limit and maximum LRV of 6.5 are used to calculate the maximum feed concentration that can be used during a challenge test, as shown in Equation 3.2 (40 CFR 141.719(b)(2)(iii)):

$$C_{f-max} = (3.16 \bullet 10^6) \bullet DL \qquad \text{Equation 3.2}$$

Where: C_{f-max} =   maximum feed concentration (number or mass / volume)

                DL   =   detection limit in the filtrate (number or mass / volume)

The coefficient in Equation 3.2 represents the antilog of 6.5, thus capping the maximum feed concentration in the test solution to allow a maximum of 6.5 log removal to be demonstrated if the challenge particulate is removed to the detection limit in the filtrate. The 6.5 log limit under the rule is intended to prevent excessive over-seeding that can result in artificially high LRVs.

The minimum required feed concentration can be calculated from the LRV_t and the detection limit using Equation 3.3:

$$C_{f-min} = 10^{LRVt} \bullet DL \qquad \text{Equation 3.3}$$

Where: C_{f-min} =   minimum feed concentration (number or mass / volume)

                LRV_t =   the target log removal value for the challenge test (dimensionless)

                DL   =   detection limit in the filtrate (number or mass / volume)

Equation 3.3 implicitly assumes complete removal of the challenge particulate by an integral membrane as a conservative means of estimating the minimum feed concentration. Note that Equations 3.2 and 3.3 result in identical feed concentrations for a LRV_t of 6.5. Equations 3.2 and 3.3 also demonstrate that the requisite feed concentration is a function of the detection limit associated with the analytical technique used to enumerate the challenge particulate in the filtrate. The detection limit is typically expressed in terms of the number of challenge particulates per unit volume, or in the case of a molecular marker, mass per unit volume. For a given analytical method with a known sensitivity, the detection limit can be reduced by increasing the sample volume analyzed. For example, if a microbiological method is capable of detecting one organism in a sample, the detection limit can be improved by an order of magnitude by increasing the volume analyzed from 100 mL to 1,000 mL; however, many methods have limitations with respect to the maximum sample volume that can be analyzed. Also, note that the feed and filtrate concentrations must be expressed in terms of equivalent volumes for the purposes of calculating log removal, even if different sample volumes are collected and analyzed during the test.

If the LRV_t selected is less than the maximum of 6.5 permitted under the LT2ESWTR, the maximum and minimum feed concentrations will be different. Note that it is desirable to use a concentration greater than the minimum since use of the minimum feed concentration would only demonstrate the LRV_t if the challenge particulate were removed to the detection limit in the filtrate. After the maximum and minimum concentrations are established using Equations 3.2 and 3.3, respectively, a convenient value between these boundaries can be selected for the challenge particulate concentration (C_{test}).

Once the required test solution volume and concentration have been determined, the total number of challenge particulates required for the test can be calculated using Equation 3.4. If a molecular marker is used, Equation 3.4 would be used to determine the total mass of the marker required.

$$TCPP = C_{test} \bullet V_{test}$$ Equation 3.4

Where: $TCPP$ = total challenge particulate population (number or mass of particles)

C_{test} = feed concentration of challenge particulate (number or mass / volume)

V_{test} = challenge test solution volume (gallons)

3.10.4 Challenge Particulate Seeding Method

There are two approaches commonly used to introduce the challenge particulate into the test solution: batch seeding and in-line injection. Batch seeding involves the introduction of the total challenge particle population (TCPP) into the entire volume of test solution followed by complete mixing to a uniform concentration. In-line injection allows for continuous or intermittent introduction of challenge particulates into the feed stream entering the membrane filtration system. The specific method used may depend on the circumstances of the particular challenge test, although either is permitted under the LT2ESWTR.

Batch seeding is simpler and requires less equipment than in-line injection, but it is only feasible when the entire test solution volume is contained in a reservoir. Furthermore, mixing the challenge particulates into large volumes of water to create a uniform concentration can be logistically problematic. For these reasons, batch seeding is typically only used in challenge studies for small-scale modules with a relatively small membrane area, for which test solution volumes are easier to handle and mix.

In-line injection is the most common seeding approach used in challenge studies, particularly for those involving full-scale modules with greater membrane area. In-line injection allows the challenge particulate to be introduced into the feed on either a continuous or intermittent basis. In general, continuous seeding is advantageous for challenge testing, although intermittent seeding may be appropriate for long-term studies in which it is only necessary to seed organisms at key times during an operational cycle. If intermittent seeding is used, it is necessary to ensure that equilibrium is achieved during each seeding event prior to collection of any feed or filtrate samples.

In-line injection requires additional equipment, such as chemical feed pumps, injection ports, and in-line mixers. These components must be properly designed and integrated into the test apparatus to ensure a consistent challenge particulate concentration in the feed. A chemical metering pump that delivers an accurate and steady flow of challenge material is recommended, while pumps that create a pulsing action should be avoided. The injection port should introduce the challenge material directly into the bulk feed stream to aid in dispersion. Examples of acceptable and unacceptable injection ports are shown in Figure 3.1. An in-line static mixer should be placed downstream of the injection port, and a feed sample tap should be located approximately ten pipe diameters downstream of the mixer.

Figure 3.1 Schematic of Acceptable and Unacceptable Injection Ports

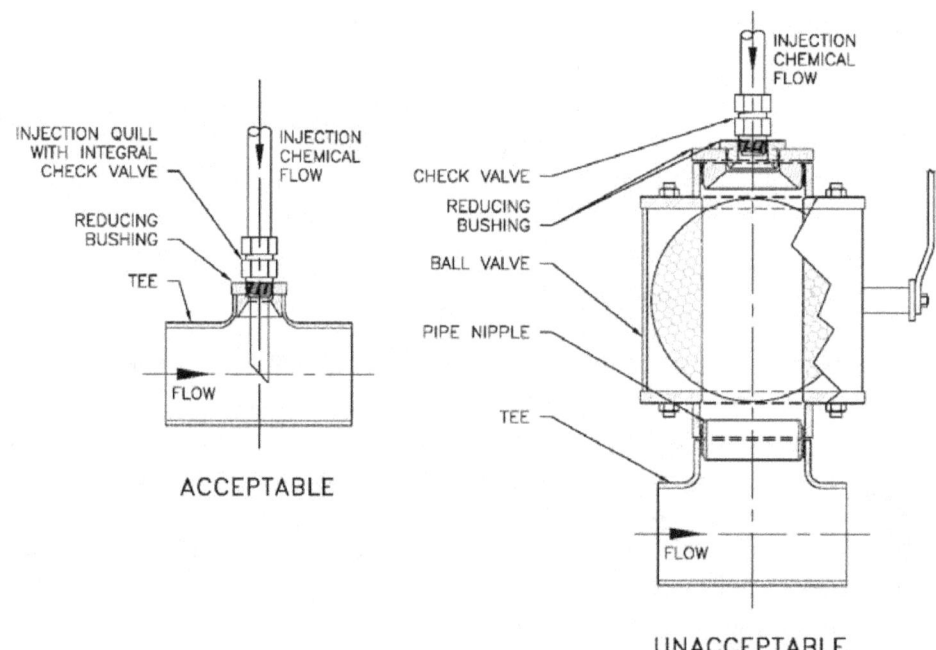

The in-line injection method of seeding delivers challenge particulates from a stock solution with a known concentration as calculated using Equation 3.5:

$$C_{ss} = \frac{TCPP}{V_{ss}}$$

Equation 3.5

Where: C_{ss} =   challenge particulate concentration in the stock solution
(number or mass / volume)

TCPP   =   total challenge particulate population
(number or mass of particles)

V_{ss} =   challenge particulate stock solution volume (gallons)

The TCPP is calculated according to Equation 3.4, while the stock solution volume, V_{ss}, can be selected for convenience. However, the V_{ss} should be between 0.5 and 2 percent of the total test solution volume, as determined using Equation 3.1.

Once the concentration of the stock solution has been determined, the stock solution delivery rate (SSDR) for the in-line injection method of seeding can be calculated using Equation 3.6:

$$SSDR = \frac{C_{test} \bullet Q_p}{C_{ss} \bullet R}$$

Equation 3.6

Where: SSDR   =   stock solution delivery rate (gpm)

C_{test} =   feed concentration of challenge particulate
(number or mass / volume)

Q_p =   filtrate flow (gpm)

C_{ss} =   challenge particulate concentration in the stock solution
(number or mass / volume)

R   =   system recovery during test (decimal percent)

The SSDR represents the setting for the chemical feed pump used to inject the challenge particulate into the feed stream.

3.10.5 <u>Example</u>: Challenge Test Solution Design

<u>Scenario</u>:

Using the methodology described in sections 3.10.1 through 3.10.4, design a challenge test solution using the following assumptions and parameters:

- The target LRV for the challenge test is 4 log.
- The membrane module has an area of 100 m^2.
- The maximum flux for the module 85 Lmh.

- The module operates in deposition mode.
- A test duration of 30 minutes is required to conduct the required sampling.
- The system has a hold-up volume of 200 L.
- The filtrate sample volume used during the challenge test is 500 mL.
- The detection limit for the filtrate sampling technique is 1 particle per 500 mL.
- Challenge particulate seeding is conducted via continuous in-line injection.

Solution:

Step 1:   Determine the required test solution volume.

$$V_{test} = \left(\frac{Q_p \bullet T_{min}}{R} + V_{hold} + V_{eq} \right) \bullet SF \qquad \text{Equation 3.1}$$

T_{min}	=	30 minutes	from given information
R	=	100 percent	standard for deposition mode hydraulic configuration
V_{hold}	=	200 L	from given information
V_{eq}	=	?	to be determined
SF	=	?	to be determined
Q_p	=	?	to be determined

Assume that assume that the equilibrium volume is equal to three times the hold-up volume, as discussed in section 3.10.2:

$$V_{eq} = 3 \bullet V_{hold}$$

$$V_{eq} = 3 \bullet (200 \text{ L})$$

$$V_{eq} = 600 \text{ L}$$

Also, as discussed in section 3.10.2, a suitable safety factor is approximately in the range of 1.1 to 1.5. Since no other information is given in this example, a value for the safety factor is arbitrarily assumed.

$$SF = 1.2$$

The filtrate flow, Q_p, can be calculated simply by multiplying the given maximum flux and the membrane area (and converting to convenient units), as shown below:

$$Q_p = \frac{85(Lmh) \bullet 100(m^2)}{60\left(\dfrac{min}{hour}\right)}$$

$$Q_p \quad = \quad 142 \text{ L/min}$$

Therefore, the required test solution volume can be calculated as follows:

$$V_{test} = \left[\frac{142\left(\dfrac{L}{min}\right) \bullet 30(min)}{1} + 200L + 600L \right] \bullet 1.2$$

$$V_{test} \quad = \quad 6,072 \text{ L}$$

Step 2:   Determine feed challenge particulate concentration within the bounds established by Equations 3.2 and 3.3.

Minimum feed concentration:

$$C_{f-min} = 10^{LRVt} \bullet DL \qquad\qquad \text{Equation 3.3}$$

$$LRV_t \ = \ 4 \text{ log} \qquad\qquad \text{from given information}$$

$$DL \quad = \ 1 \text{ particle per 500 mL} \qquad \text{from given information}$$

$$= \ 2 \text{ particles per L}$$

$$C_{f-min} = 10^4 \bullet 2 \, particles/L$$

$$C_{f-min} = 2 \bullet 10^4 \, particles/L$$

Maximum feed concentration:

$$C_{f-max} = (3.16 \bullet 10^6) \bullet DL \qquad\qquad \text{Equation 3.2}$$

$$DL = \text{1 particle per 500 mL} \qquad \text{from given information}$$

$$= \text{2 particles per L}$$

$$C_{f-\max} = (3.16 \bullet 10^6) \bullet 2 \, particles / L$$

$$C_{f-\max} = 6.32 \bullet 10^6 \, particles / L$$

Select a feed challenge particle concentration between the minimum
($2 \bullet 10^4$ particles/L) and maximum ($6.32 \bullet 10^6$ particles/L). Arbitrarily:

$$C_{\text{test}} = 5 \bullet 10^4 \text{ particles/L}$$

Step 3: Determine the total challenge particulate population required for the test solution.

$$TCPP = C_{test} \bullet V_{test} \qquad\qquad \text{Equation 3.4}$$

$$C_{\text{test}} = 5 \bullet 10^4 \text{ particles/L} \qquad \text{as calculated in step 2 above}$$

$$V_{\text{test}} = 6,072 \text{ L} \qquad\qquad \text{as calculated in step 1 above}$$

$$TCPP = (5 \bullet 10^4 \, particles / L) \bullet (6,072L)$$

$$TCPP = 3.036 \bullet 10^8 \, particles$$

Step 4: Determine the challenge particulate stock solution concentration using Equation 3.5.

$$C_{SS} = \frac{TCPP}{V_{SS}} \qquad\qquad \text{Equation 3.5}$$

$$TCPP = 3.036 \bullet 10^8 \text{ particles} \qquad \text{as calculated in step 3 above}$$

$$V_{\text{ss}} = ? \qquad\qquad \text{to be determined}$$

As discussed in section 3.10.4, select a stock solution volume that is between
0.5 and 2 percent of the total test solution volume of 6,072 L (as determined in
step 1 above). Thus:

$$(0.005 \bullet 6,072 \text{ L}) < V_{\text{ss}} < (0.02 \bullet 6,072 \text{ L})$$

$$(30.4 \text{ L}) < V_{ss} < (121.4 \text{ L})$$

$$V_{ss} = 100 \text{ L} \qquad\qquad \text{arbitrary selection}$$

Substituting values:

$$C_{ss} = \frac{3.036 \bullet 10^8 \, particles}{100L}$$

$$C_{ss} = 3.036 \bullet 10^6 \, particles \, / \, L$$

Step 5: Determine the challenge particulate stock solution delivery rate for continuous, in-line injection.

$$SSDR = \frac{C_{test} \bullet Q_p}{C_{ss} \bullet R} \qquad\qquad \text{Equation 3.6}$$

C_{test} = 5 • 10⁴ particles/L as calculated in step 2 above

Q_p = 142 L/min as calculated in step 2 above

C_{ss} = 3.036 • 10⁶ particles/L as calculated in step 4 above

R = 100 percent standard for deposition mode hydraulic configuration

Substituting values:

$$SSDR = \frac{(5 \bullet 10^4 \, particles \, / \, L) \bullet (142L \, / \, \min)}{(3.036 \bullet 10^6 \, particles \, / \, L) \bullet 1}$$

SSDR = 2.34 L/min

3.11 Challenge Test Systems

A system used for challenge testing should be carefully designed to both meet the objectives of the test and simulate full-scale operation to the greatest practical extent. Guidance for both designing an appropriate test apparatus and determining operational parameters for challenge testing is provided in the following subsections.

3.11.1 Test Apparatus

The equipment used to conduct challenge testing is product-specific to some extent, although there are some basic components that are common to all systems. In many cases, a manufacturer may maintain a special test apparatus to check individual modules as a component of its QA/QC program. Such an apparatus may be suitable for conducting challenge testing and typically includes equipment such as pumps, valves, instrumentation, and controls necessary to evaluate full-scale modules. This same type of equipment would be used in the design of systems for testing small-scale modules.

Both the seeding and sampling methods selected for challenge testing, as well as the hydraulic configuration of the system, affect the design of the test apparatus. Batch seeding requires a feed tank and mixing equipment, while continuous seeding requires a stock solution reservoir, chemical metering pump, and in-line mixers. Sampling requirements may dictate the location and design of sample taps in the system. In addition, the test apparatus should be designed to mimic the hydraulic configuration of the full-scale system as much as practical; however, the test apparatus may alternatively utilize a more conservative recovery than the full-scale system. If 100 percent recovery (i.e., the most conservative scenario) is used, a crossflow system must operate without a bleed stream such that all of the concentrate is recirculated (see section 2.5.2.2), and a deposition mode system (see section 2.5.1) must filter the entire test solution volume. Note that a full-scale crossflow system could not operate at 100 percent recovery on a sustained basis, since the feed would become increasingly concentrated. However, operation at 100 percent recovery is feasible for a short-term challenge test in the interest of generating conservative results. The test apparatus should allow the membrane module to undergo direct integrity testing both before and after the challenge test. Figures 3.2 through Figure 3.6 are schematic representations of typical apparatuses for challenge testing under various conditions. Note that ancillary equipment and operational processes (e.g., backwashing, chemical cleaning, and integrity testing) are not shown.

Figure 3.2 illustrates a pressure-driven apparatus operating in deposition mode with batch seeding and composite sampling. This type of system may be well suited for a MCF or other membrane module with limited surface area. With this apparatus the test solution is prepared as a batch and a composite filtrate sample would be generated, yielding a single data pair (i.e., a feed sample and a composite filtrate sample) for the purposes of calculating the log removal efficiency for the challenge particulate.

Figure 3.2 Schematic of a Typical Pressure-Driven System in Deposition Mode with Batch Seeding and Composite Sampling

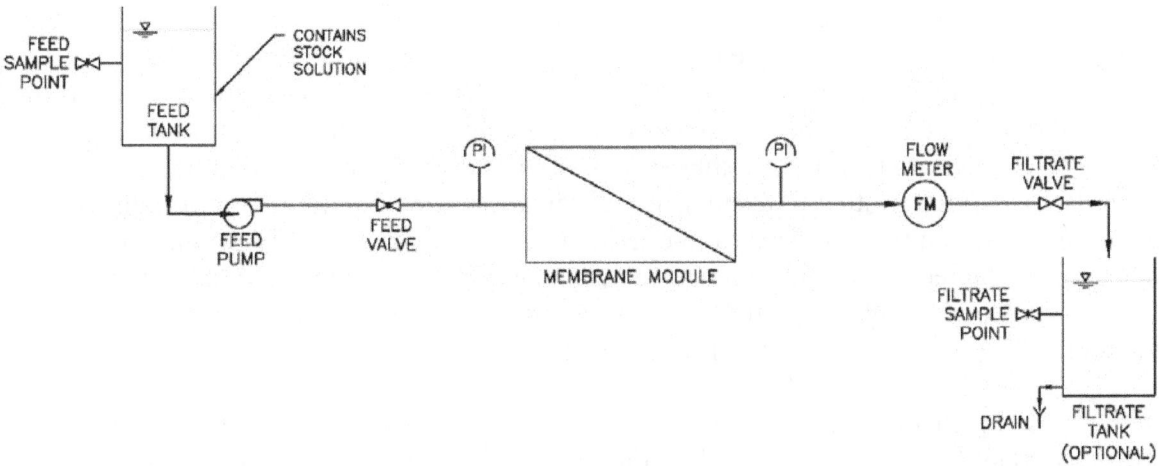

Figure 3.3 illustrates an apparatus similar to that shown in Figure 3.2, but designed for continuous in-line seeding and grab sampling rather than batch seeding and composite sampling. With this apparatus the challenge particulate is introduced from a stock solution reservoir and mixed prior to the feed sampling point. The use of grab sampling allows the collection of multiple feed and filtrate samples from a single test run.

Figure 3.3 Schematic of a Typical Pressure-Driven System in Deposition Mode with Continuous Seeding and Grab Sampling

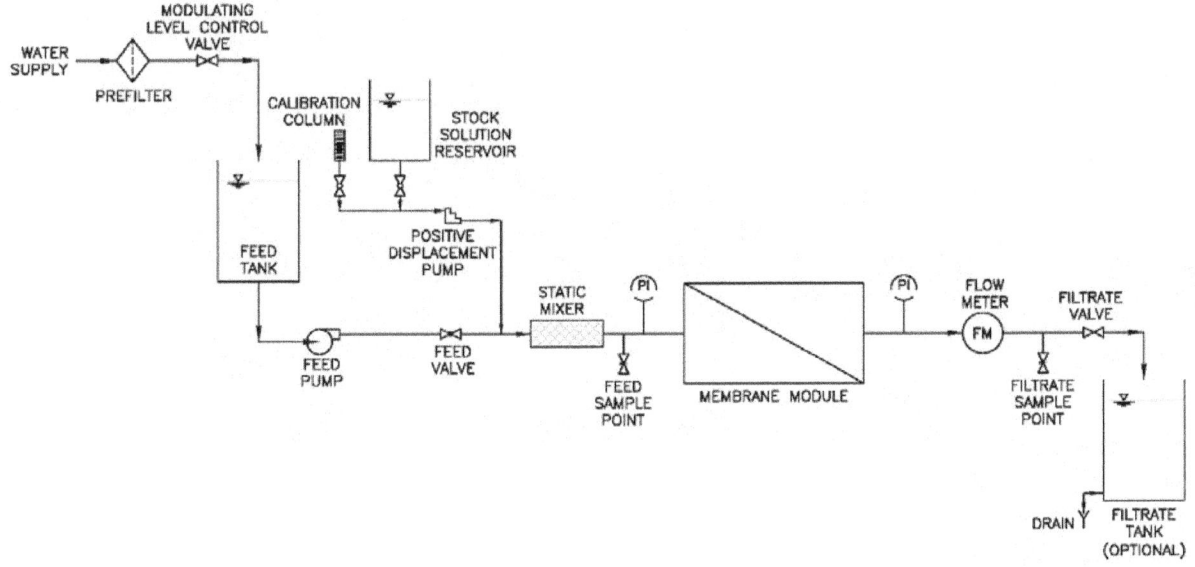

Figure 3.4 shows a schematic of a pressure-driven apparatus operated in suspension mode with continuous seeding and grab sampling. In systems that operate in suspension mode, the concentration of suspended solids increases on the feed side of the membrane, as discussed in section 2.5. While it may not be practical to accurately replicate the solids concentration profile for a membrane system in which the feed side concentration of suspended solids varies as a function of filtration time, concentrate recirculation can produce conservative feed side conditions for the purpose of challenge testing, assuming appropriate operating conditions for recovery and recycle ratio are selected. Selection of an appropriate recovery for challenge testing can be complicated by the fact that system recoveries can vary significantly in some cases (particularly for NF/RO systems). Guidance regarding the selection of an appropriate recovery for a challenge test of a membrane system operated in suspension mode is provided in section 3.11.2. The recycle ratio should be selected such that velocities across the membrane surface are high enough to keep particles in suspension. The manufacturer can typically recommend a minimum scour velocity for a crossflow system.

For systems that utilize concentrate recycle there are some additional considerations that are important to take into account regarding the feed side system volume and the location of the feed sample point. In general, larger feed side system volumes require longer system equilibration times. For example, Figure 3.4 shows the concentrate return location at the feed tank rather than directly into the module feed line, thus increasing the effective feed side system volume significantly. If such an arrangement is necessary (e.g., to provide an air break in the recirculation system), then the feed tank volume should be minimized. In a properly apparatus utilizing concentrate recycling, the feed sample point is located upstream of the return point, as shown in Figure 3.4.

Figure 3.4 Schematic of a Typical Pressure-Driven System in Suspension Mode with Continuous Seeding and Grab Sampling

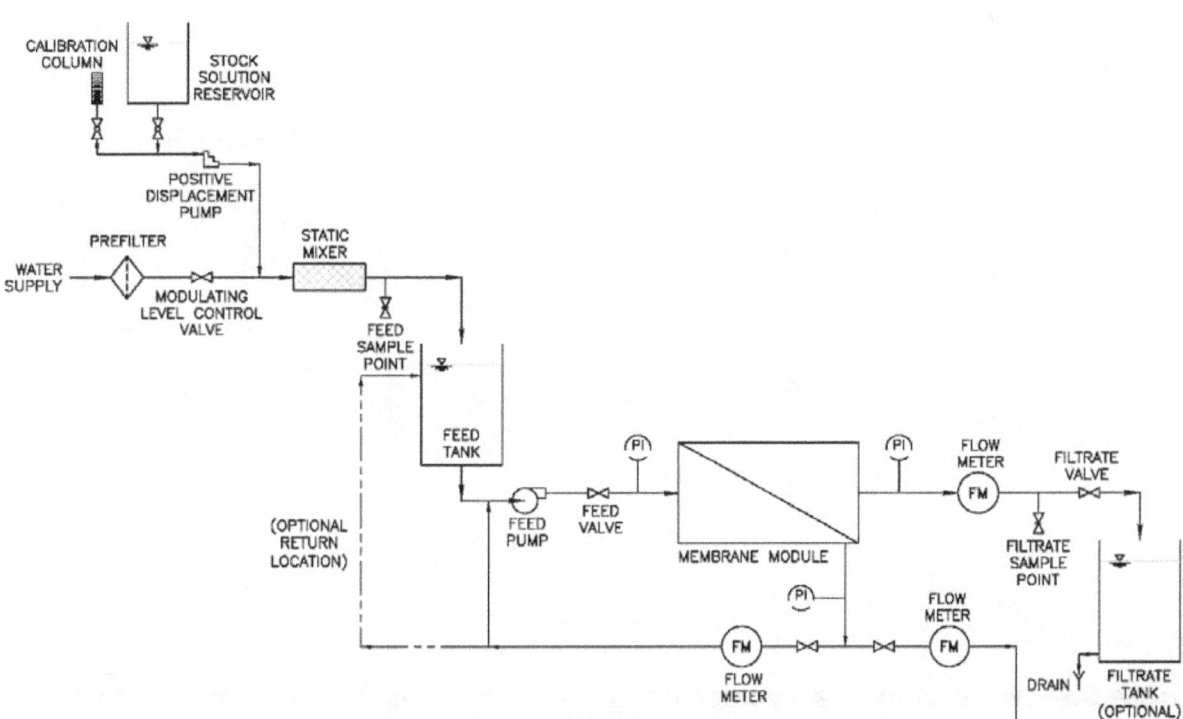

Figure 3.5 shows a typical vacuum-driven test apparatus operated in deposition mode with continuous seeding and grab sampling. Although the module is immersed in a tank, the feed water is not agitated, thus allowing particles to deposit on the membrane surface. With this apparatus, the filtrate sampling point should be located downstream of the vacuum pump.

Figure 3.6 illustrates an apparatus for a vacuum-driven system operated in suspension mode with continuous seeding and grab sampling. With this apparatus the feed tank is mechanically agitated to keep particles in suspension and the system can be modeled as a continuous stirred tank reactor (CSTR) (see section 2.5). As with the vacuum-driven apparatus shown in Figure 3.5, the filtrate sampling point should be located downstream of the vacuum pump.

Figure 3.5 Schematic of a Typical Vacuum-Driven System in Deposition Mode with Continuous Seeding and Grab Sampling

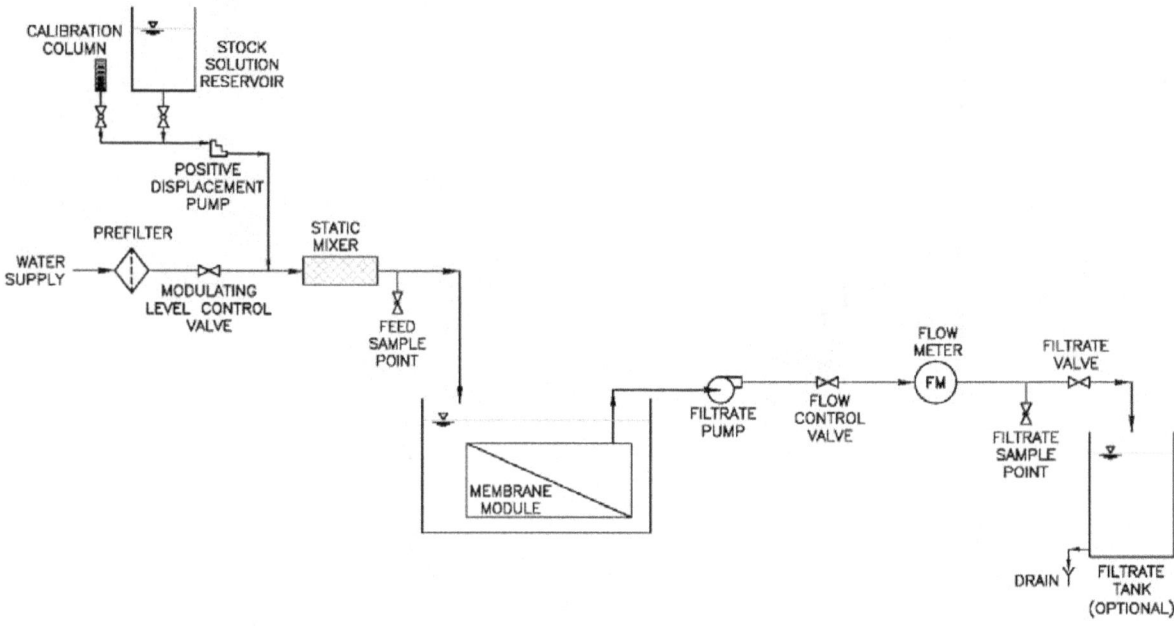

Figure 3.6 Schematic of a Typical Vacuum-Driven System in Suspension Mode with Continuous Seeding and Grab Sampling

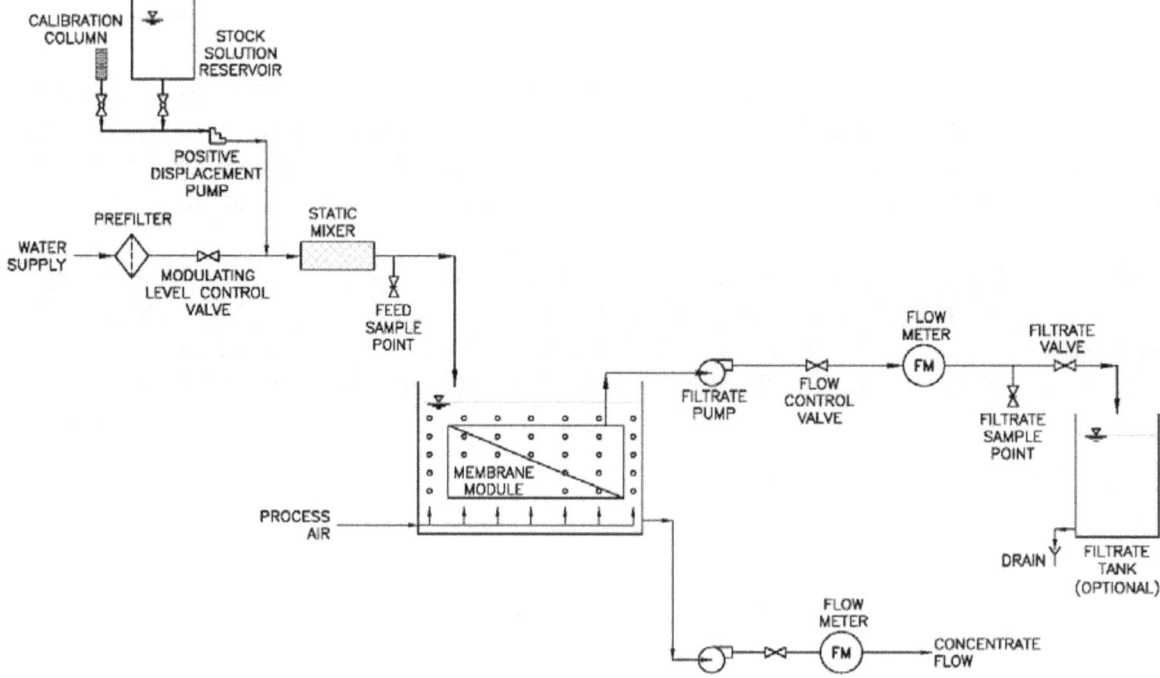

3.11.2 Test Operating Conditions

The design of a challenge test includes specifications for the following operating conditions: flux, recovery, and hydraulic configuration. The LT2ESWTR requires the challenge test to be conducted at the maximum design flux and recovery, and that the test apparatus be operated under representative hydraulic conditions (40 CFR 141.719(b)(2)(iv)). These requirements dictate the operating conditions for the test apparatus during challenge testing. Note that under the LT2ESWTR, recovery is defined as the volumetric percent of feed water that is converted to filtrate in the treatment process over the course of an uninterrupted operating cycle (i.e., excluding losses that occur due to the use of filtrate in backwashing or cleaning operations).

Testing at the maximum recovery is important to ensure that the volumetric concentration factor (VCF) simulated during challenge testing is representative of (or conservative for) full-scale system operation. For systems that operate in deposition (i.e., direct flow or "dead-end") mode, such as most MF/UF and MCF systems, the value of the VCF is one, and thus the recovery does not have a significant impact on the suspended solids concentration. However, it is still recommended that systems operating in deposition mode process at least 90 percent of the challenge test solution, thus resulting in an effective recovery of at least 90 percent. Similarly, for MF/UF systems that operate in suspension mode but without a concentrate waste stream (i.e., a bleed stream), it is also recommended that at least 90 percent of the challenge test solution be processed in order to generate an effective recovery of at least 90 percent. For MF/UF systems that operate with a concentrate waste stream, it is recommended that a recovery of at least 75 percent be utilized for challenge testing unless a more representative system recovery can be demonstrated by the manufacturer; a recovery of 100 percent would represent the most conservative case. NF/RO systems, for which the utilization of a concentrate waste stream is standard, should be operated at a recovery of at least 45 percent for the purposes of challenge testing, which is representative of a flow-weighted average recovery for a module in such a system operating at an overall recovery of greater than 90 percent for a single stage. If challenge testing is conducted on a small-scale NF/RO system (such as one using a single module), then concentrate recycle should be used to increase the recovery to at least 45 percent.

The membrane area is also typically given in the specifications, and the maximum filtrate flow can be calculated by multiplying the membrane area and the maximum flux. Although the LT2ESWTR stipulates specific requirements for challenge testing flux, the filtrate flow is also necessary for designing the challenge test solution, as described in section 3.10. Table 3.5 summarizes some typical specifications for membrane area and maximum flux associated with various types of membrane modules, as well as the corresponding filtrate flow. For the purposes of challenge testing, the membrane area exposed to the feed (i.e., as opposed to that to the filtrate) should be used in all calculations. Note that the values listed in Table 3.5 are examples only, and that particular product specifications will vary by module manufacturer.

Table 3.5 Typical Parameters for Various Types of Modules

Module		Example Membrane Area {ft^2 (m^2)}	Example Maximum Flux {gfd (Lmh)}	Filtrate Flow {gpm (L/min)}
Cartridge filter[1]		5 (0.46)	1,364 (2,318)	4.7 (17.7)
Spiral-wound	4" diameter	75 (7.0)	17.8 (30.3)	0.9 (3.5)
	8" diameter	350 (32.5)	17.8 (30.3)	4.3 (16.4)
Hollow-fiber	Outside-in	350 (32.5)	53.5 (90.8)	13.0 (49.2)
	Inside-out	1,400 (130)	107 (182)	104 (394)

1 Note that because MCF is a new concept introduced with the LT2ESWTR, the example specifications cited
   are for cartridge filters, in general, not necessarily for MCF devices

A challenge test should be designed to simulate the hydraulic configuration of full-scale system operation, since it affects the concentration of suspended solids on the feed side of the membrane and thus the removal efficiency of the process. A membrane filtration system can be operated in either suspension or deposition mode. (The various hydraulic configurations for membrane filtration systems are discussed in further detail in section 2.5.) While it is relatively straightforward to simulate a system operating in deposition mode during a challenge test, it may not be possible to simulate all variations of suspension mode operation using a single module challenge test apparatus. For example, it is not practical to simulate a plug flow reactor (PFR) configuration typical of full-scale RO systems. In such cases, the challenge test apparatus should be designed and operated as a CSTR, since a CSTR configuration generally results in the highest concentration of suspended solids on the feed side of the membrane. If the challenge test can successfully demonstrate the target LRV with a higher concentration of suspended solids on the feed side of the membrane than expected with the full-scale system, then the use of a CSTR in place of a PFR configuration would represent more conservative, and thus acceptable, challenge test conditions. A single RO module can be operated as a CSTR by including a concentrate recycle loop, as shown in Figure 3.4.

3.12 Sampling

Although the LT2ESWTR does not stipulate any particular requirements for sampling, it is an important component of a challenge test. An effective sampling program is dependent on the development of detailed and thoroughly documented sampling plan, as well as on the selection of appropriate sampling methods and locations. These critical aspects of sampling, along with the monitoring of operational parameters during the execution of the challenge test, are discussed in the following subsections.

3.12.1 Sampling Methods

The two most common approaches for sampling are the grab and composite methods. Grab sampling involves the collection of one or more aliquots from the feed or filtrate stream, while composite sampling involves collection of the entire process stream for processing and subsequent analysis. The concentration of challenge particulates in the feed solution is typically characterized through grab sampling, while the filtrate stream may be sampled using either grab or composite sampling. If grab sampling is used for both the feed and filtrate streams, the number of feed and filtrate samples does not need to be equivalent, and samples can be collected on different schedules during the challenge test. In many cases, it may be advantageous to collect more filtrate samples than feed samples, since filtrate concentrations are expected to be very low, and an error of just a few particles in a filtrate sample can have a significant impact on the demonstrated removal efficiency. Moreover, if batch seeding is used, the feed concentration should not vary significantly over the course of challenge testing, assuming appropriate feed stock mixing. However, if continuous seeding is used, paired sampling may be preferred for simplicity of data reduction.

Grab sampling of the filtrate typically involves the collection of a predetermined volume of water in an appropriate collection vessel at predetermined times, as documented in the sampling plan. Composite sampling of the filtrate is more common and may be accomplished in one of two ways. A composite filtrate sample may be collected and processed following collection of the entire volume; however, the more common approach is to pass the entire filtrate stream through an absolute filter capable of complete retention of the challenge particulate during the course of the challenge test. The challenge particulate would then be enumerated either directly from the filter media or removed from the filter for subsequent analysis.

It is important that good sampling practices be employed during challenge testing, such as flushing taps prior to sample collection (if applicable) and isolating filtrate sampling locations from feed sampling locations to prevent cross-contamination. Furthermore, appropriate QA/QC measures should be implemented during sampling, such as collection of duplicate samples and blanks.

3.12.2 Sample Port Design and Location

As with the challenge particulate injection port, the design of the feed and filtrate sample withdrawal ports should yield a uniform sample. Figure 3.7 illustrates examples of both acceptable and unacceptable sampling ports. The unacceptable apparatus requires the sample to be pulled from a large section of pipe and has an area where stagnation of sample flow may occur. By contrast, the acceptable apparatus has a quill that extends into the center of the pipe to obtain a more representative sample.

Figure 3.7 Schematic of Acceptable and Unacceptable Sampling Ports

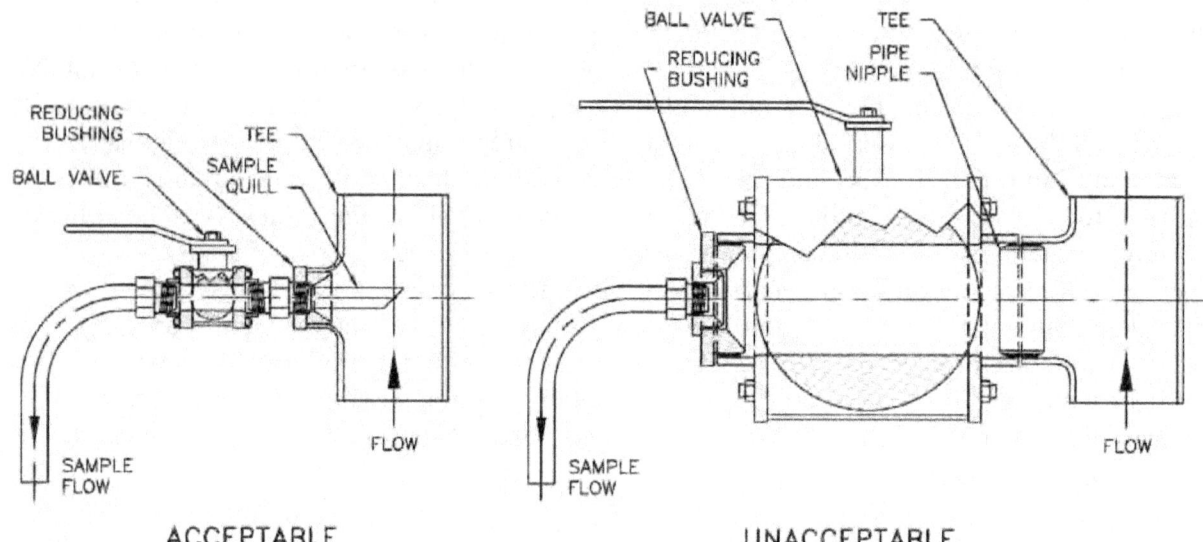

ACCEPTABLE   UNACCEPTABLE

The feed sample tap should be located at least ten pipe diameters downstream of challenge particulate injection points and in-line mixers to ensure uniform concentration. For apparatuses that utilize concentrate recycle, as illustrated in Figure 3.4, the feed sample tap should be located upstream of the T-connection where the concentrate is blended with the incoming feed water. A check valve may be used to prevent backflow of concentrate into the feed line. As a guideline, the feed sample rate should be no more than 1 percent of the flow to the membrane. Filtrate samples should be collected at a point after the filtrate passes through any filtrate side instrumentation such that any important measurements are not affected by the sampling event. In vacuum-driven apparatuses, the filtrate sample tap must be located downstream of the filtrate pump, as shown in Figures 3.5 and 3.6. Note that the filtrate sampling valve should be positioned as close as possible to the filtrate port in order to minimize error due to any potential for adhesion of the challenge particulate to the piping prior to the sampling point. In addition, if a microorganism is used as the challenge particulate, it is prudent to use a metal or heat-resistant sampling valve to allow the tips of the valve to be flame-sterilized.

3.12.3 Process Monitoring

During the challenge test it is important that the operational parameters be monitored to ensure that the test conditions remain constant. Continuous monitoring for flow (or flux) and pressure should be conducted if the apparatus is equipped with the appropriate instrumentation. If periodic monitoring must be utilized, it is important that operational parameters be checked at least both before beginning and after completing sampling for the challenge particulate. Operational data collected prior to initiating sampling should be used verify that the flux and recovery are at the required levels. It is also important that the membrane module(s) undergo direct integrity testing both before and after challenge testing to verify that the modules were integral during the test.

If the water quality of the test solution matrix is a particular concern, other water quality parameters should be sampled accordingly. Some examples of other parameters that may be important in some cases include pH, temperature, turbidity, TDS, TOC, and ionic strength. The measurement of such water quality parameters may be most relevant if a molecule marker is used as the challenge particulate, as discussed in section 3.10.1.

All aspects of process monitoring, including what parameters to monitor, how often to monitor, and the range of acceptable results should be included the sampling plan for the challenge test (see section 3.12.4). If any results are outside acceptable tolerances, the challenge test should be restarted.

3.12.4 Sample Plan Development

The primary purpose of a sampling plan is to define the samples to be collected and provide an accompanying sampling schedule for the challenge test. A sampling plan should include:

- Type of sample(s) (i.e., composite vs. grab)

- Number of feed and filtrate samples to be collected

- Sample locations

- Sampling interval

- Estimate of time required to collect each sample

- Sampling equipment required

- Sample volume(s)

- Process monitoring requirements (see section 3.12.3)

The sampling plan should also specify any particular requirements associated with the analytical technique to be employed, and well as procedures for shipping the sample(s) if they are to be analyzed at an off-site laboratory. Samples should be collected, preserved, stored, prepared, and analyzed using methods and techniques appropriate for the challenge particulate.

Sampling should not begin until the system has stabilized (i.e., reached equilibrium concentration). Most test apparatuses achieve greater than 90 percent of equilibrium concentrations after three hold-up volumes have passed through the system. Thus, the system hold-up volume and feed flow can be used to estimate the point at which the system is near equilibrium and thus the time at which sampling can begin. Both the hold-up volume and time required for system stabilization (as discussed in section 3.10.2) should be included in the sampling plan.

3.13 Analysis and Reporting of Challenge Test Results

After challenge testing is completed for a particular product, the results must be analyzed to determine the established removal efficiency of the module (i.e., LRV_{C-Test}) for the purposes of LT2ESWTR compliance. The following subsections provide guidance regarding the calculation of removal efficiency under the rule, the statistical analysis of the challenge test results, and summarizing challenge testing in a report for State review.

3.13.1 Calculation of Removal Efficiency

The removal efficiency established during challenge testing (LRV_{C-Test}) is determined from the various LRVs generated during the testing process. The LT2ESWTR requires that a single LRV be generated for each module tested for the product line under evaluation. The LRVs for each respective module tested are then combined to yield a single value of LRV_{C-Test} that is representative of the product line.

Under the LT2ESWTR, the LRV is calculated according to Equation 3.7 (40 CFR 141.719(b)(2)(v)):

$$LRV = \log(C_f) - \log(C_p)$$
Equation 3.7

Where: LRV = log removal value demonstrated during a challenge test
C_f = feed concentration of the challenge particulate (number or mass / volume)
C_p = filtrate concentration of the challenge particulate (number or mass / volume)

Note that the feed and filtrate concentrations must be expressed in identical units (i.e., based on equivalent volumes) in order for Equation 3.7 to yield a valid LRV. If the challenge particulate is not detected in the filtrate, then the term C_p is set equal to the detection limit.

The overall value of LRV_{C-Test} (i.e., the removal efficiency of the product) is based on the entire set of LRVs obtained during challenge testing, with one representative LRV established per module tested. The manner in which LRV_{C-Test} is determined from these individual LRVs depends on the number of modules tested. Under the LT2ESWTR, if fewer than 20 modules are tested, then the lowest representative LRV among the various modules tested is the LRV_{C-Test}. If 20 or more modules are tested, then the 10th percentile of the representative LRVs is the LRV_{C-Test}. The percentile is defined by [i/(n+1)] where "i" is the rank of "n" individual data points ordered from lowest to highest. It may be necessary to calculate the 10th percentile using linear interpolation (40 CFR 141.719(b)(2)(vi)).

Although the LT2ESWTR requires that one representative LRV be established per module tested, the rule does not restrict the manner in which the representative LRV for each module is calculated. Consequently, there are numerous methods that could be used to calculate the representative LRV for a module. If multiple feed/filtrate sample pairs are collected, a LRV can be calculated for each set of paired data, and the LRV for the tested module could be selected as the lowest LRV (more conservative) or the average of the LRVs (less conservative). Another approach is to average all the respective feed and filtrate concentrations from among the various samples collected and calculate a single LRV for a tested module using Equation 3.7. A more conservative approach would be to use the average feed concentration but the maximum filtrate concentration sampled, which would result in a lower representative LRV for a tested module. Likewise, a still more conservative approach would be to use the minimum feed and maximum filtrate concentrations. Note that these methods simply represent potential options; other approaches may be used for calculating a representative LRV for each module tested. Regardless of the particular method used, the range of data collected for a single module can provide some indication about the experimental error associated with the study (i.e., errors due to seeding, sampling, analysis, etc.). If a statistically valid sampling method was used to select the modules for challenge testing (as described in section 3.7), a comparison of the LRVs across the different modules tested would provide an indication of variability within the product line.

3.13.2 Statistical Analysis

If a sufficient number of modules are evaluated in the course of a challenge test to be considered a statistically significant sample of the product line, it may be useful to conduct a formal statistical analysis of the data in order to make inferences about the entire population from the sample set. Note that such a statistical analysis is not considered a substitute for the methodology for determining LRV$_{C\text{-Test}}$ required under the rule, but could provide the manufacturer with useful information. For example, if challenge testing were conducted using a random sampling of membranes from a production lot, a statistical analysis of the challenge test data (i.e., the LRV observed for each of the modules tested) would provide an estimate of the range of removal efficiency for the entire product line. This information could be used to infer the number of membrane modules in the product line that would be expected to achieve the LRV$_{C\text{-Test}}$. If it is desirable to apply a statistical analysis to the LRV data generated during challenge testing, the particular method of analysis should be considered in the design of the challenge test protocol, since the method to be employed may dictated the number of modules selected for testing.

3.13.3 Reporting

Utilities using membrane filtration for compliance with *Cryptosporidium* removal requirements under the LT2ESWTR must submit challenge testing results to the State in accordance with the rule (40 CFR 141.721(f)(10)(i)(A)). These results must be specific to the particular membrane filtration system used for compliance and must demonstrate the established removal efficiency of the process. In addition, the established removal efficiency must be equal to or greater than the *Cryptosporidium* credit proposed to be achieved with the membrane filtration system for compliance with the rule at the site-specific application. States may also have additional challenge testing reporting requirements. In order to facilitate State acceptance of challenge testing results, a sample outline of a challenge test report is provided as follows. Note that this outline should be customized to meet any particular State requirements.

1) Introduction
   a) Description of testing organization
   b) Test site
   c) Description of membrane filtration product
   d) Testing objectives (including target LRV)

2) Membrane Modules and Test Apparatus
   a) Membrane module specifications for each module evaluated
   b) Considerations for small-scale module testing (if applicable)
   c) Non-destructive performance testing
   d) Description of test apparatus

3) Challenge Test Protocol
   a) Challenge particulate (including rationale for selection)
   b) System operating conditions
   c) Challenge test solution design
   d) Seeding method
   e) Process monitoring
   f) Detailed sampling plan
   g) QA/QC procedures
   h) Data management

4) Results and Discussion
   a) Summary of measured system operating conditions
   b) Summary of LRV results for each module tested
   c) Summary of system integrity evaluation
   d) Determination of removal efficiency
   e) Summary of NDPT results for each module tested
   f) QCRV determination based on the results of non-destructive performance testing
   g) Statistical evaluation of results (if applicable)

5) Summary and Conclusions

a) Summary description of membrane filtration product
b) Summary of challenge test protocol
c) LRV demonstrated during challenge testing
d) Quality control release value for non-destructive performance testing

3.14 Re-Testing of Modified Membrane Modules

As a component of ongoing innovation and product development, manufacturers may make changes to a particular product line or its associated manufacturing process. If such a change affects the fundamental characteristics of the module, the removal efficiency, and/or the NDPT results and associated QCRV, the LT2ESWTR requires the modified product to be re-subjected to challenge testing (40 CFR 141.719(b)(2)(viii)).

Because it is not possible to develop a comprehensive listing of the potential modifications that would require re-testing, the need to re-test a modified product must be evaluated on a case-by-case basis and at the discretion of the State. Some examples of membrane properties, which if modified, might alter the fundamental removal characteristics of a module, include:

- Membrane material
  (e.g., a change in the polymer or backing material)

- Pore size (nominal and absolute)

- Porosity

- Permeability

- Membrane symmetry
  (i.e., symmetric, asymmetric, or composite)

As shown in the examples given above, most of the changes that may necessitate re-testing are modifications to the membrane material itself, such as changes to the polymer chemistry and/or pore size distribution. Even if the change to the membrane media is not intended to affect removal efficiency or NDPT parameters, it may still be necessary to re-test the module, since a modification to an intrinsic property of the membrane media could have an impact on one or both of these criteria.

Minor changes or changes that are not related to the fundamental removal characteristics of the membrane media or NDPT parameters would not require re-testing. For example, a change in the membrane area within a module would not be expected to affect either the removal efficiency or non-destructive performance testing.

Although not directly related to the product itself, any modifications to the manufacturer's NDPT may also require that the product line be re-subjected to challenge testing, since the results of the specific NDPT are used ensure that the modules produced meet the minimum requirements for achieving the demonstrated LRV. However, if the modified NDPT can be correlated to the specific test associated with product line's challenge test, additional challenge testing may not need to be conducted, at the discretion of the State.

Modifications to the hydraulic configuration of a membrane filtration system might warrant re-testing, since the concentration of suspended solids on the feed side of the membrane may be affected, in turn potentially affecting removal efficiency. In determining whether or not a modification to the hydraulic configuration would require re-testing, one significant factor is whether the hydraulic conditions of the original challenge test are conservative for the new hydraulic configuration of the system. In general, testing performed under a hydraulic configuration with a higher VCF than that associated with the modification would be considered conservative. A summary of the various hydraulic configurations and the general range of VCFs that might be expected for each based on typical system operating parameters is provided in Table 2.3, and duplicated for reference as Table 3.6, below. Note that the ranges given in Table 3.6 are intended to be illustrative only and not a substitute for more rigorous VCF determination using the theoretical models or experimental evaluation.

Table 3.6 Typical Range of VCF Values for Various Hydraulic Configurations

Hydraulic Configuration		VCF
Deposition mode	Dead-end	1
Suspension mode	PFR	3 - 20
	Crossflow[1]	4 - 20
	CSTR[2]	4 - 20

1 Encompasses both large and small volume crossflow systems
2 Encompasses CSTR systems both with and without backwash

In general, most modifications to the module itself (as opposed to the membrane media) are unlikely to affect either removal efficiency or non-destructive performance testing, although re-testing might be warranted in some cases.

The product modifications addressed in this section are provided as guidance to help assess whether or not a particular change might require a product line to undergo a new challenge test under the LT2ESWTR. This discussion is not intended to be comprehensive. In addition, the inclusion of a particular modification in this section does not necessarily imply that re-testing is required for all such modifications. Although the manufacturer can exercise its discretion in deciding the circumstances under which additional testing is necessary, the State has authority regarding whether or not to accept a particular change such that the product line would still be eligible to receive *Cryptosporidium* removal credit under the rule without re-testing.

3.15 Grandfathering Challenge Test Data From Previous Studies

As a result of the commercialization of membrane treatment processes for drinking water production, a number of States, consulting engineers, and independent testing organizations have developed programs to demonstrate the removal efficiency of membrane systems. Under these existing programs, numerous challenge studies have been conducted in which complete removal of *Cryptosporidium* was demonstrated, independent of the challenge level (USEPA 2001). The State has the discretion to accept data from a previous study provided that the testing was conducted in a manner that is consistent with LT2ESWTR requirements and demonstrates removal efficiency for *Cryptosporidium* commensurate with the treatment credit awarded to the process (40 CFR 141.719(b)(2)).

While a previously conducted challenge test may adequately demonstrate the required *Cryptosporidium* removal efficiency, it may be more difficult to correlate the results of a prior test to the NDPT currently used by the manufacturer. If the State determines that the grandfathered test does not meet the intent of the LT2ESWTR, a new challenge test that is consistent with the rule requirements must be conducted.

As a general guide, the following challenge test conditions have been identified as potentially yielding results that do not satisfy the intent of the rule:

- <u>Challenge testing conducted on obsolete products</u>: Refer to section 3.14 for guidance on the re-testing of modified membrane modules.

- <u>Challenge testing conducted on small-scale modules</u>: Small-scale module testing is permitted under the LT2ESWTR if certain criteria are met. Refer to section 3.8 for guidance regarding the testing of small-scale modules.

- <u>Challenge testing using unacceptable surrogates for *Cryptosporidium*</u>: The challenge particulate used in a grandfathered test must provide equivalent or sufficiently conservative removal efficiency relative to *Cryptosporidium* oocysts. Refer to section 3.9 regarding the selection of surrogates for use in challenge testing.

- <u>Challenge particulate enumeration using unacceptable methodology</u>: The challenge particulate must have been quantified using an acceptable method. Specifically, gross measurements are generally considered unacceptable. Refer to section 3.9 regarding methods for enumerating various challenge particulates.

- <u>Unavailable QCRV</u>: If non-destructive performance testing was not used to establish a suitable QCRV in a previous study, it may be difficult or impossible to relate the demonstrated removal efficiency to the NDPT results for untested modules that are produced.

There may also be cases in which deviations from challenge testing requirements under the LT2ESWTR may not be significant, such that additional testing would not be required. Some potential cases in which challenge test data that do not specifically comply with the rule might receive favorable consideration for grandfathering, at the discretion of the State, are listed below.

- Removal efficiency determination using different method: It may be necessary to recalculate the removal efficiency (i.e., LRV_{C-Test}) from a previous challenge test according to the requirements of the LT2ESWTR. If some moderate deficiencies exist in the data such that the LRV_{C-Test} cannot be calculated according to the procedure described in this guidance manual, the State may exercise discretion as to whether or not to approve the prior test. The State may also evaluate the impact of such deficiencies and consequently consider reducing the LRV_{C-Test} on the basis of this evaluation.

- Elevated feed challenge particulate concentration: The LT2ESTR places a limit on the feed concentration of the challenge particulate that essentially establishes a maximum LRV of 6.5 in order to prevent excessive over-seeding during challenge studies. However, exceeding this maximum feed concentration does not necessarily invalidate a prior test. In reviewing grandfathered data, it is most important to consider whether or not high LRVs were achieved through excessive over-seeding. For example, if a removal study was conducted at a challenge level of 10^8 particles/L and all filtrate concentrations were less than 10 particles/L, the process could potentially qualify for the maximum removal credit of 6.5 log. By contrast, if the same challenge level was applied to a membrane that allowed 10^3 particles/L to pass into the filtrate, States may opt not to qualify the test for a LRV of 5, since the high LRV is likely a result of over-seeding an otherwise inadequate or non-integral membrane.

- Challenge testing at a flux or recovery other than the maximum design value(s): Deviations from testing at the maximum flux and/or recovery might not exclude a prior test from consideration for grandfathering. Most available data suggests that flux and recovery will not significantly impact the removal efficiency of a membrane filtration process for *Cryptosporidium*-sized particles. If complete removal of the challenge particulate was achieved during the challenge test, then deviations from testing at maximum design flux and recovery values can probably be ignored.

- Challenge testing using a poor quality feed water matrix: Although it is recommended that challenge testing be conducted using both unfouled membranes and clean make-up water for the test solution to provide a conservative estimate of removal efficiency, prior studies conducted with waters with a greater fouling potential (e.g., untreated surface waters) can potentially be considered to meet the

LT2ESWTR challenge test requirement, since the removal of *Cryptosporidium*-sized particles using membrane filtration processes as defined under the LT2ESWTR is generally unaffected by the presence of foulants. However, recent research suggests that integrity defects on the order of 200 μm can be obscured by foulants in some cases, improving pathogen rejection (Lozier et al. 2003). Thus, prior studies conducted with poorer quality water should be considered for grandfathering on a case-by-case basis.

4.0   Direct Integrity Testing

4.1   Introduction

In order for a membrane process to be an effective barrier against pathogens and other particulate matter, the filtration system must be integral, or free of any integrity breaches. Thus, it is critical that operators are able to demonstrate the integrity of this barrier on an ongoing basis during system operation. Direct integrity testing represents the most accurate means of assessing the integrity of a membrane filtration system that is currently available.

Under the Long Term 2 Enhanced Surface Water Treatment Rule (LT2ESWTR), a direct integrity test is defined as a physical test applied to a membrane unit in order to identify and isolate integrity breaches. In order to receive *Cryptosporidium* removal credit for compliance with the rule, the removal efficiency of a membrane filtration process must be routinely verified during operation using direct integrity testing. The direct integrity test must be applied to the physical elements of the entire membrane unit, including membranes, seals, potting material, associated valves and piping, and all other components which could result in contamination of the filtrate under compromised conditions (40 CFR 141.719(b)(3)(i)).

There are two general classes of direct integrity tests that are commonly used in membrane filtration facilities: ***pressure-based tests*** and ***marker-based tests***. The pressure-based tests are based on bubble point theory (as described in Appendix B) and involve applying a pressure or vacuum (i.e., negative pressure) to one side of a membrane barrier and monitoring for parameters such as pressure loss or the displacement of air or water in order to establish whether an integrity breach is present. The various pressure-based tests include the pressure and vacuum decay tests, the diffusive airflow test, and the water displacement test. Marker-based tests utilize either a spiked particulate or molecular marker to verify membrane integrity by directly assessing removal of the marker, similar to a challenge test.

The LT2ESWTR does not require the use of a particular direct integrity test for rule compliance, but rather that any test used meet the specified performance criteria for ***resolution***, ***sensitivity***, and ***frequency***. Thus, a particular system may utilize an appropriate pressure- or marker-based test or any other method that both meets the performance criteria and is approved by the State. The performance criteria for direct integrity tests are summarized as follows:

- **Resolution**: The direct integrity test must be responsive to an integrity breach on the order of 3 μm or less (40 CFR 141.719(b)(3)(ii)).

- **Sensitivity**: The direct integrity test must be able to verify a log removal value (LRV) equal to or greater than the removal credit awarded to the membrane filtration process (40 CFR 141.719(b)(3)(iii)).

- **Frequency**: A direct integrity test must be conducted on ***each membrane unit*** at a frequency of no less than once each day that the unit is in operation. (The definition of a membrane unit under the LT2ESWTR is provided in section 1.5.) Less frequent testing may be approved by the State if supported by demonstrated process reliability, the use of multiple barriers effective for *Cryptosporidium*, or reliable process safeguards (40 CFR 141.719(b)(3)(vi)).

In addition to the performance criteria, the rule also requires the establishment of a ***control limit*** within the associated sensitivity limits for the direct integrity test that is indicative of an integral membrane unit capable of achieving the *Cryptosporidium* removal credit awarded by the State (40 CFR 141.719(b)(3)(iv)). If the results of the direct integrity test exceed this limit, the rule requires that the affected membrane unit be taken off-line for diagnostic testing and repair (40 CFR 141.719(b)(3)(v)). The unit may be returned to service only after system integrity has been confirmed by a direct integrity test.

The objective of Chapter 4 is to describe the various pressure- and marker-based direct integrity tests currently in use and how these tests can be applied to meet the performance criteria specified under the LT2ESWTR. Diagnostic tests, data collection, and reporting are also addressed.

This chapter is divided into the following sections:

Section 4.2:   Test Resolution
This section discusses the determination of pressure- and marker-based direct integrity test resolution for meeting the performance criteria required by the rule.

Section 4.3:   Test Sensitivity
This section discusses the determination of pressure- and marker-based direct integrity test sensitivity for meeting the performance criteria required by the rule, including general concepts and methods.

Section 4.4:   Test Frequency
This section reviews the direct integrity testing frequency requirements of the rule.

Section 4.5:   Establishing Control Limits
This section describes the mathematical and experimental determination of control limits for direct integrity testing.

Section 4.6:   Example: Establishing Direct Integrity Test Parameters
This section illustrates the calculation of some of the critical direct integrity test performance criteria, including test resolution, sensitivity, and control limits for an example membrane filtration system.

Section 4.7: Test Methods
This section provides an overview of the various types of pressure- and marker-based tests, including generic test protocols as well as some advantages and disadvantages of each.

Section 4.8: Diagnostic Testing
This section describes some of the diagnostic tests that are used to identify and isolate integrity breaches following a failed direct integrity test.

Section 4.9: Data Collection and Reporting
This section provides guidance on direct integrity test data collection and reviews the associated reporting requirements of the rule.

4.2 Test Resolution

Resolution is defined as the size of the smallest integrity breach that contributes to a response from a direct integrity test. Any direct integrity test applied to meet the requirements of the LT2ESWTR is required to have a resolution of 3 μm or less. This resolution criterion is based on the lower size range of *Cryptosporidium* oocysts and is intended to ensure that any integrity breach large enough to pass oocysts contributes to a response from the direct integrity test used. The manner in which the resolution criterion is met depends on whether the direct integrity test is pressure-based or marker-based, as described in the following subsections.

4.2.1 Pressure-Based Tests

In order to achieve a resolution of 3 μm with pressure-based direct integrity tests, the net pressure applied during the test must be great enough to overcome the capillary forces in a 3 μm hole, thus ensuring that any breach large enough to pass *Cryptosporidium* oocysts would also pass air during the test. The minimum applied test pressure necessary to achieve a resolution of 3 μm is calculated using Equation 4.1:

$$P_{test} = (0.193 \bullet \kappa \bullet \sigma \bullet \cos\theta) + BP_{max} \qquad \text{Equation 4.1}$$

Where:

P_{test}	=	minimum test pressure (psi)
κ	=	pore shape correction factor (dimensionless)
σ	=	surface tension at the air-liquid interface (dynes/cm)
θ	=	liquid-membrane contact angle (degrees)
BP_{max}	=	maximum backpressure on the system during the test (psi)
0.193	=	constant that includes the defect diameter (i.e., 3 μm resolution requirement) and unit conversion factors

Equation 4.1 is based on bubble point theory and is derived from the balancing of capillary static forces. Note that the constant of 0.193 accounts for the LT2ESWTR resolution requirement of responding to a defect of at least 3 µm in diameter, as well as the appropriate unit conversion factors, in order to simplify the equation for the purposes of rule compliance. The general form of Equation 4.1 includes the capillary diameter as a variable and represents an expression relating this diameter to the bubble point pressure. A discussion of bubble point theory and a derivation of Equation 4.1 is provided in Appendix B.

Values for several parameters in Equation 4.1 must be determined in order to calculate the minimum test pressure necessary to achieve a resolution of 3 µm when using a pressure-based integrity test. The parameters κ and θ are intrinsic properties of the membrane. In the absence of data supplied by the membrane manufacturer, conservative values of $\kappa = 1$ and $\theta = 0$ should be used. Appendix B provides an additional discussion of these parameters. The surface tension, σ, is inversely related to temperature; consequently, the surface tension at the coldest anticipated water temperature should be used to calculate a conservative value for the minimum required test pressure. As a point of reference, the surface tension of water at 5 °C is 74.9 dynes/cm. Substituting these three values (i.e., $\kappa = 1$, $\theta = 0$, and $\sigma = 74.9$ dynes/cm) into Equation 4.1 yields the following simplified equation:

$$P_{test} = 14.5 + BP_{max} \hspace{3cm} \text{Equation 4.2}$$

Where: P_{test} = minimum test pressure (psi)

BP_{max} = maximum backpressure on the system during the test (psi)

Equation 4.2 indicates that the minimum test pressure necessary to achieve a 3-µm resolution is 14.5 psi plus the maximum backpressure on the system during application of a pressure-based direct integrity test (at a conservative temperature of 5 °C). Ideally, there should be no hydrostatic backpressure on the system during the test. However, it is not always practical to perform the test without any hydrostatic backpressure, and in these cases the additional backpressure must be considered in establishing the minimum test pressure necessary to meet the resolution criterion. For example, there might be hydrostatic pressure on the undrained side of the membrane if a pressure-driven membrane module remains filled with water or if a vacuum-driven (i.e., immersed) membrane remains submerged under water in a basin. Thus, if the bottom of the membrane is under 7 feet of water, BP_{max} would be approximately 3 psi, yielding a P_{test} value of 17.5 psi to achieve a resolution of 3 µm.

Both Equations 4.1 and 4.2 assume that the applied pressure remains constant during the direct integrity test. However, in many cases there may be some baseline decay (e.g., that attributable to diffusion) that is measurable over the duration of the test. In this case, it is important to account for this baseline decay in the resolution calculation. Thus, in order to ensure that the resolution requirement is satisfied throughout the duration of the test, the anticipated pressure at the end of the direct integrity test should be used to calculate the resolution. This value can be estimated using the initial applied pressure, the typical rate of baseline pressure decay for a fully integral system, and the duration of the test. If the baseline decay is small enough such that the final test pressure is within approximately 5 percent of the initial applied pressure, the baseline decay can be assumed to be negligible, and the initial applied pressure may be used to calculate the test resolution.

The LT2ESWTR does not establish the minimum test pressure to be used during a pressure-based direct integrity test, but rather only requires that the test achieve a 3-μm resolution. If a membrane manufacturer has information to support the use of values other than $\kappa = 1$ and $\theta = 0$, and these less conservative values are approved by the State, then Equation 4.1 can be used to calculate the minimum required test pressure. It is essential that the use of values other than $\kappa = 1$ and $\theta = 0$ be scientifically defensible, since the use of inappropriate values could result in the use of a test pressure that does not meet the resolution criterion established by the rule. One approach for determining membrane-specific values for κ and θ is through direct experimental evaluation. Because these parameters can have a significant effect on the required direct integrity test pressure, it is strongly recommended that States require sufficient justification from a membrane manufacturer prior to approving the use of values other than $\kappa = 1$ and $\theta = 0$, such as independent third party testing results using a method accepted by the scientific community and demonstrating statistically significant data.

Although the rule does not include a frequency requirement for the recalculation of resolution, the resolution should be recalculated if the system backpressure during direct integrity testing is adjusted. Alternatively, if desired, direct integrity test instrumentation and the data recording system could be configured to calculate the resolution after each application of the direct integrity test using the applied test pressure, system backpressure, and surface tension corresponding to the temperature at which the test is conducted. Note that the liquid-membrane contact angle can also change over the life of a membrane module (e.g., as a result of the adsorption of organic matter by the membrane material), and these changes may not necessarily be uniform among the various modules in a unit (Childress et al. 1996; Jucker et al. 1994). Thus, if a value other than $\theta = 0$ (i.e., the most conservative value) is used, then it may be appropriate to periodically recalculate the resolution based on a revised estimate of the actual value of θ, an exercise that may necessitate destructive testing of a representative sample of membrane modules in the system.

4.2.2 Marker-Based Tests

A marker-based direct integrity test can be viewed as a "mini challenge study," in which a surrogate is periodically applied to the feed water in order to verify the integrity of a membrane filtration system. In order to meet the resolution criterion of the rule, the surrogate used in a marker-based test must have an effective size of 3 μm or smaller, as described in section 3.9.2. A marker-based direct integrity test can use either particulate or molecular surrogates, but in either case, it must be established that the surrogate meets the resolution criterion. Section 3.9 presents guidelines for the selection of a conservative surrogate for *Cryptosporidium* during challenge testing, and these same guidelines are applicable to selection of an appropriate surrogate for a marker-based direct integrity test. The effective size of the marker can be established through any accepted methodology such as size distribution analysis of particulate markers or estimation techniques based on the molecular weight and geometry of molecular markers.

4.3   Test Sensitivity

Sensitivity is defined as the maximum log removal value that can be reliably verified by the direct integrity test (i.e., LRV_{DIT}). The sensitivity of the direct integrity test establishes the maximum log removal credit that a membrane filtration process is eligible to receive if it is less than or equal to that demonstrated during challenge testing (i.e., LRV_{C-Test}). For example, if the challenge test demonstrated a LRV_{C-Test} of 5.5 log, and the direct integrity test is capable of verifying a LRV_{DIT} of 4.5 log, the membrane filtration process would be eligible for removal credit up to 4.5 log. Although the sensitivity of the direct integrity test should not be expected to vary significantly over time, the determination of sensitivity as described in this section is designed to produce a conservative result that would remain applicable over the life of the membrane filtration system. However, if significant changes occur in terms of operational parameters, direct integrity test conditions, or any basic assumptions that might affect the value of the direct integrity test sensitivity, it is suggested that the sensitivity be reestablished to verify that it is at least equal to the removal credit awarded to the process.

The sensitivity of a direct integrity test is logarithmic in nature. For example, a test with a LRV_{DIT} of 5 log is 100 times more sensitive than a test with a LRV_{DIT} of 3 log. Thus, when a higher sensitivity is required, the test must be capable of measuring very small changes in the direct integrity test response and distinguishing these results from background or baseline data. Data suggest that many direct integrity tests, as currently applied, have sensitivities in excess of 4 log; however, sensitivity must be determined on a case-by-case basis using the information provided by the membrane manufacturer and the guidance in this document. While determination of integrity test sensitivity can be complex, it provides a rational basis for awarding high removal credits to membrane processes that are commensurate with their abilities. As was the case with resolution, the manner in which sensitivity is determined depends on whether the type of direct integrity test used is pressure- or marker-based.

Note that for systems that utilize multiple stages of membrane filtration, the sensitivity for each stage must be determined independently. The most common example of such an application would be a case in which a second stage is used to treat the backwash water from the first stage, after which the filtrate from the two stages is blended (i.e., concentrate staging). In this case, as well as in others involving multiple stages, if the filtrate from the various stages is blended, the stage using the membrane units with the lowest sensitivity would limit the maximum log removal credit that could be awarded to the overall process. However, if a second (or subsequent) stage is used strictly for residuals management such that the filtrate is recycled to an upstream point in the overall treatment process, the LT2ESWTR would not be applicable to such ancillary stages.

4.3.1 Pressure-Based Tests

The discussion in this section regarding the calculation of sensitivity for pressure-based direct integrity tests is divided into three parts, as follow. First, the basic concepts that are applicable to all pressure-based tests are introduced. The subsequent section describes calculation of the sensitivity for pressure-based direct integrity tests based on this general conceptual framework. A third section discusses the determination of diffusive (or baseline) losses in a fully integral system during the application of a pressure-based direct integrity test.

4.3.1.1 Basic Concepts

The determination of sensitivity for pressure-based direct integrity tests is more complex than for marker-based tests. The equation used to determine the sensitivity of a pressure-based direct integrity test is specified in the LT2ESWTR (40 CFR 141.719(b)(3)(iii)(A)) and given below as Equation 4.3:

$$LRV_{DIT} = \log\left(\frac{Q_p}{VCF \bullet Q_{breach}}\right) \qquad \text{Equation 4.3}$$

Where: LRV_{DIT} = direct integrity test sensitivity in terms of LRV (dimensionless)

Q_p = membrane unit design capacity filtrate flow (L/min)

Q_{breach} = flow from the breach associated with the smallest integrity test response that can be reliably measured, referred to as the *critical breach size* (L/min)

VCF = volumetric concentration factor (dimensionless)

Equation 4.3 represents a dilution model that assumes water passing through the intact membrane barrier is free of the particulate contaminant of interest and that water flowing through an integrity breach has a particulate contaminant concentration equal to that on the high pressure side of the membrane. Under these assumptions, LRV_{DIT} is a function of the ratio of total filtrate flow to flow through the critical breach (i.e., Q_p/Q_{breach}), which quantifies the dilution of the contaminated stream passing through the breach as it mixes with treated filtrate. For a membrane unit of a given capacity (i.e., constant Q_p), LRV_{DIT} will increase as Q_{breach} decreases. This implies that a more sensitive direct integrity test capable of detecting a smaller breach can verify a higher log removal value and thus potentially increase the removal credit that a membrane filtration system is eligible to receive.

The volumetric concentration factor (VCF) is a dimensionless term that accounts for the increase in the suspended solids concentration that occurs on the feed side of the membrane for some hydraulic configurations. The VCF is important in the context of membrane integrity because the risk of filtrate contamination in the event of an integrity breach is increased for systems in which the influent suspended solids are concentrated on the feed side of the membrane. For example, for the same size integrity breach, systems with higher VCFs will allow increased passage of pathogens to the filtrate, reducing the verifiable removal efficiency. This affect is taken into account in the determination of sensitivity for pressure-based direct integrity tests in that systems with higher VCFs have proportionately lower test sensitivity. Thus, in the denominator of Equation 4.3, the term Q_{breach} is multiplied by the VCF to account for the impact of this concentration effect.

The VCF is calculated as the ratio of the concentration of suspended solids maintained on the feed side of the membrane to that in the influent feed water, as shown in Equation 4.4 (also Equation 2.22):

$$VCF = \frac{C_m}{C_f} \qquad \text{Equation 4.4 \quad (2.22)}$$

Where:

    VCF   =   volumetric concentration factor (dimensionless)

    C_m =   concentration of suspended solids maintained on the feed side of the membrane (number or mass / volume)

    C_f =   concentration of suspended solids in the influent feed water to the membrane system (number or mass / volume)

The VCF generally ranges between 1 and 20, and the value depends on the hydraulic configuration of the system. Membrane systems that operate in deposition mode do not increase the concentration of suspended solids on the feed side of the membrane and thus have a VCF equal to one. In contrast, membrane systems that operate in suspension mode, such as those that utilize a crossflow hydraulic configuration, typically have a VCF in the range of 4 to 20, representing a 4- to 20-fold increase in the suspended particle concentration on the feed side of the membrane. The methods and equations used to calculate the VCF for various hydraulic configurations are provided in section 2.5, and Table 2.4 presents equations for calculating both the average and maximum VCF for various hydraulic configurations. Alternatively, the VCF could be determined experimentally, as discussed in section 2.5.4.

Note that the LT2ESWTR does not specify use of the maximum or average VCF value in calculating sensitivity, but the rule does require that the increase in suspended solids concentration on the high-pressure side of the membrane, as occurs with some hydraulic configurations, be considered in the calculation. The maximum VCF typically ranges from 1 to 20 and provides the most conservative value for LRV_{DIT}, while the average VCF typically ranges from 1 to 7. In selecting between the maximum, average, or any other value for the VCF, consideration should be given to the concentration profile along the membrane surface in the direction of water flow and the implication of integrity breaches at various locations within the membrane unit. For example, although the maximum VCF does provide the most conservative value for LRV_{DIT}, for systems in which the concentration varies with position this value represents only a very small portion of the concentration profile and thus is only representative of breaches that occur at the extreme end of the membrane unit. Similarly, some systems exhibit a concentration profile as a function of time within a filtration cycle, and the maximum VCF only occurs at the end of the filtration cycle immediately before a backwash event; prior to this time, the VCF is significantly lower.

4.3.1.2 Calculating Sensitivity

The sensitivity of a pressure-based direct integrity test can be calculated by converting the response from a pressure-based test that measures the flow of air (e.g., the diffusive airflow test) or rate of pressure loss (e.g., the pressure decay test) to an equivalent flow of water through an integrity breach during normal operation, as described in the following subsection. The second subsection outlines a general procedure for determining the threshold response of a pressure-based direct integrity test experimentally, if this information is not available from the membrane filtration system manufacturer.

Calculating Sensitivity Using the Air-Liquid Conversion Ratio

In order to calculate the LRV_{DIT}, the flow through the critical breach for a direct integrity test (i.e., Q_{breach}) must be determined (as shown in Equation 4.3). Since most direct integrity tests do not directly measure Q_{breach}, it is necessary to establish a correlation between the direct integrity test response and the flow of water through the critical breach during system operation. In some of the most commonly used pressure-based direct integrity tests, including the pressure- or vacuum-decay test and the diffusive airflow test, air is applied to the drained side of a membrane and subsequently flows through any integrity breach that exceeds the test resolution. The response from such a test is typically measured as pressure decay or airflow. In order to relate the response from a pressure-based integrity test to Q_{breach}, it is necessary to establish a correlation between airflow and liquid flow through the critical integrity breach.

This correlation can be characterized through the air-liquid conversion ratio (ALCR), which is defined as the ratio of air that would flow through a breach during a direct integrity test to the amount of water that would flow through the breach during filtration, as defined in Equation 4.5:

$$ALCR = \left(\frac{Q_{air}}{Q_{breach}} \right)$$ Equation 4.5

Where: ALCR = air-liquid conversion ratio (dimensionless)

Q_{air} = flow of air through the critical breach during a pressure-based direct integrity test (L/min)

Q_{breach} = flow of water through the critical breach during filtration (L/min)

The ALCR can be used to express the liquid flow through a breach in terms of corresponding flow of air, as shown in Equation 4.6:

$$Q_{breach} = \left(\frac{Q_{air}}{ALCR} \right)$$

Equation 4.6

Where:

Q_{breach}	=	flow of water through the critical breach during filtration (L/min)
Q_{air}	=	flow of air through the critical breach during a pressure-based direct integrity test (L/min)
ALCR	=	air-liquid conversion ratio (dimensionless)

Substituting Equation 4.6 into the general expression for sensitivity (Equation 4.3) yields the following expression:

$$LRV_{DIT} = \log \left(\frac{Q_p \bullet ALCR}{Q_{air} \bullet VCF} \right)$$

Equation 4.7

Where:

LRV_{DIT}	=	direct integrity test sensitivity in terms of LRV (dimensionless)
Q_p	=	membrane unit design capacity filtrate flow (L/min)
ALCR	=	air-liquid conversion ratio (dimensionless)
Q_{air}	=	flow of air through the critical breach during a pressure-based direct integrity test (L/min)
VCF	=	volumetric concentration factor (dimensionless)

Equation 4.7 can be used to directly calculate the sensitivity for any pressure-based direct integrity test that is based on bubble point theory and which measures the flow of air (Q_{air}) through an integrity breach. The four parameters that need to be determined to calculate sensitivity are: Q_p, VCF, ALCR, and Q_{air}. The VCF can be established as previously described in section 4.3.1.1. Q_p is the design capacity filtrate flow approved by the State from the membrane unit to which the direct integrity test is applied. For a constant Q_{breach}, higher filtrate flows yield greater direct integrity test sensitivity. Thus, if different sizes of membrane units with varying capacity are used in the same treatment system, the sensitivity of the direct integrity test should be independently determined for each unit size, with the lowest sensitivity establishing the maximum log removal credit for the overall process. The flow of air, Q_{air}, is related to the response from the direct integrity test. For tests that measure the airflow through an integrity breach directly, such as the diffusive airflow test, Q_{air} is simply the airflow measured during application of the test. On the other hand, methods such as the pressure and vacuum decay tests yield results in terms of pressure loss per unit time, which must be converted to an equivalent flow of air using Equation 4.8:

$$Q_{air} = \frac{\Delta P_{test} \bullet V_{sys}}{P_{atm}}$$

Equation 4.8

Where:

Q_{air}	=	flow of air (L/min)
ΔP_{test}	=	rate of pressure decay during the integrity test (psi/min)
V_{sys}	=	volume of pressurized air in the system during the test (L)
P_{atm}	=	atmospheric pressure (psia)

Note that Equation 4.8 assumes that the temperature of the water and air are the same since the air temperature should rapidly equilibrate with the water temperature. V_{sys} encompasses the entire pressurized volume, including all fibers (typically the insides of the fibers are pressurized), piping, and other void space on the pressurized side.

Substituting Equation 4.8 into Equation 4.7 yields Equation 4.9, which can be used to calculate the sensitivity of a direct integrity test that measures the rate of pressure or vacuum decay.

$$LRV_{DIT} = \log\left(\frac{Q_p \bullet ALCR \bullet P_{atm}}{\Delta P_{test} \bullet V_{sys} \bullet VCF}\right)$$

Equation 4.9

Where:

LRV_{DIT}	=	direct integrity test sensitivity in terms of LRV (dimensionless)
Q_p	=	membrane unit design capacity filtrate flow (L/min)
ALCR	=	air-liquid conversion ratio (dimensionless)
P_{atm}	=	atmospheric pressure (psia)
ΔP_{test}	=	smallest rate of pressure decay that can be reliably measured and associated with a known integrity breach during the integrity test (psi/min)
V_{sys}	=	volume of pressurized air in the system during the test (L)
VCF	=	volumetric concentration factor (dimensionless)

Regardless of whether the flow of air (Q_{air}) or pressure decay rate (ΔP_{test}) is measured during the direct integrity test, the smallest response from the test that can be reliably measured and associated with an integrity breach should be used in the sensitivity calculation. This should not be confused with the baseline integrity test response from an *integral* membrane unit, since there may be a small airflow or pressure decay due to diffusion of air through water in the wetted pores and/or the membrane material, even if there are no breaches in the system. In many cases, this smallest measurable response associated with an integrity breach may be provided by the membrane filtration system manufacturer. If this information is not available from the system manufacturer, it may be determined experimentally by progressively creating small integrity breaches of a known size in an otherwise integral membrane unit in order to determine the smallest measurable response from the direct integrity test that is distinguishable from the baseline response for an integral membrane unit. A general procedure for this experimental method is described in the following subsection ("Measuring The Threshold Direct Integrity Test Response Experimentally").

The basic procedure for calculating the ALCR involves first making a reasonable assumption regarding whether the flow through the critical breach is laminar or turbulent for the particular membrane filtration system of interest. Then the ALCR can be calculated using the appropriate equation, as summarized in Table 4.1. Note that the equations in Table 4.1 assume that the flow regimes for the passage of air and water through an integrity breach are the same (i.e., both either laminar or turbulent). If this assumption is not considered appropriate for a specific system and/or application such that inaccurate estimates for direct integrity test sensitivity may result, a hybrid approach may be considered, as described in Appendix C. Other assumptions utilized in the derivation of the ALCR equations in Table 4.1 may also not be valid for some membrane filtration systems. Section C.5 provides some examples of these types of systems and along with general guidance for modifying the ALCR formula derivations to accommodate these systems.

Table 4.1 Approaches for Calculating the ALCR

Module Type	Defect Flow Regime	Model	ALCR Equation	Appendix C Equation
Hollow-fiber[1]	Turbulent[2]	Darcy pipe flow	$170 \bullet Y \bullet \sqrt{\dfrac{(P_{test} - BP) \bullet (P_{test} + P_{atm})}{(460 + T) \bullet TMP}}$	C.4
	Laminar	Hagen-Poiseuille[3]	$\dfrac{527 \bullet \Delta P_{eff} \bullet (175 - 2.71 \bullet T + 0.0137 \bullet T^2)}{TMP \bullet (460 + T)}$	C.15
Flat sheet[4]	Turbulent	Orifice	$170 \bullet Y \bullet \sqrt{\dfrac{(P_{test} - BP) \bullet (P_{test} + P_{atm})}{(460 + T) \bullet TMP}}$	C.9
	Laminar	Hagen-Poiseuille[3]	$\dfrac{527 \bullet \Delta P_{eff} \bullet (175 - 2.71 \bullet T + 0.0137 \bullet T^2)}{TMP \bullet (460 + T)}$	C.15

1 Or hollow-fine-fiber
2 Typically characteristic of larger diameter fibers and higher differential pressures
3 The binomial in the Hagen-Poiseuille equation (C.15) approximates the ratio of water viscosity to air viscosity and is valid for
     temperatures ranging from approximately 32 to 86 °F. Additional details are provided in Appendix C.
4 Includes spiral-wound and cartridge configurations

The various parameters given in the ALCR equations listed in Table 4.1 include the following:

Y	=	net expansion factor for compressible flow through a pipe to a larger area (dimensionless)
P_{test}	=	direct integrity test pressure (psi)
BP	=	backpressure on the system during the direct integrity test (psi)
P_{atm}	=	atmospheric pressure (psia)
T	=	water temperature (^{o}F)
TMP	=	transmembrane pressure (psi)
ΔP_{eff}	=	effective integrity test pressure (psi)

Additional guidance for both calculating the ALCR and determining appropriate values for the component parameters of the respective ALCR equations (as listed above) is provided in Appendix C, along with derivations of the ALCR equations. Further information regarding the net expansion factor (Y) may be found in various hydraulics references, including Crane (1988). Note that although the Darcy and orifice equations for the ALCR are identical (as shown in Table 4.1), the method for determining the net expansion factor (Y) is different for these two models, as described in Appendix C.

The ALCR can also be determined via empirical means, which would be applicable to any flow regime or configuration of membrane material and are independent of a particular hydraulic model. Some manufacturers may have developed empirical models that could be used to determine the ALCR. If an empirical approach is preferred for determining the ALCR, and a valid empirical model is not available for the system, it may be necessary to develop one. One conceptual procedure for empirically deriving the ALCR for hollow-fiber membrane filtration systems is the correlated airflow measurement (CAM) technique; the details of this procedure are presented in Appendix D.

Measuring the Threshold Direct Integrity Test Response Experimentally

The smallest measurable response of a pressure-based direct integrity test that is associated with a known breach can be evaluated experimentally if this information is not available from the membrane filtration system manufacturer. For pressure-based tests, this response corresponds to the value of ΔP_{test} that should be used in the calculation of sensitivity. This experimental evaluation involves intentionally compromising system integrity in small, discrete, and quantifiable steps and monitoring the corresponding integrity test responses. In the case of microfiltration (MF) and ultrafiltration (UF) systems, several fiber-cutting studies conducted to evaluate the threshold response of various direct integrity tests have been documented in the literature (Adham et al. 1995; Landsness 2001). In general, the procedure for measuring the threshold response experimentally involves the following steps:

1. The membrane system is determined to be integral through the application of a direct integrity test.

2. The investigator intentionally compromises a membrane to generate a known defect. Examples of such compromises include generating a hole in the membrane using a pin of a known diameter or cutting a hollow fiber at a predetermined location. In order to identify the threshold response, it is desirable to utilize a small integrity breach, such as a single cut fiber in a membrane unit.

3. After compromising the membrane, the integrity of the membrane unit is measured using the designated direct integrity test.

4. The process is repeated with additional defects of progressively increasing size or quantity until a measurable response from the direct integrity test is detected. This minimum measurable response represents ΔP_{test} for the purposes of calculating the sensitivity of a direct integrity test.

4.3.1.3 Diffusive Losses and Baseline Decay

If it is determined to be appropriate for the membrane under consideration, the diffusive losses that constitute the baseline integrity test response may be subtracted from the smallest measurable response associated with an integrity breach for the purpose of determining sensitivity. For example, if a pressure decay rate of 0.05 psi/min is typical for an integral membrane unit and the limitations of the test are such that the smallest pressure decay rate that can be reliably associated with an integrity breach is 0.12 psi/min, the incremental response associated with an integrity breach is 0.07 psi/min, and this value may be used in the sensitivity calculation, as illustrated in Equation 4.10 (a variation of Equation 4.8).

$$Q_{air} = \frac{(\Delta P_{test} - D_{base}) \bullet V_{sys}}{P_{atm}}$$

Equation 4.10

Where:

Q_{air}	=	flow of air (L/min)
ΔP_{test}	=	rate of pressure decay during the integrity test (psi/min)
D_{base}	=	baseline pressure decay (psi/min)
V_{sys}	=	volume of pressurized air in the system during the test (L)
P_{atm}	=	atmospheric pressure (psia)

In general, diffusive losses are most likely to be observed over the duration of the direct integrity test for thin-skinned, asymmetric membranes. The manufacturer should be able to provide information regarding whether or not diffusive losses are expected to be significant. If a high level of sensitivity is required for a membrane filtration system, baseline diffusive losses may be an important consideration. If diffusive losses are determined to be significant and the membrane manufacturer cannot provide a value for the baseline decay, this section describes a means to estimate that decay.

For porous MF, UF, and membrane cartridge filtration (MCF) membranes, diffusive losses occur during direct integrity testing because a certain amount of compressed air used during the test dissolves into the water in the fully-wetted pores and is transported across the membrane surface. In order to calculate the diffusive losses, it can be assumed that the water fills the pores of the membrane and forms a film of thickness z, and that diffusion directly through the membrane material itself is insignificant in comparison to the diffusion across the film of water. Using these assumptions, Equation 4.11 illustrates the relationship between diffusive losses and the various parameters that influence these losses.

$$Q_{diff} = 6 \bullet \left(\frac{D_{aw} \bullet A_m \bullet (P_{test} - BP) \bullet H \bullet \varepsilon}{z} \right) \bullet \left(\frac{R_{gas} \bullet T}{P_{atm}} \right) \qquad \text{Equation 4.11}$$

Where:

Q_{diff}	=	diffusive flow of air through the water held in the membrane pores (L/min)
D_{aw}	=	diffusion coefficient for air in water (cm^2/s)
A_m	=	total membrane surface area to which the direct integrity test is applied (m^2)
P_{test}	=	membrane test pressure (psi)
BP	=	backpressure on the system during the test (psi)
H	=	Henry's constant for air-water system (mol/psi-m^3)
ϵ	=	membrane porosity (dimensionless decimal)
z	=	membrane thickness (mm)
R_{gas}	=	universal gas constant (L-psia/mol-K)
T	=	water temperature during direct integrity test (K)
P_{atm}	=	atmospheric pressure (psia)
6	=	unit conversion factor

Note that the dimensionless porosity (ϵ) is defined as the ratio of area of pores to the total membrane area in the unit. This term should not be confused with the pore size of porous MF, UF, and MCF membranes, which is given in terms of the limiting dimension of the openings in the membrane. The porosity of the membrane material can typically be provided by the manufacturer, if necessary. In addition, the diffusion flow path, which is affected by porosity, tortuosity, and the differential pressure across the membrane, is approximated by the membrane thickness (z) in Equation 4.11. Because membrane porosity and tortuosity may be difficult to measure, thus making it problematic to accurately quantify the actual length of the diffusion flow path, a more precise empirical method accounting for these two factors as a combined term has been developed by Farahbakhsh (2003). Both the diffusion coefficient (D_{aw}) and Henry's constant (H) vary with temperature, and Henry's constant also varies somewhat with the concentration of dissolved solids in the water. However, these affects may partially offset and may not be significant. Values for D_{aw} and H as a function of these variables (as applicable) may be found in standard tables in the literature.

The parameters given above for Equation 4.11 are applicable to flat sheet porous membranes, such as those used in membrane cartridge configurations. For porous membranes in a hollow-fiber configuration, such as most MF and UF systems, the following modifications are required:

A_m = log mean total membrane area to which the direct integrity test is applied (m^2)

= $(A_2-A_1)/\ln(A_2/A_1)$

A_1 = total membrane surface area to which the direct integrity test is applied based on the ***inside*** fiber diameter

A_2 = total membrane surface area to which the direct integrity test is applied based on the ***outside*** fiber diameter

z = differential fiber radius (mm)

= $r_2- r_1$

r_1 = inside radius of the hollow fiber

r_2 = outside radius of the hollow fiber

It is generally accepted that diffusion of air across the membrane can reduce the measured LRV_{DIT}. Under most circumstances, the amount of diffusion is small in comparison to the flow of air through a breach. However, if the membrane has a propensity to diffuse a significant amount of air (e.g., if the porosity is unusually high) or if a high level of sensitivity is required, it may be necessary to account for diffusive losses. Typically, there is only limited information available regarding the amount of diffusion that occurs for membrane processes used in water treatment under production conditions; however, MF/UF membrane manufacturers can typically provide a value for the baseline decay during a pressure-based direct integrity test for their specific proprietary systems, so it is generally not necessary to explicitly calculate diffusive losses using Equation 4.11.

In the absence of information provided by the manufacturer, it may be advantageous to directly measure the baseline pressure decay on an integral membrane unit, a process that should be conducted using clean membranes to avoid the potential for fouling to artificially hinder diffusion. Because the diffusive loss is directly proportional to temperature (as shown in Equation 4.11), and the diffusion coefficient (which is also directly proportional to diffusive loss) also increases with temperature, it is also important to characterize the baseline decay for the membrane filtration system at a typical water temperature in order to generate an appropriately representative value for diffusive loss. This temperature of evaluation should be recorded for future reference.

With semi-permeable nanofiltration (NF) and reverse osmosis (RO) membranes, diffusive losses occur via the diffusion of air through the saturated membrane material itself. However, because NF and RO modules are manufactured separately from the accompanying filtration systems and inserted manually into pressure vessels, small seal leaks may occur that can be difficult to distinguish from baseline decay. Thus, it is recommended that baseline response for a pressure-based direct integrity test be evaluated for each unit in a NF/RO system on a site-specific basis.

If the membrane module manufacturer has information for typical diffusion airflow rates per unit of membrane area, the expected diffusive airflow for the entire membrane unit should be calculated and compared against the baseline observed during the unit-specific evaluation. If the observed baseline is significantly higher than the predicted diffusive losses, the result could be indicative of an integrity problem, and diagnostic testing (see section 4.8) may be necessary to identify the source of additional airflow or pressure loss. If the membrane module manufacturer is only able to provide a diffusion coefficient for air through the membrane material, Equation 4.12 may be used to estimate the diffusive airflow for a membrane unit with a known membrane area, if necessary:

$$Q_{diff} = 6 \bullet \left(\frac{D_{am} \bullet A_m \bullet (P_{test} - BP) \bullet H}{z} \right) \bullet \left(\frac{R_{gas} \bullet T}{P_{atm}} \right) \qquad \text{Equation 4.12}$$

Where:

Q_{diff} = diffusive flow of air through a saturated semi-permeable membrane (L/min)

D_{am} = diffusion coefficient for air through a saturated semi-permeable membrane material (cm^2/s)

A_m = total membrane surface area to which the direct integrity test is applied (m^2)

P_{test} = membrane test pressure (psi)

BP = backpressure on the system during the test (psi)

H = Henry's constant for air-water system (mol/psi-m^3)

z = membrane thickness (mm)

R_{gas} = universal gas constant (L-psia/mol-K)

T = water temperature during direct integrity test (K)

P_{atm} = atmospheric pressure (psia)

6 = unit conversion factor

Note that the equations for the diffusion of air through porous and semi-permeable membranes – Equations 4.11 and 4.12, respectively – are very similar. These equations differ only in that Equation 4.12 for semi-permeable membranes does not require the membrane porosity and utilizes a diffusion coefficient for a composite membrane layer consisting of both the membrane and the water of saturation bound in the microscopic interstices of the membrane material.

Because spiral-wound NF/RO membranes are typically composite structures consisting of two or more layers (as described in section 2.3.1), it is important that the membrane thickness (z) used corresponds to the layer to which the diffusion coefficient (D_{am}) provided by the membrane manufacturer applies. For example, the diffusion coefficient may apply to the thin, active, semi-permeable layer, in which case the membrane thickness would correspond to this layer only. Alternatively, if the diffusion coefficient is a composite representing the diffusion of air through all the layers of the membrane taken as whole, then the thickness used should be that of the entire membrane, including all layers.

Both Equations 4.11 and 4.12 show that diffusive airflow is directly proportional to membrane area, the applied direct integrity test pressure, and the system backpressure. As a result, the decay should be quantified for each membrane unit of different size in the system and also recalculated if either the applied test pressure or system backpressure are modified.

If the sensitivity is calculated using the ALCR approach described in section 4.3.1.2 and if diffusion is significant, Equation 4.6 can be modified to compensate for diffusive airflow as shown in Equation 4.13:

$$Q_{breach} = \frac{Q_{air} - Q_{diff}}{ALCR} \qquad \text{Equation 4.13}$$

Where:

Q_{breach}	=	flow of water through the critical breach during filtration (L/min)
Q_{air}	=	flow of air through the critical breach during a pressure-based direct integrity test (L/min)
Q_{diff}	=	diffusive flow of air (L/min)
ALCR	=	air-liquid conversion ratio (dimensionless)

Note that the flow of air through the critical breach during a pressure-based direct integrity test (Q_{air}) includes diffusive losses and thus should always be larger than the diffusive flow of air (Q_{diff}). Equation 4.13 should not be used if the baseline decay has already been accounted for by subtracting this baseline decay from the total rate of pressure decay (ΔP_{test}) observed during the integrity test using Equation 4.10, as described at the beginning of this section.

Combining Equation 4.13 with Equation 4.7 yields Equation 4.14, which enables the calculation of sensitivity using the ALCR approach, taking into account diffusive losses.

$$LRV_{DIT} = \log\left(\frac{Q_p \bullet ALCR}{VCF \bullet \left(Q_{air} - Q_{diff}\right)}\right)$$

Equation 4.14

Where:

LRV_{DIT}	=	direct integrity test sensitivity in terms of LRV (dimensionless)
Q_p	=	filtrate flow (L/min)
ALCR	=	air-liquid conversion ratio (dimensionless)
VCF	=	volumetric concentration factor (dimensionless)
Q_{air}	=	flow of air (L/min)
Q_{diff}	=	diffusive flow of air (L/min)

Note that Equations 4.13 and 4.14 are applicable to MCF, MF/UF, and NF/RO membrane filtration systems.

4.3.2 Marker-Based Tests

Sensitivity for marker-based direct integrity tests is determined via a straightforward calculation of the log removal value, similar to the determination of log removal values during a challenge study, as shown in Equation 4.15 (also Equation 3.7):

$$LRV_{DIT} = \log(C_f) - \log(C_p)$$

Equation 4.15 (3.7)

Where:

LRV_{DIT}	=	direct integrity test sensitivity in terms of LRV (dimensionless)
C_f	=	feed concentration (number or mass / volume)
C_p	=	filtrate concentration (number or mass / volume)

In using Equation 4.15 to calculate the sensitivity of a marker-based test, the LT2ESWTR specifies that the feed concentration, C_f, is the typical feed concentration of the marker used in the test, and that the filtrate concentration, C_p, is the baseline filtrate concentration of the marker from an integral membrane unit. If the marker is not detected in the filtrate, the term C_p must be set equal to the detection limit. Unlike the sensitivity calculations for pressure-based tests, Equation 4.15 does not incorporate a VCF. This term is not necessary to account for concentration effects on the feed side of the membrane in association with marker-based tests because the concentration of the marker in the filtrate is measured. This measurement would directly account for any increase in the quantity of the marker passing through an integrity breach as a result of feed side concentration effects.

Due to variability in dosing the marker, the day-to-day LRV is likely to vary during system operation. Thus, in order to establish the sensitivity of a marker-based direct integrity test using Equation 4.15, it is necessary to assume an appropriately conservative feed concentration such that the LRVs determined on a day-to-day basis meet or exceed the LRV_{DIT} unless there is an integrity breach. An example of such a conservative approach would be the use of the lower bound of the anticipated concentration range of seeded marker as the feed concentration for the purposes of calculating the sensitivity of the test. Note that a marker-based direct integrity test must utilize a feed concentration sufficient to demonstrate the required *Cryptosporidium* LRV for rule compliance. Because a typical feed water will normally not contain a sufficient number of particles at the required 3-μm resolution or else will not be sufficiently characterized to demonstrate this, seeding will usually be required for marker-based tests.

In order to optimize the sensitivity and reliability of a marker-based direct integrity test, it is important to use an accurate method for quantifying the feed and filtrate concentrations. Since the feed and filtrate concentrations will differ by orders of magnitude, an analytical method with a wide dynamic range is desired. If such a range is not available using a single device, two different instruments may need to be used to measure these respective concentrations. Regardless of the dynamic range of the instrument(s), it is likely that different analytical volumes will need to be used to deal with the different concentration ranges; however, the concentrations will have to be expressed in terms of equivalent volumes for the purpose of calculating an LRV. Some specific considerations regarding the use of particulate and molecular marker-based direct integrity tests are discussed in section 4.7.5.

4.4   Test Frequency

Most currently available direct integrity tests require the membrane unit to be taken off-line for testing and thus are conducted in a periodic manner, requiring a balance between the need to verify system integrity with the desire to minimize system downtime and lost productivity. In addition, although some marker-based tests may be conducted while the membrane unit is on-line and in production, it is generally neither practical nor cost effective to implement these tests on a continual basis. Thus, the frequency at which direct integrity testing is conducted for membrane filtration systems represents a compromise between these competing objectives.

The LT2ESWTR requires that direct integrity testing be conducted on each membrane unit at least once each day that the membrane unit is in operation for rule compliance, unless the State approves less frequent testing (40 CFR 141.719(b)(3)(vi)). This minimum test frequency is intended to balance the need to verify system integrity as often as possible with cost and production considerations and is based in part on a USEPA report that indicated daily direct integrity testing was relatively common practice at membrane filtration facilities (USEPA 2001). It is important to note that the rule requires that each unit be subjected to direct integrity testing on a daily basis, even if the unit is operational for only a fraction of each day. It is also recommended that a given unit be tested at approximately the same time each day (site-specific facility operations permitting) in order to maintain a roughly consistent time interval between applications of direct integrity testing; however, this is not specifically required by the rule.

The State may require either more or less frequent integrity testing for compliance with the rule at its discretion, although less frequent testing must be supported by demonstrated process reliability, the use of multiple barriers effective for *Cryptosporidium*, or reliable process safeguards (40 CFR 141.719(b)(3)(vi)). For example, in terms of process reliability, the State may opt to reduce the frequency of direct integrity testing if the membrane filtration system has a significant history of demonstrated operation either without the detection of an integrity breach or with only rare occurrences of integrity breaches that have never been large enough to compromise the ability of the system to achieve the awarded removal credit. Alternatively, the State may also reduce the required test frequency if the overall treatment scheme (including both membrane filtration and other processes) incorporates at least one additional process that is capable of achieving a substantial portion of the required *Cryptosporidium* treatment credit. In this case, even with a small integrity breach the State may be confident that the multiple barrier treatment process is fully capable of achieving compliance with the LT2ESWTR.

The State may also permit a reduced direct integrity test frequency if other safety factors are utilized to mitigate the risk associated with a potential integrity breach. One such strategy might be maintaining filtrate storage with a detention time equivalent to or longer than the time between direct integrity test events. If an integrity breach were detected, this storage would provide the utility with sufficient time to take the necessary mitigation measures to ensure that any contamination risks are addressed before the water enters the distribution system. Another possibility might be the use of continuous indirect integrity monitoring techniques with very high demonstrated sensitivities relative to other indirect monitoring methods. This strategy would represent a trade-off in that although the more sensitive (relative to the indirect monitoring methods) direct integrity test would be conducted less often, the ability to detect an integrity breach between applications of the direct test would be increased.

Note that any reduction in direct integrity test frequency is subject to the discretion of the State, which may utilize any of these suggested strategies as the basis for its decision or any other criteria it determines to be appropriate. Also, although unrelated to rule compliance, more frequent testing may be appropriate under certain specific circumstances, such as during initial facility start-up, as described in Chapter 8. More frequent direct integrity testing may also be advantageous for systems that rely on membrane filtration to achieve a high log removal of *Cryptosporidium*, since the risk associated with an integrity breach is greater. Conducting direct integrity testing on a membrane unit more frequently than once per day may be voluntarily implemented by a utility or required at the discretion of the State.

4.5   Establishing Control Limits

A control limit (CL) is defined as a response that, if exceeded, indicates a potential problem with the system and triggers a response. Multiple CLs can be set at different levels to indicate the severity of the potential problem. In the context of direct integrity testing, CLs are set at levels associated with various degrees of integrity loss. Under the provisions of the LT2ESWTR, a CL within the sensitivity limits of the direct integrity test must be established at the threshold test response that is indicative of an integral membrane unit capable of achieving the *Cryptosporidium* removal credit awarded by the State for rule compliance

(40 CFR 141.719(b)(3)(iv)). Because utilities or States would have the option to implement a series of tiered CLs that may represent progressively greater levels of integrity loss leading up to the specific CL required under the rule, in this guidance manual the LT2ESWTR-mandated CL is referred to as an upper control limit (UCL). If the integrity test response is below the UCL, the membrane unit should be achieving a LRV equal to or greater than the removal credit awarded to the process. Alternatively, if the UCL is exceeded, the membrane unit is required to be taken off-line for diagnostic testing (as described in section 4.8) and repair.

The same principles used to establish direct integrity test sensitivity are also used to establish the UCL. For *pressure-based tests*, the UCL may be calculated using the ALCR methodology. A modified version of Equation 4.7 yields an expression for the UCL for, as shown in Equation 4.16:

$$UCL = \frac{Q_p \bullet ALCR}{10^{LRC} \bullet VCF} \qquad \text{Equation 4.16}$$

Where:

UCL	=	upper control limit in terms of airflow (L/min)
Q_p	=	membrane unit design capacity filtrate flow (L/min)
ALCR	=	air-liquid conversion ratio (dimensionless)
LRC	=	log removal credit awarded (dimensionless)
VCF	=	volumetric concentration factor (dimensionless)

Similarly, Equation 4.9 can be rearranged to establish an expression for calculating the UCL in terms of a pressure decay rate, as shown in Equation 4.17:

$$UCL = \frac{Q_p \bullet ALCR \bullet P_{atm}}{10^{LRC} \bullet V_{sys} \bullet VCF} \qquad \text{Equation 4.17}$$

Where:

UCL	=	upper control limit in terms of pressure decay rate (psi/min)
Q_p	=	membrane unit design capacity filtrate flow (L/min)
ALCR	=	air-liquid conversion ratio (dimensionless)
P_{atm}	=	atmospheric pressure (psia)
LRC	=	log removal credit (dimensionless)
V_{sys}	=	volume of pressurized air in the system during the test (L)
VCF	=	volumetric concentration factor (dimensionless)

Values for the parameters in Equations 4.16 and 4.17 should be the same as the analogous terms used to calculate sensitivity using Equations 4.7 and 4.9, respectively. Note that to the extent possible, these values should be selected to yield a conservative result for the UCL.

Equations 4.16 and 4.17 establish the maximum direct integrity test response that can be used as a UCL for the log removal credit (LRC) awarded by the State. The LRC, in turn, must be less than or equal to the lower value of either the log removal value determined during challenge testing ($LRV_{C\text{-Test}}$) or the sensitivity of the direct integrity test used (LRV_{DIT}).

In the context of Equations 4.16 and 4.17, CLs are expressed in terms of the actual response from the direct integrity test (i.e., flow of air or pressure decay rate, respectively). In this form the CLs may be most useful for operators, since these could be directly compared to integrity test results. However, it may also be useful to calculate the corresponding LRVs for both the CL(s) and the individual direct integrity test responses using the generic forms of Equations 4.7 (for tests yielding results in terms of the flow of air) or 4.9 (for tests yielding results in terms of pressure decay). In this case these equations will simply yield a LRV corresponding to a particular direct integrity test result (i.e., a general LRV) rather than the sensitivity of the test (i.e., LRV_{DIT}). Many membrane systems have automated data acquisition equipment that could be programmed to calculate the LRV based upon the results of the most recent integrity test results and current operating conditions. These parameters may be displayed and trended to track system performance. Additional guidance on data analysis is provided in section 4.9.

For *marker-based tests*, which yield results in terms of log removal, the UCL is simply equal to the log removal credit awarded by the State.

Any CLs other than that mandated by the LT2ESWTR that are either voluntarily implemented by the utility or required by the State are referred to as lower control limits (LCLs). For example, a LCL may be established to provide operators with an indication that there may be an integrity breach before the breach becomes a compliance concern. In this scenario the LCL could be used in the context of preventative maintenance, and excursions above the LCL could prompt investigation and repair during scheduled downtime for the unit rather than require an immediate shutdown. The use of CLs and integrity testing in the context of a comprehensive integrity verification program is discussed further in Appendix A.

Unlike the UCL that is established by the log removal credit awarded to the process, a LCL can be established based on the needs and objectives of the utility. However, any LCLs should be above the baseline integrity test response for an integral membrane and below the UCL in order to be useful. The baseline integrity test value for a membrane unit can be established during the commissioning of the facility, after the membrane system has been fully wetted and determined to be integral. The baseline level is described as the normal range of direct integrity test results that would occur for an integral membrane unit. A practical lower bound for any LCLs is the sensitivity of the direct integrity test (i.e., the lowest response that can be reliably measured that is indicative of an integrity breach).

As with the determination of membrane unit sensitivity, for systems that utilize multiple stages of membrane filtration, the UCL for each stage must be determined independently. Thus, for pressure-based tests, the parameters used to calculated the UCL (e.g., log removal credit, ALCR, VCF, etc.) must be specific to the membrane units in each respective stage. This requirement is applicable to all stages to which the LT2ESWTR applies (i.e., those that produce filtrate for drinking water or post-filtration treatment rather than for recycling to an upstream point in the treatment process).

4.6   <u>Example</u>: Establishing Direct Integrity Test Parameters

<u>Scenario</u>:

A submerged, vacuum-driven hollow-fiber membrane system that operates in a deposition (i.e., dead-end) mode hydraulic configuration is required to provide a total of at least 3 log removal for *Cryptosporidium* by the State under the LT2ESWTR. A pressure decay test is applied to one of the membrane units in the system. Assume the plant is at sea level. Applicable parameters are as follows:

Operational parameters
- The permitted design capacity of the membrane unit is 1,200 gpm.
- The maximum anticipated water temperature is 75 °F (24 °C).
- The minimum anticipated water temperature is 41 °F (5 °C).
- The maximum anticipated backpressure that might be exerted on the units during direct integrity testing is 75 inches of water.
- The minimum anticipated backpressure that might be exerted on the units during direct integrity testing is 60 inches of water.
- The backpressure measured prior to the most recent pressure decay test was 65 inches of water.
- The filtrate flow measured prior to the most recent pressure decay test was 1,000 gpm.
- The TMP measured prior to the most recent pressure decay test was 10 psi.

Direct integrity test parameters
- The volume of pressurized piping during the test is 285 L.
- The initial applied test pressure is 16 psi.
- The duration of the pressure decay test is 10 minutes.
- Baseline (i.e., diffusive) decay is negligible over the duration of the test.
- The smallest verifiable rate of pressure decay under known compromised conditions is 0.10 psi/min.
- The most recent pressure decay test yielded a result of 0.13 psi/min.
- The temperature of both the water and the applied air were 68 °F (20 °C) during the most recent pressure decay test.

Unit and membrane characteristics
- The maximum rated TMP is 30 psi.
- The pore shape correction factor (κ) for the membrane material was not determined experimentally, and thus a conservative value of one is assumed.
- The membrane material is relatively hydrophilic and has a liquid-membrane contact angle (i.e., "wetting angle") of 30°, as determined using a method acceptable to the State.
- The membrane surface variation (i.e., roughness) is 0.3 μm, as provided by the manufacturer.

- The hollow-fiber lumen diameter is 500 μm.
- The depth of the membrane into the potting material is 50 mm.
- All flow through any integrity breach that may be present is assumed to be turbulent.

Calculate:

1. The minimum direct integrity test pressure commensurate with the required resolution of 3 μm for the removal of *Cryptosporidium*

2. The sensitivity of the direct integrity test

3. The UCL for this system

4. The LRV verified by the most recent direct integrity test

Solution:

1. Calculate the minimum direct integrity test pressure commensurate with the required resolution of 3 μm for the removal of *Cryptosporidium*.

$$P_{test} = (0.193 \bullet \kappa \bullet \sigma \bullet \cos\theta) + BP_{max} \qquad \text{Equation 4.1}$$

$\kappa = 1$ from given information

$\sigma = 74.9$ dynes/cm   the surface tension of water at 5 °C

$\theta = 30°$ from given information

$BP_{max} = 75$ inches (of water column)   from given information

$$P_{test} = \left(0.193\frac{psi \cdot cm}{dyne} \bullet 1.0 \bullet 74.9\frac{dynes}{cm} \bullet \cos(30°)\right) + \frac{75\ inches \cdot H_2O}{27.7\ \dfrac{inches \cdot H_2O}{psi}}$$

$$P_{test} = 12.5\ psi + 2.7\ psi$$

$$P_{test} = 15.2\ psi$$

Because the problem scenario states that baseline diffusive losses are negligible, the pressure calculated above represents the lowest permissible initial applied test pressure. If diffusive losses could not be neglected, P_{test} would represent the lower bound above which the pressure must be maintained to ensure that the resolution is maintained throughout the duration of the test. If this were the case, the applied pressure would have to be increased over P_{test} by the total anticipated pressure decay over the duration of the test. In this particular example, since the applied test pressure is given as 16 psi, the resolution requirement of the LT2ESWTR is satisfied.

2. <u>Calculate the sensitivity of the direct integrity test</u>.

$$LRV_{DIT} = \log\left(\frac{Q_p \bullet ALCR \bullet P_{atm}}{\Delta P_{test} \bullet V_{sys} \bullet VCF}\right)$$ Equation 4.9

Q_p = 1,200 gpm   design capacity filtrate flow (from given information)

P_{atm} = 14.7 psia   atmospheric pressure at sea level (from given information)

The minimum backpressure that might be exerted during direct integrity testing is used to establish a conservative value for sensitivity.

ΔP_{test} = 0.10 psi/min   smallest verifiable decay rate (from given information)

V_{sys} = 285 L   from given information

VCF = 1   standard for deposition mode hydraulic configuration

ALCR = ?   to be determined

Consult Table 4.1 and Appendix C – use Darcy pipe flow model for a hollow-fiber membrane filtration system under conditions of turbulent flow (as specified):

$$ALCR = 170 \bullet Y \bullet \sqrt{\frac{(P_{test} - BP) \bullet (P_{test} + P_{atm})}{(460 + T) \bullet TMP}}$$ Equation C.4

P_{test} = 16 psi   initial applied test pressure (from given information)

If diffusion through an integral membrane unit (i.e., baseline pressure decay was significant, the cumulative decay over the duration of the test would be subtracted from the initial test pressure before applying this parameter to Equation C.4 to yield a conservative result for the ALCR.

BP = 60 inches of water   minimum backpressure (from given information)

P_{atm} = 14.7 psia   atmospheric pressure
                                          at sea level
                                          (from given information)

T = 75 °F   maximum anticipated
                                          temperature
                                          (from given information)

TMP = 30 psi   maximum allowable TMP
                                          (from given information)

The values for system backpressure, temperature, and transmembrane pressure (TMP) were selected to establish a conservative (i.e., lower) value for the ALCR, which in turn yields a conservative value for sensitivity.

Y = ?   net expansion factor
                                          (to be determined)

As indicated in Equation C.5 in Appendix C, the gas compressibility factor (Y) is a function of the applied test pressure (P_{test}), the system backpressure during the test (BP), atmospheric pressure (P_{atm}), and a flow resistance coefficient (K), as follows:

$$ Y \propto \left[\frac{1}{\left(\dfrac{P_{test} - BP}{P_{test} + P_{atm}} \right)}, K \right] \qquad \text{Equation C.5} $$

Smaller values for Y are generated with larger values of the first parameter in Equation C.5 and smaller values for K. Thus, values for these variables should be selected to produce smaller values of Y, which in turn yield smaller values for the ALCR and a more conservative value for the test sensitivity.

Quantifying the first parameter in Equation C.5:

P_{test} = 16 psi

BP = 75 inches (of water column)

P_{atm} = 14.7 psia

$$\left(\frac{P_{test} - BP}{P_{test} + P_{atm}} \right) = \left(\frac{16\,psi - \dfrac{75\ inches \cdot H_2O}{27.7\ \dfrac{inches \cdot H_2O}{psi}}}{16\,psi + 14.7\,psi} \right) = 0.43$$

The flow resistance coefficient is defined by Equation C.6, as described in Appendix C:

$$K = f \bullet \frac{L}{d_{fiber}}$$

Equation C.6

L = 50 mm — potting depth (from given information)

d_{fiber} = 0.5 mm — fiber diameter (from given information)

f = ? — friction factor (to be determined)

The friction factor can be obtained using the iterative method described in Appendix C.

f = 0.037 — friction factor (from iterative method)

$$K = f \bullet \frac{L}{d_{fiber}} = (0.037) * \left(\frac{50mm}{0.5mm} \right) = 3.7$$

Using the appropriate chart on page A-22 from Crane (1988) with the values calculated above:

$$\left(\frac{P_{test} - BP}{P_{test} + P_{atm}} \right) = 0.43, \qquad K = 3.7$$

…yields a value for Y as shown below.

Y = 0.78

Having determined a value for Y, the ALCR can be calculated follows:

$$ALCR = 170 \bullet Y \bullet \sqrt{\frac{(P_{test} - BP) \bullet (P_{test} + P_{atm})}{(460 + T) \bullet TMP}}$$

$$ALCR = 170 \bullet (0.78) \bullet \left(\frac{\left(16\,psi - \dfrac{75\ inches \cdot H_2O}{27.7\ \dfrac{inches \cdot H_2O}{psi}} \right) \bullet (16\,psi + 14.7\,psi)}{(460^o F + 75^o F) \bullet 30\,psi} \right)^{0.5}$$

ALCR = 21.1

Substituting values into Equation 4.9 for sensitivity:

$$LRV_{DIT} = \log \left(\frac{\left(1{,}200\,gpm \bullet 3.785\,\dfrac{L}{gal} \right) \bullet 21.1 \bullet (14.7\,psi)}{0.10\,\dfrac{psi}{min} \bullet 285L \bullet 1} \right)$$

LRV$_{DIT}$ = 4.7

Therefore, the maximum removal value that this membrane filtration system is capable of verifying is 4.7 log.

3. <u>Calculate the UCL for this system.</u>

$$UCL = \frac{Q_p \bullet ALCR \bullet P_{atm}}{10^{LRC} \bullet V_{sys} \bullet VCF}$$ Equation 4.17

Q$_p$ = 1,200 gpm   design capacity filtrate flow (from given information)

ALCR = 21.1   as determined in Part 2 of this example, above

P_{atm} = 14.7 psia   atmospheric pressure at sea level (from given information)

LRC = 3   from given information

V_{sys} = 285 L   from given information

VCF = 1   standard for deposition mode hydraulic configuration

$$UCL = \frac{\left(1{,}200\,gpm \bullet 3.785\,\dfrac{L}{gal}\right) \bullet 21.1 \bullet (14.7\,psi)}{10^3 \bullet 285L \bullet 1}$$

UCL = 4.9 psi/min

4. <u>Calculate the LRV verified by the most recent integrity test.</u>

In addition to calculating the sensitivity, Equation 4.9 can also be used to determine the LRV verified by the most recent direct integrity test via applying values for the variables specific to this test event.

$$LRV = \log\left(\frac{Q_p \bullet ALCR \bullet P_{atm}}{\Delta P_{test} \bullet V_{sys} \bullet VCF}\right) \qquad \text{Equation 4.9}$$

Q_p = 1,000 gpm   flow measured prior to testing (from given information)

P_{atm} = 14.7 psia   atmospheric pressure at sea level (from given information)

ΔP_{test} = 0.13 psi/min   measured test decay rate (from given information)

V_{sys} = 285 L   from given information

VCF = 1   standard for deposition mode hydraulic configuration

ALCR = 21.1   as determined in part 2, above

Note that because the ALCR is designed to be a conservative value, it is not necessary to recalculate the ALCR for a specific pressure decay test in order to use Equation 4.9 for the purpose of determining the LRV verified by that test.

Substituting values into Equation 4.9 for sensitivity:

$$LRV = \log\left(\frac{\left(1{,}000\,gpm \bullet 3.785\dfrac{L}{gal}\right) \bullet 21.1 \bullet (14.7\,psi)}{0.13\dfrac{psi}{\min} \bullet 285L \bullet 1}\right)$$

LRV = 4.5

The above result of part 4 of this example demonstrates the effect that the membrane unit filtrate flow and integrity test response can have on the verifiable LRV at any point during operation. Although the sensitivity of this direct integrity test for this particular system under typical membrane unit operating conditions was determined to be an LRV of 4.7 in part 2, operating the unit at about 83 percent of the design flow, coupled with a measured pressure decay rate somewhat higher than the baseline detectable value under known compromised conditions, reduced the verifiable LRV to 4.5. Because this reduction still results in a verifiable LRV that is significantly higher than the required LRC of 3.0, operation under these conditions would still allow the system to maintain compliance with the LT2ESWTR.

4.7   Test Methods

The LT2ESWTR does not specify a particular type of direct integrity test for rule compliance, but instead allows the use of any test that meets the requirements of the rule (40 CFR 141.719(b)(3)). The two general classes of tests currently employed in municipal water treatment applications are pressure- and marker-based tests. Within these two categories, the particular types of tests most commonly used are described in the following subsections, including the pressure and vacuum decay tests, the diffusive airflow test, the water displacement test, and particulate and molecular marker tests. General procedures for conducting each of these tests are provided, along with a listing of some of the advantages and disadvantages of each method. The particular manner in which each of these tests is applied may vary according to manufacturer or site- or system-specific conditions.

The specific tests addressed in this guidance manual are not intended to represent a comprehensive list of all types of direct integrity tests that could be used to comply with the requirements of the LT2ESWTR. Any method that is both consistent with the definition of a direct integrity test under the rule and capable of meeting the specified resolution, sensitivity, and frequency requirements could be used for rule compliance at the discretion of the State.

4.7.1 Pressure Decay Test

The pressure-based pressure decay test is the most common direct integrity test currently in use and is generally associated with MF, UF, and MCF systems, which utilize porous membranes. In a pressure decay test, a pressure below the bubble point value of the membrane is applied, and the subsequent loss in pressure is monitored over several minutes. An integral membrane unit will maintain the initial test pressure or exhibit a very slow rate of decay. Note that the pressure decay test is applicable to currently available pressure-driven and vacuum-driven systems, since this test is conducted under positive pressure for both types of systems. A schematic illustrating a pressure decay test is shown in Figure 4.1.

Figure 4.1 Schematic Illustrating a Pressure Decay Test

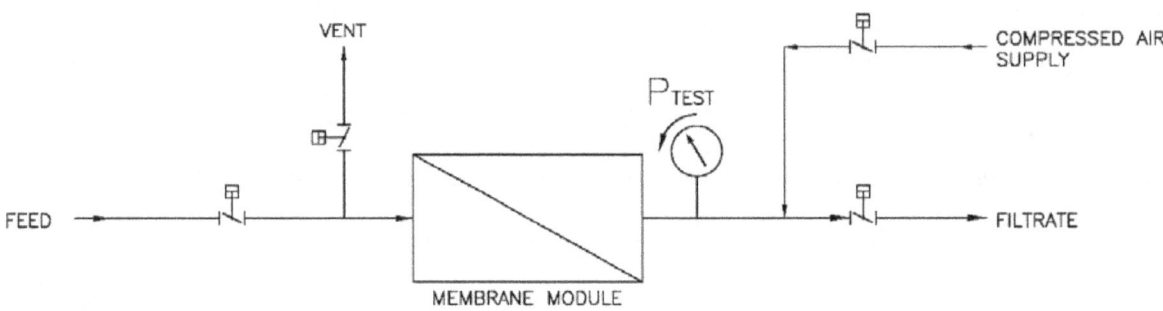

An outline of a generic protocol for a pressure decay test as is as follows:

1. <u>Drain the water from one side of the membrane</u>.
   For hollow-fiber systems, typically the inside of the fiber lumen is drained, which may represent the feed or the filtrate, depending on whether the system is operated in an "inside-out" or "outside-in" mode, respectively.

2. <u>Pressurize the drained side of the fully wetted membrane</u>.
   The applied pressure must be lower than the bubble point pressure of the membrane (i.e., the pressure required to overcome the capillary forces that hold water in the membrane pores). Pressures ranging from 4 to 30 psi are typically applied during the pressure decay test, depending on the particular system. Membrane construction may limit the pressure at which a membrane can be tested. For compliance with LT2ESWTR requirements, the applied pressure must be sufficient to meet the resolution criterion of 3 μm based on Equation 4.1. For systems that utilize membranes submerged in an open basin, the test is typically applied to the filtrate side of the membranes without draining of the basins. As a result, the hydrostatic pressure at the deepest part of the membrane unit at which feed water passes through

the membrane must be considered in determining the resolution actually achieved by the test, as discussed in section 4.2.

3. Isolate the pressure source and monitor the decay for a designated period of time. If there are no leaks in the membrane, process plumbing, or other pressurized system components, then air can only escape by diffusing through the water contained in the pores of the fully wetted membrane. Typically, this test is monitored over a period of 5 to 10 minutes such that a stable rate of decay can be determined. The rate of pressure decay should be compared to the UCL (or any LCLs that may also be established) for the test to determine what, if any, subsequent action is triggered.

Advantages of the pressure decay test include:

- Ability to meet the resolution criterion of 3 µm under most conditions

- Ability to detect integrity breaches on the order of single fiber breaks and small holes in the lumen wall of a hollow fiber, depending on test parameters and system-specific conditions

- Standard feature of most MF and UF systems

- High degree of automation

- Widespread use by utilities and acceptance by States

- Simultaneous use as a diagnostic test to isolate a compromised module in a membrane unit in some cases (see section 4.8.1)

Limitations of the pressure decay test include:

- Inability to continuously monitor integrity

- Calculation of test method sensitivity requires measurement of the volume of pressurized air in the system (see Equation 4.8)

- Potential to yield false positive results if the membrane is not fully wetted (which may occur with newly installed and hydrophobic membranes that are difficult to wet, or when the test is applied immediately after a backwash process that includes air)

- More difficult to apply to membranes that are oriented horizontally as a result of potential draining and air venting problems

In addition to the disadvantages cited above, another important concern associated with the pressure decay test is the potential for larger decay rates even within the UCL to affect the accuracy of the test. For example, if the total pressure decay over the duration of the test reduces the applied pressure on the membrane to a level below that sufficient to meet the resolution criterion, the test would not comply with the requirements of the rule. Consequently, parameters should be established so as to ensure that the pressure decay test meets the 3-μm resolution criterion throughout the duration of the test, as discussed in section 4.2.

4.7.2 Vacuum Decay Test

The vacuum decay test is analogous to the pressure decay test with the exception that the test pressure is applied by drawing a vacuum on the membrane and monitoring the rate of vacuum (as opposed to pressure) decay over a period of time. An integral membrane unit will maintain the initial test vacuum or exhibit a very slow rate of decay. This test is generally associated with spiral-wound NF and RO membranes. A schematic illustrating a vacuum decay test is shown in Figure 4.2.

An outline of a generic protocol for a vacuum decay test as is is as follows:

1. Drain the water from one side of the membrane.
   Typically, the filtrate side of a spiral-wound NF or RO membrane is drained.

2. Apply a vacuum to the drained side of the fully wetted membrane.
   A vacuum of 20 to 26 inches Hg is typically applied during the test. Membrane construction may limit the vacuum at which a membrane can be tested. For compliance with LT2ESWTR requirements, the applied vacuum must be sufficient to meet the resolution criterion of 3 μm.

3. Isolate the vacuum source and monitor the decay for a designated period of time.
   If there are no leaks in the membrane, process plumbing, or other system components under vacuum, then the vacuum should not decay over the duration of the test. Typically, this test is monitored over a period of 5 to 10 minutes to allow a stable rate of decay to be determined. The rate of pressure decay should be compared to the UCL (or any LCLs that may also be established) for the test to determine what, if any, subsequent action is triggered.

Figure 4.2 Schematic Illustrating a Vacuum Decay Test

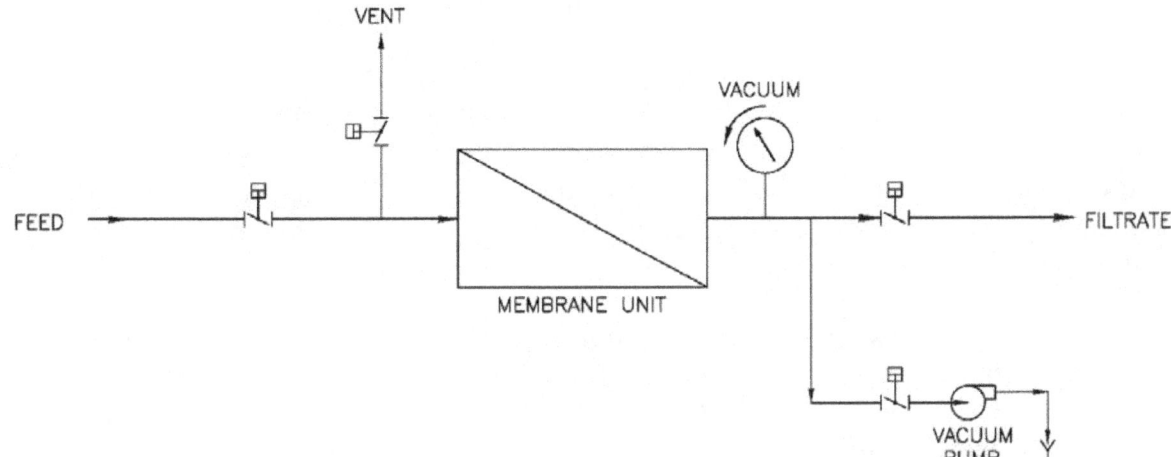

<u>Advantages</u> of the vacuum decay test include:

- Ability to test spiral-wound membranes or other systems that cannot be pressurized on the filtrate side of the membrane

- Ability to meet the resolution criterion of 3 μm under most conditions

<u>Limitations</u> of the vacuum decay test include:

- Inability to continuously monitor integrity

- Not widely used for full-scale systems in current practice

- Difficulty in removing entrained air after the test has been completed

- Calculation of test method sensitivity requires measurement of the volume of air under vacuum in the system (see Equation 4.8)

As with the pressure decay test, another important concern associated with the vacuum decay test is the potential for larger decay rates even within the UCL to affect the accuracy of the test. Thus, if the total vacuum decay over the duration of the test reduces the applied vacuum on the membrane to a level below that sufficient to meet the resolution criterion, then the test would not comply with the requirements of the rule. Consequently, test parameters should be established so as to ensure that the vacuum decay test meets the 3-μm resolution criterion throughout the duration of the test, as discussed in section 4.2.

4.7.3 Diffusive Airflow Test

The diffusive airflow test provides a direct measurement of the airflow through an integrity breach (i.e., Q_{air}). Like the pressure decay test, the diffusive airflow test is also based on bubble point theory and generally associated with MF, UF, and MCF membranes. However, instead of measuring the rate of pressure decay, the test pressure is kept constant and the airflow through a breach is measured. If there are no breaches in the system, there will typically be a small flow of air resulting from the diffusion of air through water held in the fully wetted membrane pores. This diffusive airflow represents the baseline test response that is indicative of an integral membrane.

The diffusive airflow test has not been commonly used in municipal water treatment plants, at least partially as a result of the incorporation of pressure decay test as a standard feature with most MF and UF systems. However, the diffusive airflow test has long been used for measuring the integrity of cartridge filters used in sterilization applications and has been described in the literature by a number of authors, including Meltzer (1987), Vickers (1993), and Johnson (1997), among others. Note that this test is not applicable to submerged membrane filtration systems. A schematic illustrating a diffusive airflow test is shown in Figure 4.3.

Figure 4.3 Schematic Illustrating a Diffusive Airflow Test

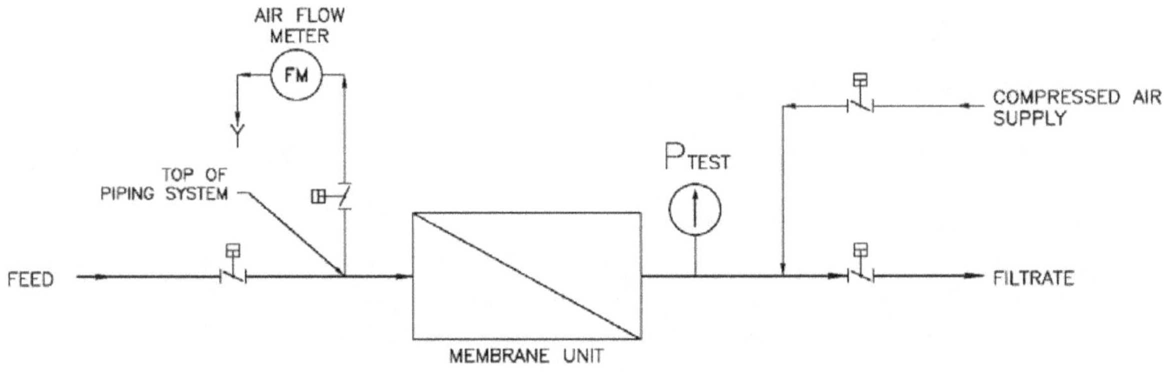

The outline of a generic protocol for a diffusive airflow test is similar to that for the pressure decay test, and is as follows:

1. <u>Drain the water from one side of the membrane.</u>
   For hollow-fiber systems, typically the inside of the fiber lumen is drained, which may represent the feed or the filtrate, depending on whether the system is operated in an "inside-out" or "outside-in" mode, respectively.

2. <u>Pressurize the drained side of the fully wetted membrane.</u>
   The applied pressure must be lower than the bubble point pressure of the membrane (i.e., the pressure required to overcome the capillary forces that hold water in the membrane pores). As with the pressure decay test, pressures ranging from 4 to 30 psi are typically applied during the diffusive airflow test, depending on the particular system. Membrane construction may limit the pressure at which a membrane can be tested. For compliance with LT2ESWTR requirements, the applied pressure must be sufficient to meet the resolution criterion of 3 μm.

3. <u>Maintain constant pressure and monitor the airflow for a designated period of time.</u>
   If there are no leaks in the membranes, process plumbing, or other pressurized system components, then air can only escape by diffusing through the water contained in the pores of the fully wetted membrane. Typically, this test is monitored over a period of 5 to 10 minutes to allow a stable airflow to be established. The flow of air should be compared to the UCL (or any LCLs that may also be established) for the test to determine what, if any, subsequent action is triggered.

<u>Advantages</u> of the diffusive airflow test include:

- Ability to directly measure the flow of air through an integrity breach

- Ability to meet the resolution criterion of 3 μm under most conditions

<u>Limitations</u> of the diffusive airflow test include:

- Inability to continuously monitor integrity

- Requires equipment to measure flow of air that is generally not a standard inclusion with a membrane filtration system

- Not widely used for full-scale systems in current practice

- Potential to yield false positive results if the membrane is not fully wetted (which may occur with newly installed and hydrophobic membranes that are difficult to wet, or when the test is applied immediately after a backwash process that includes air)

4.7.4 Water Displacement Test

The water displacement test is similar to the diffusive airflow test with the exception that the volume of water displaced by air flowing through an integrity breach is measured rather than the actual flow of air through the breach. Specifically, the air that passes across the membrane displaces a corresponding volume of water on the opposite side. This test can only be applied to membrane systems operated under positive pressure (i.e., as opposed to immersed systems that operate under vacuum). The volume of displaced water collected over a known period of time is measured and converted to a corresponding volume of air to determine the airflow (Q_{air}). If there are no breaches in the system, there will typically be a small flow of displaced water resulting from the diffusion of air. This flow represents the baseline test response that is indicative of an integral membrane. A larger volume (or flow) of displaced water would indicate an integrity breach. It is important to note that although this test measures the flow of water through an integrity breach, it does not constitute the flow of water through the critical breach (Q_{breach}) that is necessary for determining the direct integrity test sensitivity.

The water displacement test is described in an American Water Works Association Research Foundation (AWWARF) report prepared by Jacangelo et al. (1997). While this test is not commonly utilized in municipal drinking water treatment plants, it is relatively easy to use, requires minimal equipment, and can detect very small integrity breaches on the order of a single broken fiber. Since publication of the referenced AWWARF report, a membrane filtration facility in Tauranga, New Zealand has adopted the use of this integrity test. A schematic illustrating a water displacement test is shown in Figure 4.4.

Figure 4.4 Schematic Illustrating a Water Displacement Test

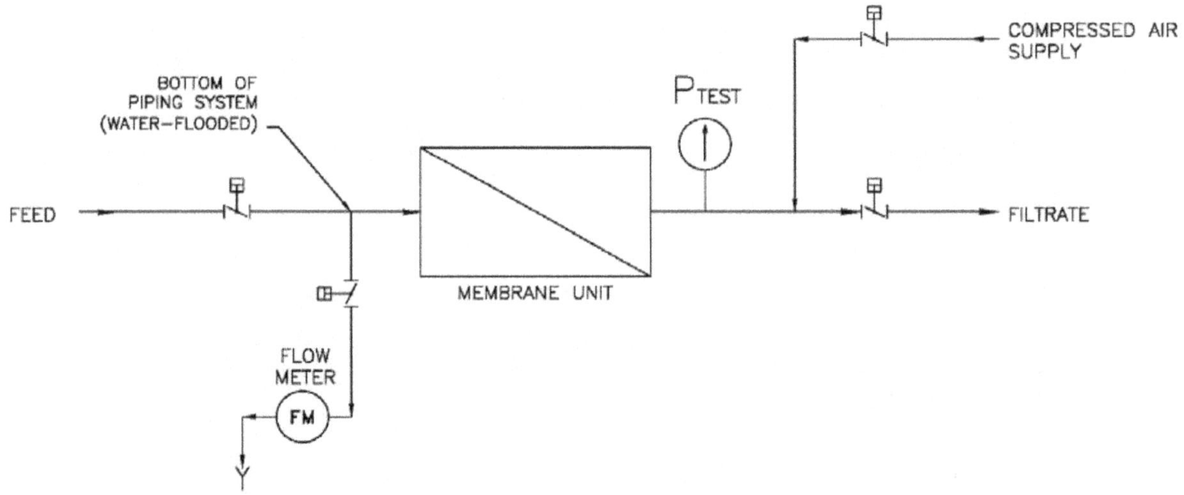

The outline of a generic protocol for a water displacement test is similar to that for the diffusive airflow test, and is as follows:

1. <u>Drain the water from one side of the membrane.</u>
   For hollow-fiber systems, typically the inside of the fiber lumen is drained, which may represent the feed or the filtrate, depending on whether the system is operated in an "inside-out" or "outside-in" mode, respectively.

2. <u>Pressurize the drained side of the fully wetted membrane.</u>
   The applied pressure must be lower than the bubble point pressure of the membrane (i.e., the pressure required to overcome the capillary forces that hold water in the membrane pores). Pressures ranging from 4 to 30 psi are typically applied during the pressure decay test, depending on the particular system. Membrane construction may limit the pressure at which a membrane can be tested. For compliance with LT2ESWTR requirements, the applied pressure must be sufficient to meet the resolution criterion of 3 μm. Since one side of the membrane must remain flooded with water during the test, there is a potential for some hydrostatic backpressure. If this pressure is determined to be significant, it must be considered in the resolution calculations as discussed in section 4.2.

3. <u>Maintain constant pressure and monitor the flow for a designated period of time.</u>
   If there are no leaks in the membrane, process plumbing, or other pressurized system components, then air can only escape by diffusing through the water contained in the pores of the fully wetted membrane. Typically, this test is monitored over a period of 5 to 10 minutes to allow a stable flow to be established. The air displaces water that in turn is measured via either a flow meter or a graduated cylinder and timing device. Since the flow of water is equivalent to the flow of air, the resulting flow should be compared to the UCL (or any LCLs that may also be established) for the test to determine what, if any, subsequent action is triggered. Note that the flow meter and associated sample line used in this test must be configured to prevent the gravity flow of water.

<u>Advantages</u> of the water displacement test include:

- Ability to directly measure the flow of air via a corresponding flow of water (provided that the backpressure on the system does not result in the compression of diffused air)

- Ability to meet the resolution criterion of 3 μm under most conditions

- Ability to detect integrity breaches on the order of single fiber breaks and small holes in the fiber lumen wall of a hollow

- Ease of measuring flow of water relative to that of air

<u>Limitations</u> of the water displacement test include:

- Inability to continuously monitor integrity

- Not widely used for full-scale systems

- Potential to yield false positive results if the membrane is not fully wetted (which may occur with newly installed and hydrophobic membranes that are difficult to wet, or when the test is applied immediately after a backwash process that includes air)

4.7.5 Marker-Based Integrity Tests

A marker-based test directly verifies the removal efficiency of a membrane filtration system by measuring the concentrations of a particulate or molecular marker in the feed and filtrate. The difference in the log of these concentrations represents the LRV demonstrated by the integrity test. It is important to note that a marker-based direct integrity test must utilize a feed concentration sufficient to demonstrate the required *Cryptosporidium* LRV for rule compliance. In addition, the full concentration of markers measured to demonstrate the required LRV must meet the 3-μm resolution criterion. Because a typical feed water will normally not contain a sufficient number of particles at the required 3-μm resolution or else will not be sufficiently characterized to demonstrate the fulfillment of this criterion, seeding will usually be required for marker-based tests.

Marker-based tests are advantageous in that they provide a direct assessment of the LRV achieved by the process. Both particulate and molecular marker-based direct integrity tests may be used to meet the requirements of the LT2ESWTR provided the specified performance criteria for resolution, sensitivity, and frequency are satisfied (40 CFR 141.719(b)(3)). Since a marker-based direct integrity test is essentially a form of challenge testing, much of the guidance provided in Chapter 3 may be useful in designing a marker-based test. Because these tests are applied to water treatment equipment in active use, the particulate or molecular marker used must be inert and suitable for use in a water treatment facility (e.g., approved by Food and Drug Administration (FDA) or certified by the National Sanitation Foundation (NSF) as an NSF-60 approved material). In addition, because MF, UF, and MCF systems utilize porous membranes that would not remove a molecular marker, only particulate marker tests should be used with these systems. (Note that one possible exception is a system utilizing UF membranes with a low molecular weight cutoff (MWCO), which could potentially remove larger molecular markers.) Conversely, because the semi-permeable NF and RO membranes are not specifically designed to accommodate large particle loads and typically cannot be backwashed to removed particulate matter from the membrane surface, only molecular markers should be used with NF and RO systems. Schematics illustrating the particulate and molecular marker tests are shown in Figures 4.5 and 4.6, respectively.

Figure 4.5 Schematic Illustrating a Particulate Marker Test

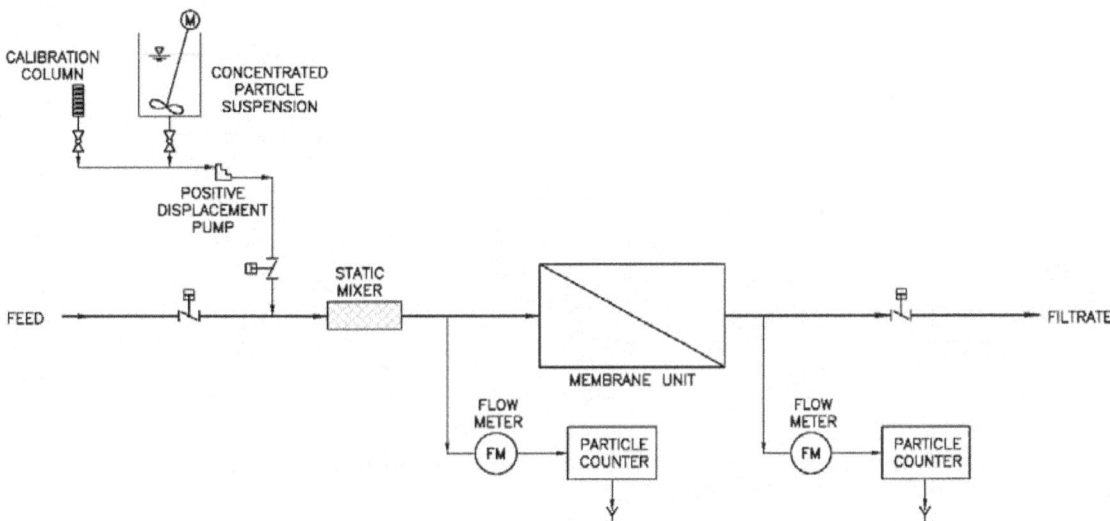

Figure 4.6 Schematic Illustrating a Molecular Marker Test

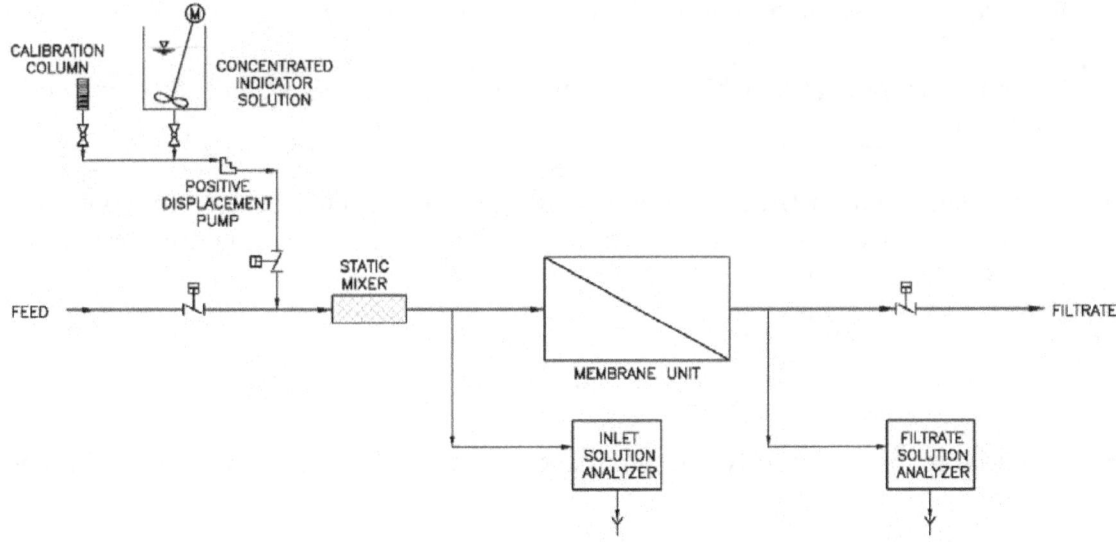

The generic protocols for both particulate and molecular marker tests are similar, and are as follows:

1. Select an appropriate marker and verify that it meets the resolution criterion. (Additional guidance is given in section 3.9.)

2. Ensure that the expected feed and filtrate sampling concentrations fall within the limits for the measuring instrumentation used.

3. Calculate the feed dosage rate and filtrate sample volume(s).
   (Additional guidance is given in section 3.10.)

4. Make sure that any operational processes such as chemical cleaning or backwashing are not initiated during the marker-based test.

5. Determine the amount of time the marker has to be applied to attain a steady state or equilibrium condition.
   (Additional guidance is given in section 3.10.)

6. Feed the marker into the system.

7. Commence filtrate sampling when equilibrium is reached.

8. Continue simultaneous feed and filtrate sampling until results are attained.

9. Discontinue dosing.

Advantages of marker-based integrity tests include:

- Provides direct evaluation of the LRV of a membrane unit

- Potential for application while unit is on-line and producing water (i.e., during normal operation)

- Instruments used for filtrate monitoring may also potentially be used for continuous indirect integrity monitoring between applications of the direct integrity test at the discretion of the State

Limitations of marker-based integrity tests include:

- Potential for an improperly selected marker to cause fouling and interfere with normal system operation

- Cost and calibration of instrumentation

- Cost of marker

- May require disposal of filtered water produced during the test

- May require special considerations for disposal of the filtered marker waste

- Measurement of marker concentrations is typically not continuous

Additional Considerations for Particulate Marker Tests

Typically particulate marker tests require the use of particle counters to measure the feed and filtrate particle concentrations during the test, thus enabling the LRV to be directly calculated. Consequently, there are some important considerations regarding particle counters to consider in implementing a particulate marker test. First, it is important that the feed and filtrate particle counters continually monitor baseline concentrations between test events so that background particle levels can be taken into account in the determination of the LRV as measured by the test during routine operation. The background levels recorded in both the feed and the filtrate should be subtracted from the respective particle count measurements during the test event in order to yield an accurate LRV. In addition, particle counters are subject to coincidence error (i.e., two or more particles passing through the sensor simultaneously that are consequently measured as a single larger particle), particularly at high particle concentrations. This potential error may limit the maximum particle concentration that can be counted, which may in some cases be problematic for measuring the feed concentration during a particulate marker test. In order to compensate for possible coincidence error, some particle counters can allow for the sample to be diluted. However, because any particles in the dilution water would introduce error into the measurement, the quality of any dilution water used should be also be taken into account in measuring the particle concentration of interest during a particulate marker test. It is also important that the particulate marker selected does not clump, which could exacerbate coincidence error or clog the instrument sensor.

All particle counters (or other instruments used to measure particulate concentrations during a particulate marker-based test) should be routinely calibrated at a frequency recommended by the manufacturer or required by the State. The calibration should be targeted toward counting particles at the specified size and concentration used in the particulate marker test. The particles used for the particulate marker test should be inert and compatible with the membrane in order to avoid irreversible fouling.

Additional Considerations for Molecular Marker Tests

Although the ambient levels of a molecular marker in the feed and filtrate may be negligible in many cases, these background concentrations should nevertheless still be measured and accounted for in the LRV calculation. The background concentrations of the molecular marker in the feed and filtrate should be subtracted from the respective measurements collected during the molecular marker-based direct integrity test in order to accurately quantify the LRV of the membrane unit. In addition, as with particle counters, all instruments that may be used to measure molecular marker concentrations should be calibrated regularly at a frequency recommended by the manufacturer or required by the State. Another important consideration is the potential for a molecular marker to adsorb onto the membrane surface or other system components, resulting in an inaccurate LRV determination; thus, any prospective molecular marker should be known to have negligible adsorption affinity for the membrane and other materials in use (Lozier et al. 2003).

4.8   Diagnostic Testing

Diagnostic tests are types of direct integrity tests that are specifically used to identify and isolate any integrity breaches that are detected, thus supplementing primary methods such as the pressure- or marker-based tests described in section 4.7, which simply determine whether or a not a leak is present in a membrane unit. Note that there are no sensitivity or resolution requirements for diagnostic tests under the LT2ESWTR, since the objective of these tests is simply to isolate a particular compromised module and/or broken fiber. However, it is important that a diagnostic test have sufficient resolution to enable the detection of the integrity breach that caused the UCL to be exceeded.

Many diagnostic tests are most useful for identifying a particular compromised module within a unit; however, for hollow-fiber MF/UF systems some diagnostic tests may be used locate a specific broken fiber in a module. The LT2ESWTR requires that if the results of a direct integrity test exceed the UCL, the affected unit be immediately taken off-line for diagnostic testing and subsequent repair (40 CFR 141.719(b)(3)(v)). Although many different types of diagnostic tests may be available, four of the tests most commonly used are described in the following subsections, including visual inspection, bubble testing, sonic testing, conductivity profiling, and single module testing. Note that not all of these tests are applicable to every type of membrane filtration system. Visual inspection, bubble testing, and sonic testing are generally applicable to MF, UF, and MCF systems, while conductivity profiling is used with NF and RO systems. Single module testing may be applicable to all types of membrane filtration systems.

4.8.1 Visual Inspection

Visually inspecting a membrane unit for leaks is the simplest form of diagnostic testing and is generally applicable to MF, UF, and MCF systems. Since many pressure-based direct integrity tests apply air to one side of the membrane while maintaining water on the other side, it may be possible to see air bubbles form in a compromised module. If a pressure-based direct integrity is used, visual inspection may be able to be conducted simultaneously with the direct integrity test to identify a compromised module. In order to perform a visual inspection, some component of the module housing must be transparent; accordingly, some pressure-driven membrane systems have inspection ports or clear tubing at the top of the membrane module to allow an operator to identify the particular compromised module. For vacuum-driven membrane systems submerged in basins, this test simply involves observing the water surface to identify the module that is the source of the bubbles rising to the surface.

Advantages of visual inspection include:

- Ability to identify specific compromised fibers and leaking seals

- Easy to use and interpret results

- No additional equipment required

- Equally applicable to systems using enclosed pressure vessels or immersed modules

- Does not require membrane modules to be removed from the unit

Limitations of visual inspection include:

- Manual application

- Requires system design considerations for implementation (e.g., sight tubes, removable plates or carriages, etc.)

4.8.2 Bubble Testing

The bubble test is based on bubble point theory (as described in Appendix B) and is applicable to MF, UF, and MCF membranes. In conducting a bubble test, the module of interest is generally first removed from the membrane unit. The external shell of the module is then drained and pressurized to a level below the bubble point of the membrane but higher than the pressure required to achieve the required 3-μm resolution for compliance with the LT2ESWTR. The pressures applied in a bubble test are generally similar to those applied for the pressure-based direct integrity tests. The end cap is removed and a dilute surfactant solution is applied to the open ends of the membrane fibers at the end of the module. The formation of bubbles in the surfactant solution can be used pinpoint specific leaking fibers. The defective fiber may be repaired by plugging the lumen with a stainless steel pin or epoxy adhesive dispensed from a syringe. The membrane module manufacturer should be consulted for specific recommendations regarding module repair. Note that the bubble test is distinct from the bubble point test described in section 3.6. The bubble test described in this section is applied to identify leaks, and thus the pressure must be kept below the bubble point of the porous membrane. By contrast, the bubble point test is applied for the specific purpose of determining the bubble point of the membrane, and thus the pressure must be gradually increased until the bubble point is achieved.

Advantages of the bubble test include:

- Ability to identify specific compromised fibers and leaking seals

- Easy to use and interpret results

- Equally applicable to systems using enclosed pressure vessels or immersed modules

Limitations of the bubble test include:

- Manual application

- May require removal of a membrane module from the rack

4.8.3 Sonic Testing

The principle underlying sonic testing is that water passing through broken fibers or other damaged system components will make a unique sound that can be detected using specialized equipment. The analysis is typically conducted by manually applying an accelerometer (an instrument used to detect vibrations) to one or more locations on each membrane module. Using headphones, the operator listens for vibrations generated by leaking air. Since the test is applied to modules on an individual basis, it is useful for identifying the module that has a potential integrity breach when the direct integrity test indicates that there may be a problem in a membrane unit. Sonic testing is generally applicable to MF, UF, and MCF systems.

A sonic test is most effectively administered by a skilled and experienced operator, particularly since the results are more subjective than the other forms of integrity testing, either direct or indirect. Nonetheless, it is a useful diagnostic tool that can help isolate a compromised membrane module. Adham et al. (1995) reported that sonic testing was able to detect a breach as small as a 0.6 mm needle puncture in the wall of one out of over 22,000 hollow fibers.

An automated sonic testing system could have the potential to eliminate the subjectivity of this test and serve as an on-line, continuous, and direct means of integrity testing. The early-stages of development and testing of such an automated system was described in a paper by Glucina et al. (1999). This acoustic monitoring system utilizes a sensor on each membrane module to detect changes in noise caused by pressure fluctuations in any compromised fibers. Test results indicated that the system was capable of detecting a single compromised fiber, although performance was affected by the level of background mechanical noise associated with the membrane filtration system. The system described by Glucina et al. has not been fully developed to date.

Advantages of sonic testing include:

- Ability to identify a specific compromised module within a membrane unit

- Ease of use (assuming the test is conducted by a trained operator)

Limitations of sonic testing include:

- Manual application

- Potential for subjective interpretations of results

- Requires the purchase of additional equipment

- Not feasible for immersed membrane systems

4.8.4 Conductivity Profiling

Conductivity profiling is a common practice associated with NF and RO systems to identify leaks in modules, o-rings, and seals. A long sample tube is inserted into the permeate port and used to withdraw a sample. The conductivity of the sample is monitored at various points among the modules in a pressure vessel and then indexed along the length of the permeate tube. Any integrity breaches are identified by significant changes in conductivity. The most common locations of leaks in a NF or RO system are at the module filtrate tube collector and end adaptor o-ring seals.

Advantages of conductivity profiling include:

- Ability to identify a specific compromised module within a pressure vessel

Limitations of conductivity profiling include:

- Manual application

- Requires operator skill for indexing probe

- Potential for subjective interpretations of results

- Requires the purchase of additional equipment

4.8.5 Single Module Testing

Single module testing may be applicable to all types of membrane filtration systems and involves removing potentially compromised modules from a membrane unit that is known to contain an integrity breach and testing each module on an individual basis. Although a number of different methods of integrity testing could potentially be utilized to screen individual modules with this diagnostic technique, single module testing generally refers to the use of a small apparatus to conduct a pressure or vacuum decay test on one module at a time. One advantage of this diagnostic test is that the pressure or vacuum decay associated with an integrity breach may be much more pronounced in a single compromised module at a given test pressure than in one compromised module in an entire membrane unit that is otherwise integral, enabling a breach to be more readily detected. Thus, while the process of single module testing is labor-intensive, it may be especially useful for isolating a compromised module in cases in which other methods of diagnostic testing have not been successful.

Advantages of the single module testing include:

- Increased ability to detect compromised modules compared to in-situ testing applied to an entire membrane unit

- May be able to isolate compromised modules in cases in which other diagnostic tests have been unsuccessful

- Generally applicable to MF, UF, NF, RO, and MCF systems

Limitations of single module testing include:

- Manual application

- Labor-intensive testing process

- May require modules to be removed from the membrane unit

- May require separate testing apparatus

4.9   Data Collection and Reporting

The LT2ESWTR requires that direct integrity testing be conducted on each membrane unit at a frequency of at least once each day of operation unless the State approves less frequent testing based on demonstrated process reliability, the use of multiple barriers effective for *Cryptosporidium*, or reliable process safeguards (40 CFR 141.719(b)(3)(vi)). At a minimum, the direct integrity test results that exceed the UCL and result in a membrane unit being taken off line must be reported, as well as the corrective action taken as a result of the UCL exceedance. This information must be reported to the State within 10 days following the end of each monthly monitoring cycle (40 CFR 141.721(f)(10)(ii)(A)). The State may exercise its discretion as to whether or not a report is required for a monthly monitoring period in which no UCL exceedances have occurred. Routine direct integrity test results that do not exceed the UCL are not required to be reported under the LT2ESWTR; however, the State may require that additional results be reported at its discretion. The State may also require that direct integrity test results be reported as measured (e.g., the rate of pressure decay rate for the pressure decay test or the flow of air for a diffusive airflow test) and/or as converted to equivalent LRVs, as calculated using the methodology described under section 4.5. Note that it is often most advantageous for a utility to record the actual process value(s) measured (e.g., the rate of pressure decay) to track the results of the direct integrity test over time. Additional guidance regarding data collection and the use of this data for system optimization is provided in Appendix A in the context of developing a comprehensive integrity verification program.

All data collected in the process of conducting direct integrity testing as required under the rule, must be kept by utilities for a minimum of three years (40 CFR 141.722(c)). This data includes all direct integrity test results, as well as UCL exceedances and any corrective action taken.

A sample summary report form for a hollow-fiber system using the pressure decay test (the most common type of direct integrity test) is provided in Figure 4.8. The form includes the following components:

- Facility information, membrane unit number, and date (i.e., month and year)

- System parameters and test constants

- Test conditions and results

- UCL exceedances

The sample summary report form also contains a column for recording the LRV that is verified for each particular daily application of the direct integrity test, as well as for the flow, TMP, and ALCR parameters that are required to calculate the LRV. Note that these data are not required for reporting purposes under the LT2ESWTR, but are included to underscore the utility of these values for tracking overall unit performance over time.

The LT2ESWTR also requires that utilities submit to the State the resolution, sensitivity, frequency, control limit, and associated baseline response (i.e., for an integral membrane filtration system) for the direct integrity test proposed to be used on the full-scale facility (40 CFR 141.721(f)(10)(i)(B)). Because the resolution and sensitivity are site- and system-specific, it will generally not be possible to accurately quantify these parameters until the full-scale system is fully installed and operational. Consequently, utilities could submit to the State estimates for these parameters based on information available from the membrane system manufacturer, and then subsequently submit more accurate, field-verified data after the full-system is operational. However, the LT2ESWTR only specifies that this data must be submitted; the specifics of the submittal process, including the timing and procedures, are subject to the discretion of the State.

Figure 4.7 Sample Summary Report Form for Pressure Decay Testing

Month		Utility	
Year		Facility Name	
Membrane Unit No.		Test Duration	min
Volume of System (V_{sys})	L	VCF	
UCL		Total No. of UCL Violations	
Signed		Dated	

Day	Pressure (psi) Initial	Final	ΔP_{test} (psi/min)	Within UCL?	Corrective Action Taken (if required)	Filtrate Flow (gpm)	TMP (psi)	ALCR	LRV Verified
1									
2									
3									
4									
5									
6									
7									
8									
9									
10									
11									
12									
13									
14									
15									
16									
17									
18									
19									
20									
21									
22									
23									
24									
25									
26									
27									
28									
29									
30									
31									
Min									
Max									
Ave									

5.0 Continuous Indirect Integrity Monitoring

5.1   Introduction

As the name suggests, the various indirect integrity monitoring methods are not physical tests applied specifically to a membrane module or membrane unit, but instead involve monitoring some aspect of filtrate water quality as a surrogate measure of membrane integrity. Because the quality of membrane filtrate is very consistent and largely independent of fluctuations in feed water turbidity or particle levels, a marked decline in filtrate quality may indicate an integrity problem. Although indirect integrity monitoring is not as sensitive as direct testing for detecting integrity breaches, the indirect methods offer some significant benefits that make them an important and useful tool in an overall integrity verification strategy. These benefits include both the ability to be operated in a continuous, on-line mode and the non-proprietary nature of the indirect techniques, such that the same indirect method is similarly applicable to any membrane filtration system independent of manufacturer, configuration, or other system parameters.

Chapter 4 discussed the use of direct integrity testing as a means of verifying the removal efficiency of a membrane filtration process at a level commensurate with the removal credit awarded by the State. However, while direct integrity tests can be extremely sensitive and thus can potentially verify high log removal values, most direct test methods require that the membranes be taken off-line (i.e., out of production mode) to undergo testing. Thus, these direct tests are limited to periodic application in order to minimize system down time. A failed periodic direct integrity test indicates that an integrity breach occurred at some time between the most recent direct test in which integrity has been verified and the failed test, but indicates nothing about integrity over the period between direct test applications. Continuous monitoring using indirect methods does provide a real-time indication of membrane integrity, albeit with generally less sensitivity. Consequently, the advantages of the direct and indirect integrity monitoring approaches are complementary, and both are critical elements of a comprehensive integrity verification program (IVP). (The development of a comprehensive IVP, including the complementary value of direct and indirect methods for assessing system integrity, is discussed in Appendix A.)

The Long Term 2 Enhanced Surface Water Treatment Rule (LT2ESWTR) requires continuous indirect integrity monitoring in the absence of a continuous direct integrity test method (40 CFR 141.719(b)(4)). Currently, continuous direct integrity monitoring techniques that meet the performance criteria of the rule are not available, and thus it is anticipated that utilities employing membrane filtration to comply with the LT2ESWTR requirements will have to implement both periodic direct integrity testing and continuous indirect integrity monitoring at the time of rule promulgation. However, the development of a sufficiently sensitive and reliable continuous direct integrity test is potentially feasible, and implementation of such a test would preclude the requirement to conduct indirect integrity monitoring, additional State integrity monitoring requirements notwithstanding.

"Continuous" indirect integrity monitoring is defined as monitoring some aspect or component of filtrate water quality that is indicative of the removal of particulate matter at a frequency of no less than once every fifteen minutes (40 CFR 141.719(b)(4)(ii)). The LT2ESWTR specifies turbidity monitoring on the filtrate of each membrane unit as the default methodology for continuous indirect integrity monitoring, and that two consecutive filtrate turbidity readings above 0.15 NTU trigger direct integrity testing (40 CFR 141.719(b)(4)(iii), 40 CFR 141.719(b)(4)(i), and 40 CFR 141.719(b)(4)(iv), respectively). Turbidity was selected as the default indirect integrity monitoring technique because it is an accepted monitoring technology within the water treatment industry, it is used as both a relative and an absolute indicator of water quality, and it is required for plants employing both conventional and alternative technologies under current surface water treatment regulations. However, because several other techniques are generally accepted for indirect integrity monitoring in the water treatment industry, a number of which are more sensitive than turbidity monitoring, the LT2ESWTR contains a provision to allow States to approve alternative methods for continuous indirect integrity monitoring.

Independent of the method used (i.e., turbidity monitoring or an alternative method approved by the State), the LT2ESWTR specifies the following requirements for indirect integrity monitoring:

1. The filtrate from *each membrane unit* of the filtration system must be monitored independently (40 CFR 141.719(b)(4)(iii)). (The definition of a membrane unit under the LT2ESWTR is provided in section 1.5.)

2. The filtrate must be monitored *continuously* from each membrane unit. For the purposes of indirect integrity monitoring under the LT2ESWTR, continuous monitoring is defined as taking measurements at least once every 15 minutes (40 CFR 141.719(b)(4)(ii)).

3. A *performance-based control limit* must be established such that readings exceeding the control limit for a period of greater than 15 minutes (or two consecutive 15-minute readings exceeding the control limit) would immediately trigger direct integrity testing (40 CFR 141.719(b)(4)(iv) and 40 CFR 141.719(b)(4)(v)). (The establishment of control limits is further discussed in the context of each indirect integrity monitoring method addressed in Chapter 5. Additional information regarding control limits and their role in a comprehensive IVP is provided in Appendix A.)

4. Utilities must *report* a summary of all excursions above the established control limit that trigger direct integrity testing to the State on a monthly basis (40 CFR 141.721(f)(10)(ii)(B)).

Aside from turbidity monitoring, particle counting may be the most commonly employed means of indirect integrity monitoring for compliance with the LT2ESWTR. Both of these techniques are widely used throughout the water treatment industry. Particle monitoring is similar to particle counting, although the collected data are presented in a relative manner rather than as absolute values. The tracking of particles and turbidity in the filtrate is based on the passage of particulate matter through a membrane and is thus appropriate for all types of membrane filtration systems. However, other surrogate measures of integrity that monitor dissolved solids or other miscible water quality parameters such as conductivity may also satisfy the indirect integrity monitoring requirements of the LT2ESWTR for nanofiltration (NF) and reverse osmosis (RO) membranes capable of affecting the levels of these water quality indicators. Still, it is important to note that any indirect integrity monitoring method used for the purposes of LT2ESWTR compliance other than turbidity monitoring must be approved by the State.

Under the LT2ESWTR, resolution and sensitivity are both defined with respect to their applicability to direct integrity testing, as described in Chapter 4. However, unlike the direct methods, there are no requirements relating to test sensitivity or resolution with respect to the indirect integrity monitoring methods under the LT2ESWTR. Nevertheless, these two concepts, where applicable, may be useful tools for optimizing the ability of an indirect test to yield meaningful information about potential integrity breaches. Within the context of continuous indirect integrity monitoring, resolution and sensitivity may be described as specified in Chapter 4, although without the references to direct integrity testing, as specified in the rule language. Thus, resolution may be described as the smallest leak that contributes to a response to an indirect integrity monitoring method, and sensitivity may be expressed as the maximum log removal value that can be reliably verified by an indirect integrity monitoring method. Resolution and sensitivity are discussed in association with each respective indirect integrity monitoring method addressed in Chapter 5, noting the extent to which each concept applies to each method.

The purpose of Chapter 5 is to describe the most commonly used techniques for indirect integrity monitoring and discuss their use for compliance with the requirements of the LT2ESWTR. Considerations for selecting other indirect integrity monitoring methods are also addressed.

This chapter is divided into the following sections:

Section 5.2: Turbidity Monitoring
This section discusses the use of turbidity monitoring as the default indirect integrity monitoring technique under the LT2ESWR, including methods of application, control limits, and advantages and limitations.

Section 5.3: Particle Counting and Particle Monitoring
This section describes the use of particle counting or particle monitoring as a potential alternative to turbidity monitoring for indirect integrity monitoring, including methods of application, control limits, and advantages and limitations.

Section 5.4:   Other Indirect Monitoring Surrogates
This section discusses some general considerations for selecting an alternative indirect integrity monitoring method not specifically addressed in this guidance.

Section 5.5:   Data Analysis and Reporting
This section summarizes the reporting requirements for continuous indirect integrity monitoring results and provides guidance on the analysis and interpretation of the results.

Note that a non-continuous indirect method (e.g., a silt density index (SDI) test) has limited value for integrity monitoring, offering neither the ability to directly test the membranes nor the advantage of continuous, on-line monitoring. As a result, non-continuous indirect methods do not satisfy the indirect integrity monitoring requirements of the LT2ESWTR and are not addressed in this chapter.

5.2   Turbidity Monitoring

Conventional turbidimeters (also referred to as "turbidity meters") detect the intensity of light scattered at one or more angles to an incident beam of light. The angular distribution of scattered light depends on a number of conditions, including the wavelength of the incident light, as well as particle size, shape, and composition. Consequently, it is difficult to correlate the turbidity with the number or mass concentration of particles in suspension. However, turbidimeters are in widespread use throughout the water industry, and the turbidity data generated by these instruments is broadly recognized as a meaningful gauge of water quality. As such, turbidity measurements have been used as an indicator of finished water quality for previous surface water regulations.

Under the LT2ESWTR, turbidity monitoring is specified as the default method for continuous indirect integrity monitoring unless the State approves an alternate approach (40 CFR 141.719(b)(4)(i)). Using the default approach specified in the rule, filtrate turbidity must be monitored on each discrete membrane unit at a minimum frequency of once every 15 minutes (40 CFR 141.719(b)(4)(iii) and 40 CFR 141.719(b)(4)(ii), respectively). Two consecutive readings that exceed a control limit of 0.15 NTU for any one unit would require that unit to immediately undergo direct integrity testing and subsequent diagnostic testing, as necessary (40 CFR 141.719(b)(4)(iv)). Any such events that trigger direct integrity testing must be reported to the State on a monthly basis (40 CFR 141.721(f)(10)(ii)(B)). Further information regarding data collection, analysis, and reporting is provided in section 5.5.

Because turbidimeters do not yield any information about particulate size, these instruments cannot provide any specific indication about the size of an integrity breach through which turbidity may have passed. Similar turbidity readings may result from either significant particulate material passing through a small breach, or a smaller amount of larger particulate matter passing through a larger breach. As a result, the concept of test method resolution as defined under the LT2ESWTR is not applicable to turbidity monitoring. However, the concept of method sensitivity can apply to turbidity monitoring, even though there is no formal sensitivity requirement for continuous indirect integrity monitoring methods, as fiber cutting studies may be conducted to correlate a particular reduction in log removal capability of the membrane system to approximate increases in filtrate turbidity. The sensitivity of turbidity monitoring and the use of this technique to detect integrity breaches are further discussed in the context of establishing control limits in section 5.2.2.

The following sections describe turbidity monitoring methods, approaches for establishing control limits, and advantages and limitations, respectively.

5.2.1 Methods

There are two basic types of on-line turbidimeters that may have the capability to provide continuous indirect integrity monitoring as required under the LT2ESWTR: conventional turbidimeters and laser turbidimeters. The primary difference between these two types of instruments lies in the type of light source used. Conventional turbidimeters typically utilize a tungsten lamp or other light-emitting diode (LED) as a light source, whereas laser turbidimeters employ a laser light source.

Because laser turbidimetry is a relatively new technique, its effectiveness as a monitoring tool is still being evaluated within the water industry. However, recent research generally indicates that laser turbidimeters are more sensitive than conventional turbidimeters and may perform comparably to particle counters for detecting membrane integrity breaches (Banerjee et al. 2000; Colvin et al. 2001). Available manufacturer specifications indicate that laser turbidimeters may have increased sensitivity in excess of two orders of magnitude over conventional turbidimeters (in terms of measuring turbidity, not necessarily with respect to the ability of the technique to measure reductions in log removal efficiency resulting from membrane integrity breaches). Additional research indicates laser turbidimeters can be optimized to measure very low turbidities, in the range of 0 to 1 NTU (Banerjee et al. 1999a; 1999b; and 2000). Since most microfiltration (MF) and ultrafiltration (UF) systems produce filtrate water consistently in the range of 0.03 to 0.07 NTU as measured by conventional turbidimeters, laser turbidimeters may be better suited to monitor membrane filtrate.

Currently, there are four USEPA-approved analytical methods for the measurement of turbidity (USEPA 1999). These are as follows:

- USEPA Method 180.1, Determination of Turbidity by Nephelometry (USEPA 1993)

- Standard Method 2130B, Turbidity – Nephelometric Method (American Public Health Association (APHA) 1998)

- Great Lakes Instrument Method 2 (GLI 2) (USEPA 1999)

- Hach FilterTrak Method 10133

Systems must utilize turbidimeters that conform to one of these four methods for compliance monitoring purposes. Additional guidance on the installation, calibration, operation, and maintenance of on-line turbidimeters is provided in *Guidance Manual for Compliance with the Interim Enhanced Surface Water Treatment Rule: Turbidity Provisions* (USEPA 1999).

5.2.2 Control Limits

A control limit (CL) is defined as a threshold response from an integrity test that triggers a specific action. The LT2ESWTR requires CLs to be established for both direct integrity tests and indirect integrity monitoring. In the context of this guidance manual the CLs required under the rule are referenced as upper control limits (UCLs), signifying that these limits serve as indicators of minimally acceptable system performance. For direct integrity tests, the UCL corresponds to the smallest integrity breach that would compromise the log removal credit awarded by the State. However, for indirect integrity monitoring methods the UCL is simply designed to serve as a general indication that a system integrity breach may have occurred.

Unless the State approves an alternate approach (e.g., particle counting, particle monitoring, or other method), the LT2ESWTR requires continuous filtrate turbidity monitoring on each membrane unit with a control limit of 0.15 NTU (40 CFR 141.719(b)(4)(iv)). Because most membrane filtration systems consistently produce filtrate well below 0.15 NTU, a sustained high turbidity event with filtrate readings above 0.15 NTU may suggest a potentially serious integrity problem. Consequently, the LT2ESWTR requires that two consecutive filtrate turbidity readings above 0.15 NTU on any membrane unit trigger immediate direct integrity testing on that unit, as described in Chapter 4. If the unit in question passes the triggered direct integrity test, the unit may continue in production. However, if the unit fails the direct integrity test, further diagnostic testing and repair of any integrity breach(es) would be required. The unit may only be returned to service upon passing a direct integrity test.

Note that a more stringent UCL may be either voluntarily implemented by a utility or mandated at the discretion of State. For turbidity monitoring, an example of a more stringent UCL might be 0.10 NTU. A more stringent UCL might be advantageous for systems that rely on membrane filtration to achieve a high log removal of *Cryptosporidium*, since the risk associated with an integrity breach is greater. Alternatively, a more stringent UCL may also be warranted for a membrane filtration system that consistently produces filtrate with very low turbidity and which exhibits almost no spikes in the data.

The LT2ESWTR defines continuous monitoring as one reading at least every 15 minutes (40 CFR 141.719(b)(4)(ii)). The rule is designed to allow States latitude for interpreting how the continuous indirect integrity monitoring data collected by a utility would translate into a measurement for the purposes of reporting and rule compliance. Some potential strategies that States may adopt for data collection and reporting in interpretation of this requirement are further discussed in both section 5.5 and Appendix A. In the absence of any particular State requirements, a utility may adopt a data interpretation strategy that is most appropriate for its system.

While only the UCL is required under the LT2ESWTR, some utilities may choose to establish a tiered approach to control limits with one or more lower control limits (LCLs), in which the level of corrective action increases with each successive CL that is further removed from normal system performance. For example, a utility may opt to implement a LCL in which a filtrate turbidity reading exceeding 0.10 NTU triggers increased monitoring frequency. Alternatively, with hollow-fiber membrane filtration systems, a utility may choose to conduct fiber-cutting studies to correlate a quantifiable loss of log removal capability with a certain level of increased filtrate turbidity. Research has indicated that turbidity monitoring may be sensitive to some smaller integrity breaches, although this sensitivity is dependent on the type membrane system, the number of modules linked to a single instrument, the number of fibers per module, the hydraulic configuration, and other system-specific parameters (Adham et al. 1995). One study conducted by the Wisconsin Department of Natural Resources demonstrated that for one system in Kenosha, WI turbidity monitoring was unable to detect any loss of system integrity even with a substantial number of cut fibers (Landsness 2001). Consequently, as this study indicates, for some systems turbidity monitoring may not be sufficiently sensitive to allow a meaningful correlation with integrity loss for the purposes of establishing LCLs. In general, it is suggested that a utility work with both the State and the manufacturer of its membrane filtration system to establish any additional CLs and associated corrective actions that may be appropriate.

5.2.3 Advantages and Limitations

One of the most significant advantages of tracking turbidity as a means of indirect integrity testing is that system operators throughout the water treatment industry are familiar with the use of conventional turbidimeters. Because filtrate turbidity monitoring has been required for media filtration under previous federal regulations for surface water treatment, conventional turbidimeters are widely used in water treatment plants across the country.

Many of the other advantages of both conventional and laser turbidimeters are relative to other continuous indirect integrity monitoring methods. For example, neither conventional nor laser turbidimeters are subject to the same accuracy and precision inconsistencies that can be problematic for particle counters. While two well-calibrated turbidimeters are likely to yield similar data for the same water, achieving such consistency from two separate particle counters may be more difficult. In addition, turbidimeters have the advantages of measuring a parameter that is absolute (as opposed to a relative measure, as with particle monitors), broad acceptance as a meaningful gauge of water quality, and application as a water quality benchmark in other federal surface water treatment regulations.

Conventional turbidimeters are also significantly less expensive than particle counters. Although laser turbidimeters are typically more expensive than conventional turbidimeters, the ability to multiplex these instruments may help to minimize the associated cost of utilizing laser turbidimetry (Banerjee et al. 2001). In multiplexing, multiple sensors are connected to a single laser light source, detector, and control system via fiber optics. Sensors can be attached to monitor the filtrate from each membrane unit or individual membrane module, if desired. At the time of publication, however, multiplexing laser turbidimetry systems are still under development and not yet commercially available.

The most significant limitation of on-line turbidimeters is insensitivity to small changes in water quality. Since membrane systems produce a high quality filtrate, a turbidimeter may be unable to detect minor changes in filtrate water quality as a result of a small integrity breach. Laser turbidimeters are more sensitive than conventional turbidimeters, but this sensitivity is still subject to variation based on the number of modules per sensor, the system hydraulic configuration, and other system-specific parameters.

Turbidimeters are also subject to air entrainment error. Any air bubbles introduced into the system either during production, backwashing, chemical cleaning, or integrity testing may be falsely detected as particulate matter, artificially increasing the turbidity reading. Consequently, after a backwash cycle or chemical cleaning (particularly if air is utilized in the process), turbidity measurements may not be representative of filtrate quality until any entrained air is purged from the system. This purge time will vary between different membrane filtration systems and their respective backwash or chemical cleaning practices. Typically, bubble traps are employed with turbidimeters (both conventional and laser) to minimize or eliminate this error.

Advantages of turbidity monitoring include:

- Widely used in surface water treatment plants throughout the country

- Less expensive instrumentation (conventional turbidimeters)

- Consistent, precise measurements

- Absolute (as opposed to relative) measure of water quality

Limitations of turbidity monitoring include:

- Less sensitive to smaller integrity breaches than particle counters or particle monitors (conventional turbidimeters)

- Susceptible to air entrainment error

5.3   Particle Counting and Particle Monitoring

Particle counters use a laser-based light scattering technique to count particles and group them according to size. Particle monitors also operate on the principle of light obstruction; however, rather than counting particles and grouping them by size, particle monitors measure particulate water quality on a dimensionless scale relative to an established baseline.

Although neither particle counting nor particle monitoring are specified under the LT2ESWTR, either may potentially be approved by a State as an alternative technique for satisfying the continuous indirect integrity monitoring requirement of the rule. However, any approved technique must still be utilized in such a way as to satisfy the four basic criteria for indirect integrity monitoring, as specified in section 5.1. In summary, these criteria are that the filtrate of each membrane unit must be monitored independently (40 CFR 141.719(b)(4)(iii); indirect integrity monitoring on the filtrate of each membrane unit must be continuous (i.e., at a frequency no less than once every 15 minutes) (40 CFR 141.719(b)(4)(ii); two consecutive excursions above a pre-established, performance-based upper control limit must trigger direct integrity testing (40 CFR 141.719(b)(4)(iv) and 40 CFR 141.719(b)(4)(v)); and all excursions above the control limit that trigger direct integrity testing must be reported to the State on a monthly basis (40 CFR 141.721(f)(10)(ii)(B)).

The ability of particle counters to convey information about particle size renders the concept of resolution (as defined under the LT2ESWTR) applicable to this method of indirect integrity monitoring. Although there are no requirements for continuous indirect integrity monitoring method resolution (or sensitivity) under the LT2ESWTR, the ability of particle counters to yield resolution information may help to optimize the usefulness of this technique for detecting potential integrity breaches. Any significant increase in the number of particles exceeding 3 μm in size may indicate that a breach allowing the passage of *Cryptosporidium*-sized particles may have occurred. Any particle counters that are used for the purpose of membrane filtrate monitoring to satisfy the continuous indirect integrity monitoring requirement of the LT2ESWTR should be well calibrated to detect particles in the size range of *Cryptosporidium* oocysts (i.e., 3 to 7 μm). Conversely, because particle monitoring instruments do not measure particulate size, there is no specific resolution associated with the particle monitoring technique.

As with turbidity monitoring, the concept of test method sensitivity (as defined under the LT2ESWTR) applies to both particle counting and particle monitoring in that fiber cutting studies may be conducted to correlate a particular reduction in log removal capability to increases in the quantity of particles in the filtrate. Although Adham et al. (1995) determined that particle counting was the most sensitive of the three common methods of continuous indirect integrity monitoring (i.e., conventional turbidity monitoring, particle counting, and particle monitoring), particle counting instruments have a number of well-established operational problems that can potentially distort both the accuracy and precision of their measurements. Particle monitoring devices, which are more sensitive then conventional turbidimeters but less sensitive than particle counters, share many of the same limitations of particle counters (Adham et al. 1995). The advantages and limitations of both particle counters and particle monitors are discussed in section 5.3.3. The sensitivity of particle counting and particle monitoring methods for the detection of integrity breaches is further discussed in section 5.3.2 in the context of establishing control limits for these indirect integrity monitoring techniques.

5.3.1 Methods

USEPA does not currently approve particle counting and particle monitoring as compliance monitoring methods. However, either of these methods may be used for compliance with the continuous indirect integrity monitoring requirements of the LT2ESWTR if approval is granted by the State. Both the International Organization for Standardization (ISO) and the American Society for Testing and Materials (ASTM) have published standards relating to particle counting techniques, as follows:

- ISO 11943 – Hydraulic fluid power – On-line automatic particle counting systems for liquids – Methods of calibration and validation (1999)

- ASTM F658-00a – Standard Practice for Calibration of a Liquid-Borne Particle Counter Using an Optical System Based Upon Light Extinction (2000)

In addition, there are some relevant references on the use of particle counters in water treatment applications that may serve as a useful source of additional information:

- *Particle Count Method Development for Concentration Standards and Sample Stabilization.* American Water Works Association Research Foundation (AWWARF) – Chowdhury et al., 2000.

- *Fundamentals of Drinking Water Particle Counting.* AWWARF – Hargesheimer et al., 2000.

Currently there are no standards for particle monitoring, and little information has been published regarding the use of particle monitors in potable water treatment applications.

5.3.2 Control Limits

As discussed in section 5.2.2 (and throughout Chapter 4), a CL is a response to an integrity test that results in some measure of corrective action. UCLs refer to the required limits mandated under the LT2ESWTR, while LCLs are voluntary or additional State-mandated limits. In the context of continuous indirect integrity monitoring, the UCL is a response to the results of an indirect integrity test method that triggers direct integrity testing. The UCL for the default method of turbidity monitoring is defined in the LT2ESWTR; however, the rule does not establish control limits for other continuous indirect integrity monitoring methods (e.g., particle counting or particle monitoring).

Because particle count and particle monitoring data can vary significantly between two different instruments, even between adjacent membrane units applied to the same source water and within the same filtration system, site-specific CLs must be established when particle counting or particle monitoring is used as an alternative method of continuous indirect integrity monitoring. Moreover, it is recommended for these two techniques that CLs be developed on a membrane-unit-specific basis. One such approach that has been used by the Texas Commission on Environmental Quality (TCEQ) is establishing a CL for particle counting at the 95-percent confidence interval for all the data collected by a particular instrument monitoring the filtrate of a particular membrane unit over the previous month. However, because the particle counts could slowly increase over time without exceeding the 95-percent confidence interval (either as a result of slow membrane degradation or the instrument slipping out of calibration), this approach also incorporates an additional, higher control limit that is absolute (i.e., at a fixed number of particles per unit volume) (Schulze 2001). The magnitude of this UCL would be sufficient to indicate a probable membrane integrity breach independent of variations in particle count readings between different instruments. Note that if the 95-percent confidence interval of the data collected over the previous month is used to establish a new CL on an ongoing basis, it is important to exclude any data corresponding to a known integrity breach. Otherwise, a small amount of abnormally elevated data that are not indicative of typical membrane unit performance could skew the CL for the following month to an unacceptably high level, thereby diminishing the ability to detect small integrity breaches.

Absolute CLs may also be used without a lower, relative CL (e.g., the 95-percent confidence interval of the previous month's data) provided the absolute CLs are sufficiently conservative and established using a scientific methodology that is approved by the State. For example, studies may be conducted prior to placing the membrane filtration system into service in order to establish baseline particle count or particle monitoring data levels for each membrane unit. Building on the example of the framework developed by the TCEQ, the 95-percent confidence interval of this fixed data set could then represent an absolute CL. Although this method (and other membrane-unit-specific methods) may result in different CLs for each membrane unit, if studies indicate that baseline filtrate particle count or particle monitoring data are similar among all the units (such that the control limits would also be similar), the most conservative (i.e., lowest) CL may be applied to all the units. One advantage of using a statistical method such as the 95-percent confidence interval for setting CLs is that it minimizes the possibility that normal fluctuations and noise in the data that are unrelated to an integrity problem will exceed the control limit, triggering additional testing and resulting in lost unit productivity.

For hollow-fiber membrane filtration systems, fiber cutting studies may also be used to establish control limits for particle counting or particle monitoring by correlating quantifiable losses of log removal capability to specific increases in the number of particles. However, while Adham et al. (1995) and Sethi et al. (2004) demonstrated that particle counting and particle monitoring were both more sensitive than turbidity monitoring (and that particle counting was more sensitive than particle monitoring) for detecting integrity breaches, a study conducted by Landsness (2001) indicated that both particle counting and turbidity monitoring may be unable to reliably detect an integrity breach of fewer than the equivalent of several hundred fibers in a membrane unit. As with turbidity monitoring, however, the sensitivity of both particle counters and particle monitors is dependent on the type of membrane system, the number of modules linked to a single instrument, the number of fibers per module, the hydraulic configuration, and other system-specific parameters.

For particle counting, particle monitoring, or any approved method of continuous indirect integrity monitoring, one reading must be taken on the filtrate of each membrane unit at least every 15 minutes. If the readings exceed the UCL for a period of fifteen minutes (or, in the case that readings are taken only at 15 minute intervals, if two consecutive 15 minute readings exceed the UCL), then direct integrity testing is immediately required (40 CFR 141.719(b)(4)(v)). Any membrane unit taken off-line for further integrity testing (i.e., direct and potentially diagnostic testing) may be returned to service upon passing a direct integrity test. All excursions beyond the UCL that trigger direct integrity testing must be reported to the State on a monthly basis (40 CFR 141.721(f)(10)(ii)(B)). Also, although not required under the LT2ESWTR, utilities may also voluntarily choose to establish one or more LCLs that would trigger increased data collection frequency or require an operator to monitor the membrane performance for a specified period of time. Data analysis and reporting for continuous indirect integrity monitoring are further discussed in section 5.5, and guidance for the development of a comprehensive integrity verification program is provided in Appendix A.

5.3.3 Advantages and Limitations

The primary advantage associated with particle counters and particle monitors is the sensitivity of the instruments relative to conventional turbidimeters. Both have been shown to be more sensitive to breaches in membrane integrity than conventional turbidimeters, with particle counters being the most sensitive of the three instruments (Adham et al. 1995; Jacangelo et al. 1997).

However, particle counters and particle monitors have a number of well-known potential operational problems that may distort the accuracy and precision of their measurements. First, two separate but otherwise identical particle counters or particle monitors applied to the filtrate of the same membrane unit may yield different responses, the magnitude of which may be particularly significant at the very low particle concentrations that are typical of membrane filtrate quality. This inconsistency could potentially make it necessary to establish CLs on a membrane-unit-specific basis, since the magnitude of a particle count or particle monitor reading that may begin to suggest a small integrity breach may not be the same for two different instruments applied to the filtrate of different membrane units. In addition, once baseline filtrate data are established for each membrane unit, this inconsistency could complicate changing instruments, if necessary, since new baseline data may need to be generated. The impact of this potential problem may not be significant, however, if the relative differences between instruments are negligible in comparison to the differences between the baseline filtrate quality and that associated with an integrity breach of sufficient magnitude to exceed the UCL.

The variation between the readings of different instruments at low particle concentrations may be particularly problematic for using particle counters to verify log removal capabilities in association with challenge testing or a particulate-based direct integrity test. For example, the difference between 1.0 particles/mL and 0.1 particles/mL, which is within the range of observed variance for different particle counters at very low particulate concentrations, translates into a 1 log discrepancy in terms of removal capability. This variation must be considered when using particle counters to verify log removals. Considerations and guidance for challenge testing are addressed in Chapter 3, and particulate-based direct integrity tests are discussed in section 4.7.5.

Another disadvantage associated with particle counters and particle monitors is the potential for air entrainment to cause the instruments to overestimate filtrate particle concentrations. For example, in hollow-fiber membrane filtration systems, a significant amount of air may be introduced into the piping during the backwash process, particularly if the process utilizes air for scouring or pulsing the membranes. Any air bubbles remaining in the system after the backwash cycle is complete when the unit is returned to production may be falsely detected as particles. Thus, anomalous particle concentration spikes that may indicate an integrity problem could be masked until the air is purged from the system. As a result, the effective use of particle counters and particle monitors as an integrity monitoring tool is somewhat diminished after backwashing until the air is expelled and the particle concentrations return to baseline levels. The time required for this air purge varies with the particular membrane system and the backwash operating parameters. As with turbidimeters, bubble traps can be used in some cases to minimize or eliminate this potential source of error.

In terms of particle detection, the use of particle counters and particle monitors could also be potentially problematic when applied to detect pathogens with translucent cell walls that allow some level of light passage. In such cases there may be some potential for these devices to inaccurately report particle concentrations. This phenomenon may be more applicable to particle counters, which classify particles according to size, since it is more likely that some level of pathogen translucence will diminish the amount of light scatter (resulting in the apparent detection of a smaller particle) associated with that pathogen, rather than allowing the pathogen to pass across the sensor undetected.

Particle counters and particle monitors are also susceptible to coincidence error and sensor clogging, both of which can cause particle concentrations to be underestimated. However, this may not represent a significant problem for these instruments when applied to monitor membrane filtrate. Because membrane filtrate consistently has a very low concentration of particulate matter, even under compromised conditions, coincidence or clogging errors are much less likely with membrane filtrate monitoring than for most other applications in potable water treatment. Nevertheless, these errors can be problematic when using feed and filtrate particle counters in combination to determine a log removal value, such as in the application of particular marker-based direct integrity test.

Advantages of particle counting and particle monitoring are generally similar, and include the following:

- More sensitive to smaller integrity breaches than conventional turbidimeters

- Widely used in surface water treatment plants (particle counters)

- Absolute (as opposed to relative) measure of water quality (particle counters)

- Ability to yield information regarding test resolution (particle counters)

<u>Limitations</u> of particle counting and particle monitoring include:

- Imprecision between instruments at low particulate concentrations

- Susceptible to air entrainment error

- Susceptible to coincidence and clogging error at higher particle concentrations

- More expensive instrumentation (particle counters)

- Relative measure of water quality (particle monitors)

- More operation and maintenance support

- May misrepresent concentrations of particles that allow some level of light passage

5.4   Other Indirect Monitoring Methods

In addition to the three methods of continuous indirect integrity monitoring described in this chapter – turbidity monitoring, particle counting, and particle monitoring – other methods may also be used, including multiplexing and other means of characterizing the particulate matter in the filtrate. Note that turbidity monitoring is the only method specified under the LT2ESWTR for compliance with the rule, although other methods (including both particle counting and particle monitoring) may be used with State approval. All three of the methods described in this chapter relate to monitoring particulate matter in the filtrate, as these methods are applicable to all types of membrane filtration systems – MF/UF, NF/RO, and membrane cartridge filtration (MCF). However, because NF/RO processes also remove dissolved constituents, methods that monitor these constituents in the filtrate also have the potential to be used for continuous indirect integrity monitoring with NF/RO systems.

An example of one potential method that could be used for continuous indirect integrity monitoring with NF/RO systems is conductivity monitoring. Because monitoring filtrate (i.e., permeate) conductivity is routinely used as a means of assessing semi-permeable membrane performance, it may be advantageous for utilities using NF/RO systems for compliance with the LT2ESWTR to employ this technique for continuous indirect integrity monitoring, if approved by the State. However, there are some limitations with conductivity monitoring that are also applicable to other indirect methods that monitor dissolved constituents. First, increased salt passage over time may be a function of either an uncontrolled increase in membrane permeability or a planned increase in system recovery or flow, parameters unrelated to the physical integrity of the membranes and their ability to serve as a barrier to particulate matter such as *Cryptosporidium*. Although increases in membrane permeability that do not represent integrity breaches can manifest over longer periods of time and may not occur over the 15-

minute intervals at which continuous indirect integrity monitoring must be conducted, this is a consideration that should be taken into account by utilities. In addition, filtrate conductivity (and other dissolved constituents) may vary with water quality parameters such as pH and temperature, factors that are likewise unrelated to membrane integrity. While fluctuations in conductivity with these factors may potentially be minimal relative to those that might occur as a result of an integrity breach, it is nevertheless important that a NF/RO system be operated under consistent water quality and operational conditions for conductivity to yield reliable and meaningful results relative to membrane integrity.

Whether using a technique that measures particulate matter or dissolved constituents, there are several general criteria that should be considered when selecting an alternative method for continuous indirect integrity monitoring:

- The monitoring technique should be approved for use in water treatment applications by the State.

- For systems that apply membrane filtration specifically for compliance with the LT2ESWTR, the technique must be approved by the State specifically for compliance with the rule.

- The monitored parameter must be able to be measured on-line.
  (i.e., while the membrane unit is in production)

- The parameter must be able to be monitored continuously.
  (i.e., at a frequency of at least one measurement every 15 minutes for the purposes of compliance with the LT2ESWTR)

- The parameter must be monitored on the filtrate of each membrane unit.

- A significant change in the levels of the parameter should generally be indicative of an integrity breach.

- The instrumentation used to measure the parameter should be sufficiently sensitive such that relatively small integrity breaches can be detected, thus enabling the establishment of a meaningful UCL. (Note that there are no specific sensitivity requirements for continuous indirect integrity monitoring under the LT2ESWTR.)

5.5   Data Collection and Reporting

The LT2ESWTR requires utilities both to record a continuous indirect integrity monitoring reading for each membrane unit at least once every 15 minutes and to report any sequence of two consecutive excursions above the UCL (which would trigger direct integrity testing) to the State within 10 days following the end of each monthly monitoring cycle (40 CFR 141.719(b)(4)(iv), 40 CFR 141.719(b)(4)(v), and 40 CFR 141.721(f)(10)(ii)(B)). Within this regulatory framework, utilities and/or States have some latitude with respect to the specifics of how continuous indirect integrity monitoring data are collected and reported. For example, provided the basic requirements specified in the rule are satisfied, compliance strategies may vary in regard to the following:

- Frequency of data collection

- Method of data reduction

- Monitoring location

Some utilities may desire to collect data more frequently than once every 15 minutes, particularly since most of the instruments associated with the various methods of continuous indirect integrity monitoring are capable of recording data at much shorter intervals. Even if a utility is not required by the State to report data any more frequently than at 15-minute intervals, the additional data may be useful for monitoring system performance with greater precision, identifying trends and patterns, and ultimately optimizing system operation. If a utility chooses to measure and record indirect integrity monitoring data at a frequency greater than every 15 minutes (or is required to do so by the State), the State may require that data collected at intervals shorter than 15 minutes be compared to the UCL for purposes of potentially triggering direct integrity testing and compliance reporting.

If a utility does opt to collect continuous indirect integrity monitoring data more frequently than every 15 minutes, the LT2ESWTR allows some flexibility regarding how the collected data are utilized for compliance purposes. For example, if a utility records one turbidity reading each minute, there are several methods in which the data collected over a 15-minute period may be reduced. One option would be to average the data to yield a single statistic that is representative of the performance of each membrane unit for a given 15-minute interval. Two consecutive 15-minute averages exceeding the UCL would trigger direct integrity testing and subsequent State notification in the next monthly report. Another option for this example would be to make the maximum value of the 15 measurements the representative statistic. Likewise, any two consecutive 15-minute intervals with maximum values exceeding the UCL would trigger direct integrity testing and require the excursion to be reported. Alternatively, independent of how often data are collected, a utility may opt to use only the reading recorded at the 15-minute mark for comparison with the UCL. Note that while the rule allows for some flexibility for data collection and reduction, the State may require a specific methodology at its discretion.

The critical interval of 15 minutes cited in the LT2ESWTR may also be shortened at the discretion of the utility if a more conservative approach to monitoring membrane integrity is desired. For example, a utility that records a particle count measurement once every five minutes may configure its system to trigger direct integrity testing if any two consecutive 5-minute readings exceed the UCL. A more conservative approach to the location of indirect integrity monitoring may be adopted, as well, if desired. Although monitoring is required for the filtrate of every membrane unit, additional monitoring may be conducted on one or more individual membrane modules, pressure vessels, or stages within a membrane unit. Monitoring smaller incremental membrane process units, where applicable and cost-effective, can increase method sensitivity and may help to isolate any integrity breaches that are detected, as well as provide additional data for process optimization and control.

The LT2ESWTR does not require utilities to report all excursions above the UCL, but rather only those sequences of two consecutive UCL excursions that would trigger direct integrity testing. While single excursions may be anomalous measurements and thus not be indicative of an integrity breach, two consecutive measurements exceeding the UCL represent more sustained evidence of an integrity problem that warrants further testing and subsequent reporting. In conjunction with these exceedances, the rule also requires a corresponding summary of any corrective action taken in each case, including direct integrity testing and subsequent measures for addressing a detected integrity problem (40 CFR 141.721(f)(10)(ii)(B)). The ongoing record of these responses over time should also help the utility identify and streamline any appropriate future corrective action that may be necessary. Note that the State may require more detailed reporting at its discretion.

In general, the LT2ESWTR allows for some flexibility in compliance strategy with respect to continuous indirect integrity monitoring data collection and reporting to accommodate site-specific considerations, provided the basic requirements of both the rule and those of the State are satisfied. Additional guidance for data collection and analysis strategies, as well as for using this data for system optimization, is provided in Appendix A in the context of developing a comprehensive integrity verification program.

All data collected in the process of conducting indirect integrity monitoring, as required under the rule, must be kept by utilities for a minimum of three years (40 CFR 141.722(c)). This data includes all indirect integrity monitoring results, as well as UCL exceedances and any corrective action taken. If indirect integrity monitoring data are collected more frequently than at the required minimum of once every 15 minutes, only the representative statistic that is used for rule compliance by comparison with the UCL each 15-minute interval must be retained to meet the recordkeeping requirements of the rule relative to continuous indirect integrity monitoring.

6.0 Pilot Testing

6.1 Introduction

Pilot testing a membrane filtration process is often conducted as part of the design effort for a membrane treatment facility. The purpose of this chapter is to describe the considerations that are associated with pilot testing. It should be noted that the Long Term 2 Enhanced Surface Water Treatment Rule (LT2ESWTR) does not contain any requirements for pilot testing membrane filtration systems; thus, this chapter is simply intended to provide general guidance in terms of widely recognized industry practices. However, any particular State requirements would supersede the guidance provided in this chapter. This chapter is not intended to serve as a comprehensive guide, but rather to highlight some of the important benefits and considerations associated with pilot testing.

Pilot testing should not be confused with challenge testing. Challenge testing, as described in Chapter 3, is conducted on a product-specific basis under the LT2ESWTR and characterizes a membrane filtration module in terms of removal efficiency. Pilot testing, by contrast, is conducted on a site- and system-specific basis and is used to collect information for full-scale facility design. Pilot testing need not include challenge testing unless specifically required by the State. In cases in which a particular membrane product that has not undergone challenge testing is proposed for compliance with LT2ESWTR requirements, it may be convenient to include a challenge test in the scope of pilot testing.

The primary goal of a membrane pilot study is to obtain information such as treated water quality (e.g., turbidity) and operating parameters (e.g., flux) that are necessary for the design of a membrane filtration facility. The treated water quality data provides assurance that the treatment objectives can be achieved, while the pre-determination of operating parameters allows for proper sizing of the membrane facility and minimizes the uncertainties regarding footprint and utility requirements. In general, the use of pilot data helps account for unforeseen conditions that may otherwise have gone undetected. Pilot testing also helps to familiarize operators with the membrane treatment equipment.

This chapter is divided into the following sections:

Section 6.2: Planning
This section discusses important considerations that should be taken into account before conducting a pilot test.

Section 6.3: Testing Objectives
This section provides an overview of testing objectives such as the optimization of membrane flux, backwashing, and chemical cleaning, as well as balancing the flux with backwashing and chemical cleaning frequencies to maximize operational efficiency.

Section 6.4:   Testing and Monitoring
This section discusses important data that are typically collected during pilot testing, including operational, water quality, microbial, and integrity testing parameters.

Section 6.5:   Report Development
This section outlines considerations for preparing a summary report after the pilot testing is completed.

Most of the information contained in this chapter is generally applicable to both microfiltration (MF)/ultrafiltration (UF) and nanofiltration (NF)/reverse osmosis (RO) systems. Distinctions between the two systems are noted where differences occur. Pilot testing for membrane cartridge filtration (MCF) systems is typically simpler relative to testing other types of membrane filtration and is often performed simply to verify the replacement frequency and productivity for the filters. Consequently, pilot testing MCF systems is not specifically addressed in this chapter. However, MCF piloting guidelines are generally similar to those for MF/UF with the exception of references to cleaning intervals, since membrane cartridges are not typically cleaned and instead disposed of when fouled.

6.2   Planning

The planning phase prior to implementing pilot testing is an important component of an overall pilot test program. Careful planning helps ensure that all the pilot test objectives are achieved both efficiently and economically without unexpected delays. The most important element of the planning process is the development of a comprehensive pilot test protocol specifying how the testing should be conducted. This protocol should include not only instructions for carrying out the testing, but also specific testing objectives and strategies for optimizing performance in terms of flux, backwashing, and chemical cleaning (see section 6.3). A plan for collecting water quality, integrity test, and microbial (where applicable) data should also be incorporated in the test protocol (see section 6.4). In addition, it is recommended that the pilot protocol be developed in conjunction with the State to ensure that any particular State requirements are satisfied.

In addition to the test protocol, a number of other planning considerations should be taken into account. For example, it should be determined whether the pilot test units can be provided by the membrane manufacturers at no cost, rented, or custom-fabricated, as budget constraints allow. Also, once the particular membrane filtration systems to be tested are selected, the utility requirements (i.e., water, electricity, and/or air) for the pilot units must be accommodated at the test site. Appropriately sized plumbing connections should be provided for the feed, filtrate, and concentrate streams (where applicable), and provisions must be made for the disposal of backwash (for MF/UF) and chemical cleaning residuals. Labor requirements for operating the pilot units should be estimated, and it is also important to ensure that the designated testing site has sufficient area to accommodate all the units to be tested. One consideration that is sometimes overlooked is shipping the pilot units; the proper equipment for loading and unloading the pilot units must be available at the testing site or be provided with the unit(s).

Special attention in the planning phase should be given to process considerations such as scalability, screening appropriate membrane filtration systems to test, and test scheduling. These three aspects of pilot test planning are discussed in sections 6.2.1, 6.2.2, and 6.2.3, respectively.

6.2.1 Process Considerations

It is important that the pilot process be representative of the full-scale system. For example, if pretreatment is part of the planned full-scale process, then the pilot process should incorporate similar pretreatment. Likewise, piloting should also accommodate production (e.g., filtration) and intermittent (e.g., backwashing) design considerations to the extent that these parameters have been defined prior to the pilot testing phase. Thus, if it has been determined that the membrane filtration system(s) that may be used at full-scale (e.g., if an MF/UF system has been pre-selected) requires air at a specified flow rate and duration during a backwash process, then the pilot system should be designed and operated in a similar manner. The design of the membrane system should also mimic the hydraulic configuration of the full-scale system, as described in section 2.4. However, it should be recognized that it might not be possible to design a pilot system using the same hydraulic configuration used at full scale for all membrane processes. For example, it may not be economically feasible to design a NF/RO pilot unit with hydraulic characteristics identical to the full-scale system. In this case, the methodology of conversion from pilot-scale data to full-scale design should be included in the development of the pilot test protocol.

Membrane modules should also be scaleable with respect to appropriate operational characteristics. Section 3.8 includes a discussion pertaining to the use of a small-scale membrane module, which may be appropriate under some circumstances. Although the discussion in section 3.8 is presented in the context of challenge testing, the concepts are also applicable to piloting, and small-scale module testing may be viable for pilot testing depending on the objectives of the study

6.2.2 Screening and System Selection

Spiral-wound NF and RO membrane modules are standardized such that the membranes from different manufacturers are interchangeable and system design is somewhat uniform. Thus, in terms of screening for appropriate membranes, the primary consideration involves the selection of a membrane material that provides desirable productivity, resistance to fouling, and removal characteristics.

Screening is somewhat more complex for MF/UF systems, which are largely proprietary in design. The various commercially available systems may use either pressure or vacuum as the driving force and can be designed to filter from the inside-out or outside-in direction relative to the fiber lumen. In addition, the various membrane materials have differences that may be important, including removal efficiency and pH and oxidant tolerance. Membranes of different materials also have varying degrees of compatibility with water treatment chemicals such as coagulants and powdered activated carbon (PAC) that may affect performance and cost. It is important to consider such differences in MF/UF membranes and membrane systems and how these may impact system selection. For example, if the source water were periodically pre-chlorinated, a membrane that is not compatible with chlorine would be undesirable.

A list of some questions to consider when screening membranes for a pilot study is provided below:

- What are the treatment objectives of the application?

- What operational constraints/goals are to be considered in membrane selection?

- Has the membrane been used on similar waters at other sites?

- What are the pH and oxidant tolerances of the membrane? Are these compatible with the application?

- Is the membrane compatible with any pretreatment chemicals aside from oxidants that may be in use, such as alum, ferric chloride, PAC, and polymers?

- Is this membrane compatible with solids and total organic carbon (TOC) levels in the feed water?

- Does the membrane have prior applicable State regulatory approval and any required certifications (e.g., NSF 61)?

- Does the supplier have experience with full-scale operation for a facility of this size and for treating similar water with the same configuration to be used in the pilot?

- Does the system require proprietary items such as spare parts or cleaning chemicals?

- Are there any unusual operational considerations associated with the membrane filtration system, such as significant power requirements, frequent membrane replacement, or substantial or undesirable chemical use?

These types of questions should be considered as a part of the screening process before selecting any membrane filtration systems to pilot. An additional consideration for MF/UF systems is that because these systems are proprietary, the membrane modules vary in design and are not presently interchangeable. Thus, a utility is limited to obtaining replacement membranes from only one supplier after a system has been installed.

In some cases, screening tools for MF/UF membranes may be limited to experience at other sites or chemical incompatibility. Therefore, pilot testing is an important consideration. For NF/RO, proprietary software programs that are available from the various membrane manufacturers can predict membrane system removal performance with a fairly good degree of accuracy for a particular membrane product. Small-scale module testing (see section 3.8) may be an option for screening some MF/UF membranes. Similarly, NF/RO membranes could be screened for some parameters using flat-sheet studies or single element tests.

6.2.3 Scheduling

Scheduling is another important aspect of pilot test planning. One factor to consider when scheduling a pilot test for surface waters is seasonal variations in water quality, since turbidity, temperature, algae content, taste and odor, and other parameters can potentially vary significantly throughout the year. A description of typical seasonal variation in surface water quality is summarized below. Note that not all of the seasonal variations described will be applicable for every site, or necessarily pertinent for all types of membrane filtration systems.

Summer – Because user demand is typically highest in the summer, the filtration system may also have to produce more water during this season. However, greater production is facilitated by warmer water temperature. The degree to which a membrane process is affected by a change in water temperature is related to the viscosity of the water. Since the viscosity of the water is lower at higher temperatures, the membrane flux will likely be at its peak in this season (water quality variations notwithstanding). (The relationship between water temperature and viscosity as it relates to flux is discussed in further detail in Chapter 7.) Warmer water also facilitates enhanced algae growth that may be problematic for membrane system operation. Taste and odor events may also manifest during the summer months.

Autumn – In areas with hardwood cover in the watershed, autumn months typically bring an increase in the organic content of surface water resulting from the decay of fallen leaves. Turbidity may sharply increase, as well. Furthermore, cooler air temperatures and wind cause surface waters in reservoirs to "turnover," bringing deeper, more anaerobic water to the surface and thus creating the potential for both taste and odor and iron and manganese problems.

Winter – Although winter months yield the coldest water temperatures and thus typically the lowest membrane fluxes, the demand for water is typically at its lowest, as well. Both the cold temperature and decrease in demand are likely to minimize required membrane flux in this season. (As previously noted, the relationship between water temperature and viscosity as it relates to flux is discussed in Chapter 7.) Most membrane manufacturers have membrane-specific temperature compensation factors for cold-water operations.

Spring – Spring months usually yield increases in water temperature, as well as in the potential for turbidity spikes related to run-off caused from snow melt. There is also potential for springtime turnover of a reservoir, resulting in subsequent degradation in feed water quality.

Ideally, the pilot study should be conducted during the time of year yielding the most difficult water quality to treat, so that design parameters resulting from the study, such as flux and chemical cleaning frequency, would be conservative for year round operation. Some water quality effects can be accurately modeled, such as that of temperature on flux and transmembrane pressure (TMP). Multi-season piloting is advantageous if scheduling and cost constraints will allow. In fact, some States require multi-season piloting for membrane treatment facilities.

The duration of the pilot study is also an important scheduling consideration. In general, piloting through at least three cleaning cycles is recommended practice. The target cleaning frequency for hollow-fiber MF/UF systems is typically at least 30 days of continuous operation. Through the first cleaning cycle the membrane flux and backwash frequency are usually targeted to provide for 30 days of operation before cleaning is required. The second cycle provides an opportunity for optimization and operational improvement. The third cycle establishes repeatability if the operating conditions remain the same. Note that because new membranes typically perform better than membranes that have been previously fouled and subsequently cleaned, it may be beneficial to add an extra cycle to study the effects of repeated fouling and cleaning on membrane performance. Because this strategy represents three operational cycles of at least 30 days, a pilot test duration of 90 days (or 3 months) is recommended, if possible. For spiral-wound NF/RO membrane systems, longer cleaning intervals are desirable, which results in fewer operational cycles than for a MF/UF pilot of similar duration.

For more thorough MF/UF pilot studies, approximately 4 to 7 months (3,000 to 5,000 hours) of cumulative operational time is usually recommended. A longer pilot study may be appropriate for newer, less proven membrane filtration systems or for applications in which the water quality is extremely variable. NF/RO pilot studies generally range from about 2 to 7 months (1,500 to 5,000 hours) of cumulative operational time, with longer studies used for waters of variable quality. The State may also have minimum requirements for the duration of pilot studies.

6.3   Testing Objectives

Membrane flux and system productivity are typically the most important design parameters to optimize, as these dictate the number of membranes (and hence a large portion of the capital cost) required for the full-scale system. Because these two parameters are inversely related to a certain extent, the pilot testing process may help to establish the optimum balance. For example, typically operating at higher fluxes increases the rate of fouling, in turn requiring more frequent backwashing and chemical cleaning. However, the system productivity is limited by the backwashing (where applicable) and chemical cleaning frequency. Backwashing and chemical cleaning not only use filtrate as process water, but also represent time during which filtrate cannot be produced (thus affecting overall system productivity). The effect of more frequent backwashes and chemical cleanings on the system productivity may be an important consideration if the source water is limited, or alternatively if waste disposal is problematic. In addition, chemical cleaning may be a labor-intensive operation that requires the handling of harsh chemicals and produces a waste stream that may be difficult to dispose. However, chemical cleaning is a necessary process associated with membrane filtration, and establishing operating practices that extend the time between cleanings is one objective of a membrane pilot study.

This section provides guidelines for balancing flux, productivity, backwash frequency (where applicable), and chemical cleaning intervals during pilot testing. It is important to note that in order for cause and effect to be analyzed properly, only one process variable should be changed at a time during the pilot study. Economic and time constraints often dictate the duration of the pilot study and may not allow complete optimization of each of these parameters. Therefore, it is important to understand which of these parameters is most important for a particular application of membrane filtration and to structure the pilot test protocol accordingly. Thus, the result of a pilot study is not necessarily the "optimum" design, but rather a set of operational conditions that will result in feasible and economic water treatment over the anticipated range of operating and source water conditions.

6.3.1 Membrane Flux Optimization

Membrane manufacturers can recommend fluxes for particular applications and a given water quality. Table 6.1 lists some of the important water quality data that should be provided to membrane suppliers to facilitate a fairly accurate initial estimation of anticipated membrane productivity.

If scheduling permits, it is typically advantageous to begin piloting MF/UF systems at a conservative flux and then increase it based upon the rate of fouling observed. The flux may be increased either after a chemical cleaning or during a filter run if the pilot unit has undergone sustained operation with only a nominal increase in fouling. In subsequent filter runs (i.e., between chemical cleanings) the pilot unit flux may be either increased if the fouling rate of the previous run was still within acceptable tolerances or decreased if the fouling occurred at an unacceptably high rate. Note that the backwash frequency may also be adjusted either during a filter run or between filter runs to minimize fouling at a particular flux (see section 6.3.2). This fine-tuning process of adjusting the flux and/or backwashing interval (if possible, only one parameter should be adjusted at a time) between filter runs may be repeated to the extent that budget and scheduling constraints allow, or to the point at which such adjustments reach the point of diminishing returns. NF/RO performance can typically be accurately gauged by the manufacturers based on water quality models, although it is often beneficial to adjust the flux over a series of filter runs to optimize productivity based on pilot test results. The pilot test protocol should include the expected range of membrane fluxes and guidelines on how these fluxes should be adjusted based upon the results of early testing. Models may also indicate chemical pretreatment requirements necessary to control scaling on NF/RO membranes and thus to help maintain acceptable fluxes.

Table 6.1 Water Quality Parameters to Measure Prior to Piloting

Parameter	MF/UF	NF/RO
Cations		
Aluminum		X
Ammonia		X
Barium		X
Calcium	X	X
Iron	X	X
Magnesium	X	X
Manganese	X	X
Potassium		X
Sodium		X
Strontium		X
Anions		
Chloride		X
Fluoride		X
Nitrate		X
Silica	X	X
Sulfate		X
Other Chemical / Physical Parameters		
Algae	X	
Alkalinity		X
pH	X	X
SDI		X
TDS		X
TOC	X	X
TSS	X	
Turbidity	X	X
UV-254	X	X

6.3.2 Backwash Optimization

Backwashing is a periodic reverse flow process used with many hollow-fiber MF/UF systems to remove accumulated contaminants from the membrane surface, thus maintaining sufficient flux and minimizing the rate of long-term, irreversible fouling. A MF/UF membrane manufacturer can provide information on the backwash protocol required for its system, as well as a recommended backwash frequency. It is important to understand the following issues regarding the backwash process, which are applicable to both piloting and full-scale operation:

- Large water and/or air flows over short periods of time may be included as a part of the backwash process. The flows of air and water vary depending on the system manufacturer.

- Backwash effluent water contains approximately 10 to 20 times the concentration of the feed contaminants.

- Many manufacturers add free chlorine to the backwash process water to reduce membrane fouling via disinfection (a process often referred to as a "chemically enhanced backwash"). Other manufacturers may include acid or caustic solution to remove inorganic or organic foulants, respectively. As a result, the backwash effluent may contain solids and/or chemical residuals.

- In locations where the disposal of backwash water may be problematic, consideration should be given to the volume of backwash water generated. Backwash water treatment may also be considered in order to maximize system productivity and minimize residuals.

- Increased backwash efficiency can significantly enhance system performance, allowing higher fluxes and recoveries.

Variations in the backwash frequency will influence the waste flow, as well as the productivity of the system. If higher residual flows can be easily accommodated and productivity is not a critical design consideration, increasing the backwashing frequency may be a viable strategy for achieving longer filter run times or higher membrane fluxes.

Based upon the water quality data provided, MF/UF manufacturers can estimate the initial membrane flux and backwash frequency for the pilot test. The backwash frequency can be optimized during pilot testing to provide the most economical process. The potential for increased flux and useful membrane life that may be commensurate with increased backwash frequency should be balanced against decreased productivity, increased waste volume, and higher chemical usage (if applicable).

6.3.3 Chemical Cleaning Optimization

The chemical cleaning of a membrane filtration system is a necessary process that results in lost production time, produces chemical waste, and requires operator attention. Consequently, pilot testing should be used to minimize chemical cleaning frequency for both MF/UF and NF/RO systems, while maintaining acceptable productivity and controlling long-term fouling. The typical membrane cleaning cycle consists of recirculating a heated cleaning solution for a period of several hours. Some manufacturers require that the cleaning solution be prepared with softened or demineralized water, which should be accounted for during the pilot test. The specific cleaning regime will depend upon the feed water quality and the particular membrane system. Some membranes may be degraded by excessive cleaning frequencies, shortening membrane life. For MF/UF systems the typical chemical cleaning frequency assumed for design purposes is approximately once every 30 to 60 days. NF/RO systems are generally designed with significantly longer cleaning intervals, typically in the range of 3 months to 1 year. It is important for a utility to understand the impacts of chemical cleaning on the system and to determine whether to increase the flux at the expense of the consequent increase in chemical cleaning frequency. One of the objectives of a pilot test should be to determine a chemical cleaning strategy that restores membrane permeability without damaging the membrane integrity. Thus, the pilot test should be designed to include at least one chemical cleaning, even if the pilot schedule is very restrictive. If necessary, the flux can be significantly increased to induce membrane fouling for the purposes of conducting chemical cleaning.

If sufficient time is available, it may be advantageous to operate the pilot test through at least two chemical cleaning intervals, especially if the membranes used in the pilot unit are new. New membranes may exhibit slightly different performance, and thus the data collected prior to the first chemical cleaning may not always be representative of typical operation.

It is important to demonstrate the effectiveness of chemical cleaning during the pilot study. Because membrane cleaning is a time-, labor-, and chemical-intensive process, it is important to avoid experimentation on the full-scale plant. The best way to evaluate the success of a particular chemical cleaning scheme is to conduct the cleaning in a methodical manner and evaluate the performance of each step in the process. Before and after each step, a plot of the TMP as a function of flux (or filtrate flow) should be developed. This method of evaluation is normally described as a "clean water flux test." This test requires that the membrane unit be placed into filtration mode in between each of the various steps in the chemical cleaning process in order to observe the effect on the flux (or filtrate flow). (Note that filtrate produced during this process should be considered waste.) For example, if a chemical cleaning process consists of recirculating citric acid and caustic in succession, three plots of TMP versus flux (or filtrate flow) should be generated. Figure 6.1 illustrates the clean water flux test technique for this example. The three applicable plots shown in Figure 6.1 are as follows:

1. TMP vs. flux (or filtrate flow) for the fouled membrane prior to cleaning

2. TMP vs. flux (or filtrate flow) after citric acid recirculation / cleaning

3. TMP vs. flux (or filtrate flow) after caustic recirculation / cleaning
   (i.e., after all cleaning steps are completed)

Figure 6.1 Sample Chemical Cleaning Test Profile

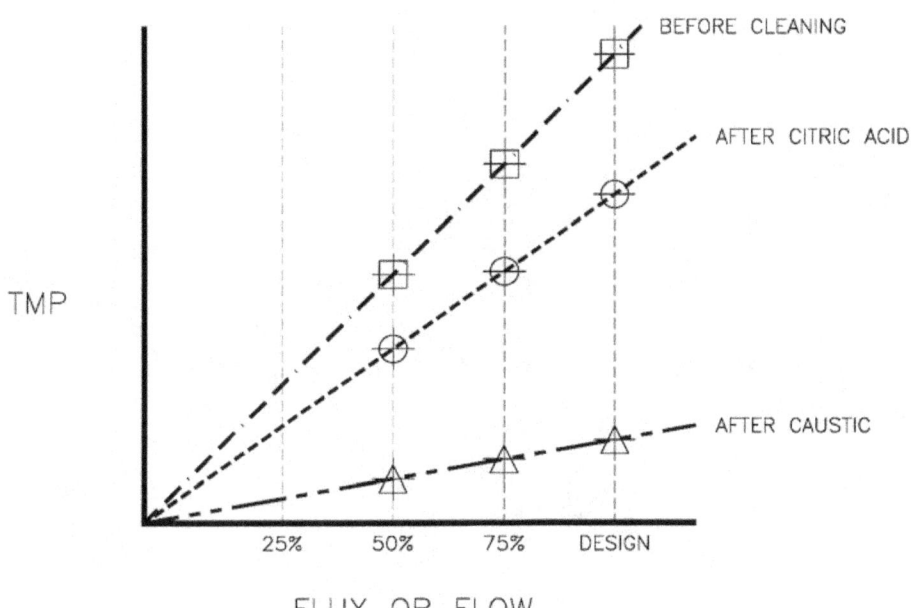

In an ideal cleaning, the final plot (i.e., after caustic) of TMP versus flux (or filtrate flow) would be similar to that after the previous cleaning event such that the plots generated after each successive chemical cleaning (i.e., the final plot of each cleaning) would overlap. Thus, the plots established during a clean water flux test facilitate this comparison and provide an indication as to the effectiveness of the cleaning regimen. This is illustrated in Figure 6.1 with plots showing the increase in TMP per incremental change in flux (or filtrate flow). Smaller slopes indicate lower pressure requirements for operation at a given flux (or filtrate flow), and thus a more effective step in the chemical cleaning process.

After cleaning, a direct integrity test should be conducted before the pilot unit is returned to service. If the membrane is integral and an acceptable flux has been restored, then the cleaning is deemed successful. If not, alternate cleaning strategies should be considered. A flow chart outlining a general pilot study sequence of events is shown in Figure 6.2, illustrating a typical series of test runs and subsequent chemical cleaning events.

Figure 6.2 Sample Pilot Study Sequence Overview

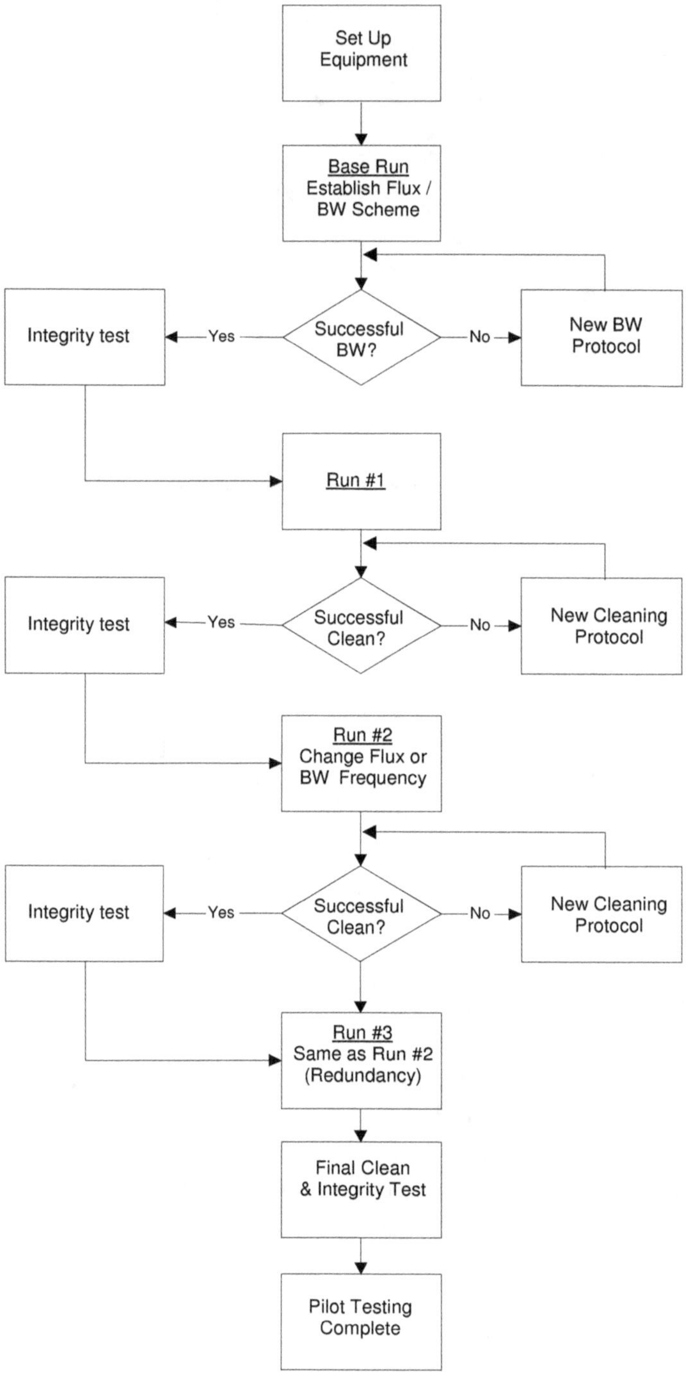

6.4   Testing and Monitoring

A thorough and carefully developed testing and monitoring plan is a critical component of a pilot test program, as this is the means by which system performance is assessed. There are several categories of testing and monitoring that should be addressed in a pilot test protocol, including operational parameters, water quality, microbial monitoring (where applicable), and integrity testing. Each of these four categories is discussed in further detail in the following subsections.

6.4.1 Operational Parameter Monitoring

Monitoring pilot unit operational parameters is an important means of assessing system performance and tracking the rate of membrane fouling. The following operational parameters should be monitored continuously, if possible:

1. Elapsed run time

2. Pressure (feed, filtrate, concentrate)

3. Flow (feed, filtrate, concentrate)

4. Temperature (feed or filtrate)

Another important consideration in the evaluation of a membrane process is operational data collection for the intermittent process sequences, such as backwashing. For these intermittent processes, the design parameters (e.g., flows, times, and volumes) associated with air, water, or chemical usage during the sequence should be established and verified during pilot testing. It is not uncommon for the pilot unit to operate under different parameters than those that would be considered appropriate for a full-scale unit. For example, a pilot system generally has faster pneumatic valve actuation times and a shorter overall backwash sequence than a full-scale unit. Some of these discrepancies are unavoidable; however, these scale-up issues should be noted for consideration prior to initiating pilot testing.

6.4.2 Water Quality Monitoring

The particular water quality data collected during a pilot study will depend upon the type of membrane filtration system, site-specific treatment objectives, and foulants of concern. In general, the following table can be used as a guideline:

Table 6.2 Suggested Water Quality Sampling Schedule for Membrane Piloting [1]

Parameter	MF/UF			NF/RO		
	Feed	Concentrate[2]	Filtrate	Feed	Concentrate	Filtrate
General Water Quality Parameters						
Algae	W		W	W		
Color	W	M	W	W	M	W
HPC	M		M	M		M
Particle counts	C		C			
pH	D	W	W	C	W	W
SDI				W		
Taste & odor	M	M	M	M	M	M
Temperature	C			C		
TOC	W	M	W	W	M	W
Total coliform	W		W	W		
TSS	M	M	M			
Turbidity	C	M	C	C		W
UV-254	W	M	W	W		W
Dissolved Solids						
Alkalinity	W	M	W	W	W	W
Barium				W	W	W
Hardness			M	W	W	W
Iron	W			W	W	W
Manganese	W			W	W	W
Silica				W	W	W
Sulfate	W			W	W	W
TDS			M	C	W	C
Simulated Distribution System Disinfection Byproducts						
HAAs / HAA5s			M			M
TTHMs			M			M

1 Generic recommendations only; the specific applicable parameters will vary with each application
2 Where applicable, based on particular hydraulic configuration used

Key:
C = Continuous
D = Daily
W = Weekly
M = Monthly

In addition to those flows listed in Table 6.2, it is recommended that the MF/UF backwash flow be checked weekly for total suspended solids (TSS) and turbidity. Note that the recommended parameters and sampling frequencies should be modified to meet the requirements and objectives of each particular site-specific pilot study. The sampling frequencies may also be modified during the course of a pilot study if conditions warrant.

6.4.3 Microbial Monitoring

One advantage of membrane filtration is that it provides a physical barrier through which water must pass, virtually eliminating pathogens larger than the membrane exclusion characteristic. As with the full-scale facility, it is important that the pilot test demonstrate that this barrier is intact and is rejecting pathogens. A site-specific microbial challenge test may not be required by the State during the pilot study for a membrane filtration system that has received prior regulatory approval. For the purposes of LT2ESWTR compliance, product-specific demonstration of *Cryptosporidium* removal efficiency is accomplished during challenge testing, although microbial monitoring may be conducted during a pilot study for additional verification of membrane performance at the discretion of the utility or if required by the State. Although additional microbial testing can be conducted for any pathogen of concern during piloting, typically coliform bacteria, and sometimes heterotrophic plate count (HPC) are used as an indicator for microbial removal efficiency.

6.4.4 Integrity Testing

The importance of integrity testing under the LT2ESWTR necessitates that greater emphasis be placed upon integrity testing during piloting for applications of membrane filtration intended for LT2ESWTR compliance. For direct integrity testing, the membrane manufacturer should use a method approved or mandated by the State. The standard direct integrity test sequences intended for the full-scale system should be incorporated into the pilot unit. Direct integrity testing should be conducted at least as frequently as required by the State for the full-scale facility. In the absence of a particular State requirement, it is generally recommended that direct integrity testing be conducted on a daily basis, a frequency consistent with the LT2ESWTR requirements. If an integrity breach occurs during the piloting testing, then diagnostic testing and membrane repair techniques may also be practiced. Continuous indirect integrity monitoring can usually be accomplished using data that are collected during a typical pilot test program (e.g., turbidity or particle count data). However, any State requirements regarding data that must be collected for the purposes of continuous indirect integrity monitoring for the full-scale facility should also be implemented during the pilot test, and may be required by the State in any case.

Note that because the number of membrane modules to which an integrity test is applied affects the sensitivity of the test (for both direct integrity testing and continuous indirect integrity monitoring), any control limits established for a full-scale system will generally not be applicable to a pilot unit, which typically utilizes only a small fraction of the number of modules in a full-scale membrane unit. Guidance on establishing full-scale system control limits for both direct integrity testing and continuous indirect integrity monitoring for the purposes of compliance with the LT2ESWTR are provided in Chapters 4 and 5, respectively, and also in Appendix A – Development of a Comprehensive Integrity Verification Program.

6.5 Report Development

After the pilot test is complete, a report should be prepared to summarize the procedures and the test results. The pilot study report should contain sufficient detail to establish design parameters for the full-scale plant to the extent for which the testing was intended. Flux, chemical cleaning, backwashing, and integrity testing should be addressed, where applicable. Collected water quality data should also be included with emphasis given to unanticipated test results. Details on the operational parameters for each filter run (i.e., between chemical cleanings) should be summarized. The State may also have specific requirements for the report if piloting is required.

7.0 Implementation Considerations

7.1 Introduction

The successful implementation of a membrane filtration system requires a general understanding of the major considerations that influence system design and operation. Although there are many such considerations, this chapter focuses on the most significant of these issues, categorized as operational unit processes, system design considerations, and residuals treatment and disposal. The purpose of this chapter is to present a general overview of some of these critical considerations, highlighting their respective roles in the implementation of the technology. Note that this chapter is intended to provide information and recommendations regarding industry practices and is not directly related to compliance with the Long Term 2 Enhanced Surface Water Treatment Rule (LT2ESWTR).

This chapter is divided into the following sections:

Section 7.2:   Operational Unit Processes
This section reviews the purpose(s) and practices associated with the major operational unit processes of membrane filtration systems, including pretreatment, backwashing, chemical cleaning, integrity testing, and post-treatment.

Section 7.3:   System Design Considerations
This section discusses some of the most significant conceptual design issues associated with membrane filtration systems, including flux, water quality, temperature compensation, cross-connection control, and system reliability.

Section 7.4:   Residuals Treatment and Disposal
This section outlines management practices for the various waste streams produced by membrane filtration systems, such as backwash and chemical cleaning residuals, as well as concentrate.

7.2 Operational Unit Processes

All membrane filtration systems have associated operational unit processes that are essential for maintaining and optimizing system performance and therefore critical to the successful implementation of the technology. These operational processes include backwashing, chemical cleaning, and integrity testing. For the purposes of this discussion, pretreatment and post-treatment are also considered operational unit processes associated with membrane filtration. Each of these processes and its role in the operation of a membrane filtration system is described in the following sections. Although not every membrane filtration system utilizes all of these processes, many utilize each process to some degree. The optimization of these operational unit processes can be addressed during pilot testing, as discussed in Chapter 6.

7.2.1 Pretreatment

Pretreatment is typically applied to the feed water prior to entering the membrane system in order to minimize membrane fouling, but in some cases may be used to address other water quality concerns or treatment objectives. Pretreatment is most often utilized to remove foulants, optimize recovery and system productivity, and extend membrane life. Pretreatment may also be used to prevent physical damage to the membranes. Different types of pretreatment can be used in conjunction with any given membrane filtration system, as determined by site-specific conditions and treatment objectives. Pilot testing can be used to compare various pretreatment options, optimize pretreatment, and/or demonstrate pretreatment performance. Several different types of commonly used pretreatment for membrane filtration systems are discussed in the following subsections.

7.2.1.1   Prefiltration

Prefiltration, including screening or coarse filtration, is a common means of pretreatment for membrane filtration systems that is designed to remove large particles and debris. Prefiltration can either be applied to the membrane filtration system as a whole or to each membrane unit separately. The particular pore size associated with the prefiltration process (where applicable) varies depending on the type of membrane filtration system and the feed water quality. For example, although hollow-fiber microfiltration (MF) and ultrafiltration (UF) systems are designed specifically to remove suspended solids, large particulate matter can damage or plug the membranes fibers. For these types of systems the pore size / micron rating of the selected prefiltration process may range from as small 100 μm to as large as 3,000 μm or higher, depending on the influent water quality and manufacturer specifications. Generally, hollow-fiber MF/UF systems that are operated in an inside-out mode are more susceptible to fiber plugging and thus may require finer prefiltration.

Because nanofiltration (NF) and reverse osmosis (RO) utilize non-porous semi-permeable membranes that cannot be backwashed and are almost exclusively designed in a spiral-wound configuration for municipal water treatment applications, these systems must utilize much finer prefiltration in order to minimize exposure of the membranes to particulate matter of any size. Spiral-wound modules are highly susceptible to particulate fouling, which can reduce system productivity, create operational problems, reduce membrane life, or in some cases damage or destroy the membranes. If the feed water has a turbidity less than approximately 1 NTU or a silt density index (SDI) less than approximately 5, cartridge filters with ratings ranging from about 5 to 20 μm are commonly used NF/RO prefiltration. However, if the feed water turbidity or SDI exceeds these values, a more rigorous method of particulate removal, such as conventional treatment (including media filtration) or MF/UF membranes, is recommended as pretreatment for NF/RO.

In conventional applications of cartridge filtration technology, high quality source waters are treated, and thus additional prefiltration is typically not required. However, membrane cartridge filtration (MCF) systems may warrant some level of prefiltration, possibly using more conventional cartridge or bag filters, to protect and extend the life of the more effective and more expensive MCF, particularly if the MCF system is applied to surface waters specifically for the purposes of LT2ESWTR compliance. Since MCF is a new class of technology defined under the LT2ESWTR, pretreatment practices for MCF will likely evolve as this technology is applied and may be similar to those for MF/UF.

A summary of the typical prefiltration requirements associated with the various types of membrane filtration is presented in Table 7.1.

Table 7.1 Typical Membrane System Prefiltration Requirements

Membrane System		Prefiltration Requirements	
Classification	Configuration	Size (μm)	Type(s)
Membrane Cartridge Filtration (MCF)[1]	Cartridge	300 - 3,000	Strainers; Bag Filters
Microfiltration (MF) / Ultrafiltration (UF)	Hollow-Fiber, Inside-Out	100 - 300	Strainers; Bag Filters
	Hollow-Fiber, Outside-In	300 - 3,000	Strainers; Bag Filters
Nanofiltration (NF) / Reverse Osmosis (RO)	Spiral-Wound	5 - 20	Cartridge Filters

1 Prefiltration is not necessarily required for MCF systems

In some cases, one type of membrane filtration may be used as prefiltration for another. This type of treatment scheme is commonly known as an integrated membrane system (IMS). Typically, this involves the use of MF/UF as pretreatment for NF/RO in applications that require the removal of particulate matter and microorganisms as well as some dissolved contaminants such as hardness, iron and manganese, or disinfection byproduct (DBP) precursors. One of the most significant advantages of an IMS treatment scheme is that the MF/UF filtrate is of consistently high quality with respect to particulate matter and often allows the NF/RO system to maintain stable operation by reducing the rate of membrane fouling.

7.2.1.2   Chemical Conditioning

Chemical conditioning may be used for a number of pretreatment purposes, including pH adjustment, disinfection, biofouling control, scale inhibition, or coagulation. Some type of chemical conditioning is almost always used with NF/RO systems, most often the addition of an acid (to reduce the pH) or a proprietary scale inhibitor recommended by the membrane manufacturer to prevent the precipitation of sparingly soluble salts such as calcium carbonate ($CaCO_3$), barium sulfate ($BaSO_4$), strontium sulfate ($SrSO_4$), or silica species (e.g., SiO_2).

Software programs that simulate NF/RO scaling potential based on feed water quality are available from the various membrane manufactures. In some cases, such as for those NF and RO membranes manufactured from cellulose acetate, the feed water pH must also be adjusted to maintain the pH within an acceptable operating range to minimize the hydrolysis (i.e., chemical deterioration) of the membrane. The addition of chlorine or other disinfectants may also be used as pretreatment for primary disinfection or to control biofouling. However, because some NF/RO membrane materials are readily damaged by oxidants, it is important that any disinfectants added upstream be quenched prior to contact with such membranes.

A number of different chemicals may be added as pretreatment for MF or UF, depending on the treatment objectives for the system. For example, lime may be added for softening applications; coagulants may be added to enhance removal of total organic carbon (TOC) with the intent of minimizing the formation of DBPs or increasing particulate removal; disinfectants may be applied for either primary disinfection or biofouling control; and various oxidants can be used to precipitate metals such iron and manganese for subsequent filtration. However, as with NF and RO membranes, it is important to ensure that the applied pretreatment chemicals are compatible with the particular membrane material used. As with conventional media filters, pre-settling may be used in conjunction with pretreatment processes such as coagulation and lime softening. While a MF/UF system may be able to operate efficiently with the in-line addition of lime or coagulants, pre-settling in association with these pretreatment processes can enhance membrane flux and increase system productivity by reducing the solids loading, thus minimizing backwashing and chemical cleaning frequency.

Although chemical pretreatment is generally not associated with cartridge filters, if the filters are compatible, disinfectants may be added upstream of MCF systems to maximize the time for primary disinfection (i.e., CT). Because cartridge filters are designed to be disposable and are generally not backwashed, pretreatment chemicals such as lime or coagulants could rapidly foul the cartridges and thus are not applied upstream these systems.

With any form of chemical pretreatment, it is very important to understand whether any chemical under consideration for use is compatible with the membrane material. In addition to irreversible fouling and/or physical damage to the membranes, the use of an incompatible chemical may void the manufacturer's warranty. Some chemicals such as oxidants can be quenched upstream, while others such as coagulants and lime cannot be counteracted prior to membrane exposure. In general, most NF/RO membranes and some MF/UF membranes are not compatible with disinfectants and other oxidants. However, some MF/UF membranes that are incompatible with stronger oxidants such as chlorine may have a greater tolerance for weaker disinfectants such as chloramines, which may allow for a measure of biofouling control without damaging the membranes. Certain types of both MF/UF and NF/RO membranes require operation within a certain pH range. Coagulants and lime are incompatible with many NF/RO membranes but are typically compatible with most types of MF/UF membranes. Polymers are incompatible with NF/RO membranes, and generally not compatible with MF/UF membranes, as well, although this depends to some degree on the charge of the polymer relative to the charge associated with the membrane. A polymer with a charge opposite to that of the membrane is likely to cause rapid and potentially irreversible fouling. Chemical compatibility with various types of membrane materials is briefly discussed in section 2.3.1; however, it is critical to consult with the membrane manufacturer prior to implementing any form of chemical pretreatment.

7.2.2 Backwashing

The backwash process for membrane filtration systems is similar in principle to that for conventional media filters and is designed to remove contaminants accumulated on the membrane surface. Each membrane unit is backwashed separately and in a staggered pattern so as to minimize the number of units in simultaneous backwash at any given time. During a backwash cycle, the direction of flow is reversed for a period ranging from about 30 seconds to 3 minutes. The force and direction of the flow dislodge the contaminants at the membrane surface and wash accumulated solids out through the discharge line. Membrane filtration systems are generally backwashed more frequently than conventional media filters, with intervals of approximately 15 to 60 minutes between backwash events. Typically, the membrane backwash process reduces system productivity in the range of 5 to 10 percent due to the volume of filtrate used during the backwash operation.

Of the various types of membrane filtration systems, backwashing is almost exclusively associated with hollow-fiber MF and UF processes. Some MCF systems may also be backwashed, depending on how the system is designed and configured; however it is more common to replace the cartridge filters, which are manufactured to be disposable, when they become fouled.

Backwashing is conducted periodically according to manufacturer specifications and site-specific considerations. Although more frequent backwashing allows for higher fluxes, this benefit is counterbalanced by the decrease in system productivity. In general, a backwash cycle is triggered when a performance-based benchmark is exceeded, such as a threshold for operating time, volumetric throughput, increase in transmembrane pressure (TMP), and/or flux decline. Ideally, the backwash process restores the TMP to its baseline (i.e., clean) level; however, most membranes exhibit a gradual increase in the TMP that is observed after each backwash, indicating the accumulation of foulants that cannot be removed by the backwash process alone. These foulants are addressed through chemical cleaning (see section 7.2.3).

Some systems also utilize pressurized air and/or chlorine (if a compatible membrane material is used) in combination with filtered water to remove solids, provide a measure of pathogen inactivation and biofouling control, and improve overall backwash effectiveness. A disinfectant such as chlorine may be added at a frequency ranging from every backwash to about once per day. If chloramines are added upstream of the membrane process, the benefits of adding chlorine during the backwash process should weighed against the cost of breakpoint chlorination to achieve a free chlorine residual and possible subsequent ammonia addition to regenerate a chloramine residual for secondary disinfection in the distribution system.

Some manufacturers have implemented backwash strategies for their proprietary MF/UF systems that involve the use of chemicals other than chlorine (such as acids, bases, surfactants, or other proprietary chemicals). These strategies are used as a method to enhance membrane flux and extend intervals between chemical cleanings, thus lowering the cost of operation. In these cases State regulators may require enhanced cross-connection control measures for the backwash piping similar to those used for chemical cleaning systems (see section 7.3.4), depending on the backwash frequency and the type(s) and concentration(s) of chemical(s) used. Special provisions for rinsing the membranes at the end of the backwash process may also be required.

Because the design of spiral-wound membranes generally does not permit reverse flow, NF and RO membrane systems are not backwashed. For these systems, membrane fouling is controlled primarily with chemical cleaning, as well as through flux control and crossflow velocity. The inability of spiral-wound membranes to be backwashed is one reason that NF and RO membranes are seldom applied to directly treat water with high turbidity and/or suspended solids.

7.2.3 Chemical Cleaning

Chemical cleaning is another means of controlling membrane fouling, particularly those foulants such as inorganic scaling and some forms of organic and biofouling that are not removed via the backwash process. As with backwashing, chemical cleaning is conducted for each membrane unit separately and is typically staggered to minimize the number of units undergoing cleaning at any time. While chemical cleaning is conducted on both MF/UF and NF/RO systems, because non-porous, semi-permeable membranes cannot be backwashed, chemical cleaning represents the primary means of removing foulants in NF/RO systems. Although cleaning intervals may vary widely on a system-by-system basis, the gradual accumulation of foulants makes eventual chemical cleaning virtually inevitable. Membrane cartridge filters are an exception, however, in that cartridge filters are usually designed to be disposable and thus are typically not subject to chemical cleaning.

As with backwashing, the goal of chemical cleaning is to restore the TMP of the system to its baseline (i.e., clean) level. Any foulant that is removed by either the backwash or chemical cleaning process is known as reversible fouling. Over time, membrane processes will also typically experience some degree of irreversible fouling which cannot be removed through either chemical cleaning or backwashing. Irreversible fouling occurs in all virtually membrane systems, albeit over a wide range of rates, and eventually necessitates membrane replacement.

There are a variety of different chemicals that may be used for membrane cleaning, and each is generally targeted to remove a specific form of fouling. For example, citric acid is commonly used to dissolve inorganic scaling, and other acids may be used for this purpose as well. Strong bases such as caustic are typically employed to dissolve organic material. Detergents and surfactants may also be used to remove organic and particulate foulants, particularly those that are difficult to dissolve. Chemical cleaning may also utilize concentrated disinfectants such as a strong chlorine solution to control biofouling. Due to the variety of foulants that are present in many source waters, it is often necessary to use a combination of different chemicals in series to address multiple types of fouling. The various types of chemical cleaning agents used are summarized in Table 7.2.

Table 7.2 Chemical Cleaning Agents

Category	Chemicals Commonly Used	Typical Target Contaminant(s)
Acid	• Citric Acid ($C_6H_8O_7$) • Hydrochloric Acid (HCl)	Inorganic scale
Base	• Caustic (NaOH)	Organics
Oxidants / Disinfectants	• Sodium Hypochlorite (NaOCl) • Chlorine (Cl_2) Gas • Hydrogen Peroxide (H_2O_2)	Organics; Biofilms
Surfactants	• Various	Organics; Inert particles

Numerous proprietary cleaning chemicals are also available, and these specialty cleaning agents may be useful in cases in which more conventional chemicals are ineffective. For example, under some circumstances enzymatic cleaners have been found to be effective at dissolving organic contaminants. Chemical cleaning options are more limited for membranes that cannot tolerate oxidants and/or extreme pH levels. A chemical cleaning regimen may be specified by the manufacturer or identified based on site-specific pilot testing and source water quality analyses to determine the prevalent form(s) of fouling experienced at a particular facility.

The term clean-in-place (CIP) is often used to describe the chemical cleaning process, since it is typically conducted while the membrane modules remain in the membrane unit (i.e., in-situ). The cleaning process generally involves recirculating a cleaning solution through the membrane system at high velocities (to generate scouring action) and elevated temperature (to enhance the solubility of the foulants). A soak cycle follows the recirculation phase. After the soak cycle is completed, the membrane system is flushed to remove residual traces of the cleaning solution(s). The process may be repeated using a different cleaning solution to target different types of foulants until the membranes have been successfully cleaned. Under some circumstances the use of softened or demineralized water may be required for the cleaning solution or as rinse water.

While backwashing may be conducted at more regular intervals or on a routine basis, chemical cleaning is typically conducted only when necessary. A chemical cleaning is generally necessary for MF and UF systems when the ability of periodic backwashing to restore system productivity (i.e., increase flux and reduce TMP) reaches a point of diminishing returns. For NF and RO systems, a 10 to 15 percent decline in temperature- and pressure-normalized flux or about a 50 percent increase in differential pressure may indicate the need for chemical cleaning. Delaying necessary chemical cleaning can accelerate irreversible fouling, reduce production capacity, and shorten membrane life. A benchmark of 30 days is commonly used as a minimum required interval between chemical cleanings for MF/UF systems, although it is not uncommon for a well-designed system to operate for much longer between cleanings, particularly when applied to feed waters with lower fouling potential. NF/RO systems are normally designed to operate for much longer periods between chemical cleanings, with intervals typically ranging from 3 months to 1 year.

In addition to a thorough CIP as necessitated when a pre-established point of diminishing productivity is reached, some MF/UF membrane system manufacturers recommend a routine, short duration chemical cleaning to minimize the accumulation of foulants. These processes are sometimes referred to "chemical washes" or "maintenance cleans," and the frequency at which they are implemented is typically not based on a benchmark for performance decline (as with a more thorough CIP), but rather on preset intervals ranging from several times per day to once every several days, depending on the propensity of the water to cause membrane fouling.

Isolating the cleaning chemicals from the treated (i.e., filtered) water is an important consideration for membrane filtration systems. Guidelines for cross-connection control are discussed in section 7.3.4. In addition, it is important to properly flush the membrane unit after the cleaning process and before restarting the filtration cycle. The flushed water should be diverted to waste until filtrate water quality parameters (such as turbidity for MF/UF systems and pH for NF/RO systems) return to normal production mode levels. Note that the volume of flushed water can be significant in cases in which surfactants are used.

For MF/UF systems, it is common to recycle as much as 90 percent of the cleaning chemicals for reuse, thus reducing the volume of chemical waste as well as the cost associated with cleaning. Recycling cleaning solutions is less common with NF/RO systems, since the used cleaning solutions accumulate dissolved constituents with repeated use, diminishing the effectiveness of the cleaning agents. The treatment and disposal of spent chemical cleaning waste is discussed in section 7.4.2.

7.2.4 Integrity Testing

Integrity testing is a means of determining whether or not a membrane system is "integral," or free of any breaches, leaks, or defects that might allow unfiltered water to bypass the membrane barrier, passing contaminants that are normally removed. As with both backwashing and chemical cleaning, integrity testing is conducted on each membrane unit (or smaller system subdivision in some cases) separately and is typically staggered to minimize the number of units simultaneously undergoing testing. The use of periodic or continuous integrity testing and monitoring methods allows ongoing operational verification that the membranes are performing as expected based on their established exclusion characteristics. This verification is an essential component of any membrane filtration system, particularly when the constituents of concern are pathogenic microorganisms. Integrity testing and monitoring are described in detail in Chapters 4 and 5 in the context of applying membrane filtration for compliance with the LT2ESWTR. The discussion in this section is intended to be a general overview of integrity testing as an important operational unit process for membrane filtration systems.

There are a number of potential modes of failure associated with membrane filtration systems that would result in an integrity breach. For example, membranes may become damaged via exposure to oxidants, pH levels outside the recommended range, or other chemicals or operating conditions to which the membranes are sensitive. In addition, membranes may break or puncture as a result of extreme pressure, scratches or abrasions, or operational fatigue over time. Spiral-wound membranes can be damaged at glue lines if the pressure on the filtrate side of the membrane exceeds that on the feed side. Factory imperfections such as glue line gaps or potting defects may cause integrity breaches, as well. Improper installation of membrane modules can also create integrity problems at o-rings or interconnections.

The various types of integrity tests are generally divided into two categories: direct and indirect methods. Direct methods are physical tests that are applied specifically to the membrane unit to detect integrity breaches and/or determine their sources. Indirect methods are surrogate measures of integrity that involve filtrate water quality monitoring such that a significant decline in filtrate quality may indicate an integrity problem. Both direct integrity testing and continuous indirect integrity monitoring are required for compliance with the LT2ESWTR.

Direct test methods are non-destructive techniques that can be used to identify and/or isolate leaks. While these methods yield direct information about membrane integrity, they cannot be conducted continuously while the membrane filtration system is in operation. Thus, the longer and more often that a direct test is conducted, the greater the impact on the overall system productivity. The minimum frequency requirements for direct testing vary among different States, although daily testing is commonly recommended as good practice and is required for those utilities using membrane filtration for LT2ESWTR compliance, unless the State approves less frequent testing.

Although there are a number of types of direct integrity tests, the most common method is the pressure decay test, which measures the rate of pressure loss across the membrane relative to a maximum acceptable threshold. Almost all currently available proprietary MF and UF systems are designed with the capability to conduct this test automatically at regular intervals. MCF systems may also be equipped to conduct automatic pressure decay tests. A similar test using a vacuum can also be conducted on NF and RO membranes, although currently spiral-wound systems are generally not designed to conduct this testing automatically during the course of normal operation. Other on-line direct test methods measure the flow of air or water from integrity breaches or the concentration of either spiked inert particles or a molecular marker. Additional information regarding direct integrity testing, including common methods and their application for facilitating compliance with the LT2ESWTR, is provided in Chapter 4.

The various indirect methods consist primarily of water quality monitoring practices that are common throughout the water treatment industry, such as turbidity monitoring and particle counting. Particulate-based indirect monitoring techniques are applicable to all membrane classes used for filtration. For NF/RO membranes capable of removing dissolved contaminants, other parameters such as conductivity or sulfide may also potentially be used as surrogate measures of membrane integrity. The effectiveness of indirect methods is a function of the ability of membrane filtration to produce very consistent, high quality filtrate, such that a marked decline in filtrate quality is likely to indicate an integrity problem. Although indirect methods are not as sensitive as the direct methods for detecting integrity breaches, the primary advantage of indirect methods is that they allow continuous monitoring, a capability not possible with currently available direct testing methods. Typically, if an indirect method indicates a potential integrity problem, a direct test will be conducted to determine more conclusively whether or not a breach has occurred. Indirect monitoring methods are further addressed in Chapter 5, including advantages and disadvantages of the various techniques, as well as requirements for utilizing these tests in the course of applying membrane filtration for LT2ESWTR compliance.

Guidance for the development of comprehensive integrity verification program, including the use of both direct and indirect methods, is included in Appendix A.

7.2.5 Post-Treatment

Post-treatment for membrane filtration systems typically consists of chemical conditioning and/or disinfection and is typically applied to combined filtrate. Most chemical conditioning is associated with NF and RO systems, because the removal of dissolved constituents that is achieved by these processes has a more significant impact on water chemistry than the filtering of suspended solids alone. For example, because NF and RO pretreatment often includes acid addition to lower the pH and, consequently, increase the solubility of potential inorganic foulants, a portion of the carbonate and bicarbonate alkalinity in the water is converted to aqueous carbon dioxide, which is not rejected by the membranes. The resulting filtrate can thus be corrosive given the combination of a low pH, elevated carbon dioxide levels, and minimal buffering capacity of the filtrate. Other dissolved gases such as hydrogen sulfide, will also readily pass through the semi-permeable membranes, further augmenting the corrosivity of the filtrate and potentially causing turbidity and taste and odor problems. As a result, the primary goal of chemical conditioning is the stabilization of NF/RO filtrate with respect to pH, buffering capacity, and dissolved gases.

Degasification is commonly achieved via packed tower aeration (i.e., air stripping). Air stripping also increases dissolved oxygen levels, which may be very low in the case of an anaerobic ground water source. The pH of the water may subsequently be readjusted to typical finished water levels (i.e., approximately 6.5 - 8.5) by adding a base such as lime or caustic. Alkalinity (e.g., in the form of sodium bicarbonate) may also be added, if necessary, to increase the buffering capacity of the water. Alternatively, if the pH is raised prior to degasification (thus converting the dissolved carbon dioxide to bicarbonate), much of the alkalinity may be recovered. However, this post-treatment strategy also converts any dissolved hydrogen sulfide gas into dissociated sulfide, which may readily react with other dissolved species to produce sparingly soluble sulfide compounds that may precipitate. Because MF, UF, and MCF systems do not directly affect the pH or remove alkalinity, these processes do not generally require chemical conditioning to stabilize the filtrate.

While the use of membrane filtration does not specifically necessitate disinfection post-treatment as a result of process considerations, the need for post-disinfection is generally required by regulation for primary and/or secondary disinfection. However, in some States the use of membrane filtration may reduce primary disinfection (i.e., CT) requirements, thus helping to control DBP formation. Because membrane filtration is often the last major process in the treatment scheme, it is common to apply a disinfectant to the filtrate prior to entry into a clearwell and/or the distribution system. This application is particularly important if disinfectants were either neutralized or not added at all prior to the membrane filtration process to avoid damaging oxidant-intolerant membranes.

For NF and RO membrane processes, if the disinfectant is applied prior to filtrate pH adjustment, post-disinfection may have the additional benefit of oxidizing sulfide to sulfate, thus reducing the potential for both sulfide precipitation and taste and odor concerns. Corrosion inhibitors may also be added prior to distribution, particularly for NF and RO systems that produce more corrosive water.

7.3   System Design Considerations

In the process of planning and implementing a membrane filtration system, there are several important issues that can have a particularly significant impact on system design and operation, thus warranting special consideration. These issues include membrane flux, water quality, temperature compensation, cross-connection control, and system reliability. Each of these subjects is discussed briefly in the following subsections.

7.3.1 Membrane Flux

The flux – the flow per unit of membrane area, as defined in section 2.4.1 – is one of the most fundamental considerations in the design of a membrane filtration system, since this parameter dictates the amount of membrane area necessary to achieve the desired system capacity and thus the number of membrane modules required. Because the membrane modules represent a substantial component of the capital cost of a membrane filtration system, considerable attention is given to maximizing the membrane flux without inducing excessive reversible fouling, thereby minimizing the number of modules required.

Typically, the maximum flux associated with a particular membrane filtration system is determined during pilot testing, mandated by the State, or established via a combination of these two in cases in which the State specifies a maximum operating flux based on the pilot results. Independent of maximum flux, pilot testing is also commonly used to determine a reasonable operating range that balances flux with backwash and chemical cleaning frequencies. Because higher fluxes accelerate fouling, backwashing and chemical cleaning must usually be conducted more frequently at higher fluxes. The process of using pilot testing to optimize the flux relative to the backwash and chemical cleaning frequencies is described in section 6.3. The upper bound of the range of acceptable operating fluxes (provided this bound does not exceed the State-mandate maximum) is sometimes called the "critical" flux, or the point at which a small increase in flux results in a significant decrease in the run time between chemical cleanings. A membrane filtration system should operate below this critical flux to avoid excessive downtime for cleaning and the consequent wear on the membranes over time due to increased chemical exposure.

The flux through a membrane is influenced by a number of factors, including pore size (for MF, UF, and MCF membranes), module type (i.e., cartridge, hollow-fiber, spiral-wound, etc.), membrane material, and water quality. However, it is important to note that higher fluxes do not necessarily indicate that one membrane is better than another for a particular application. Factors such as estimated membrane life, fouling potential, frequency and effectiveness of chemical cleaning, chemical use, and energy requirements to maintain a given flux should also be considered.

7.3.2 Water Quality

Because water quality can have a significant impact on membrane flux, feed water quality is also a primary design consideration for membrane filtration systems. Poorer water quality will require lower fluxes, which in turn increase the necessary membrane area and required number of modules, augmenting both the cost and the size of the system. However, pretreatment (as described in section 7.2.1) can often improve feed water quality at a lower cost than additional membrane area. Conversely, better water quality will allow higher fluxes, reducing the required membrane area, the size of the system, and the capital cost. Typically, membrane flux is determined through pilot testing, as described in Chapter 6. In the absence of pilot test data, it is important to have some understanding of how critical water quality parameters such as SDI, turbidity, organic carbon, and dissolved solids affect the flux. The influence of each of these parameters on flux is briefly described in this section. Temperature also has a significant impact on membrane flux, and this relationship is discussed separately in section 7.3.3.

Silt Density Index

The silt density index, or SDI, is an empirical, dimensionless measure of particulate matter in water and is generally useful as a rough gauge of the suitability of a source water for efficient treatment using NF/RO processes. The American Society for Testing and Materials (ASTM) Standard D 4189-95: *Standard Test Method for Silt Density Index (SDI) of Water* details the procedure for determining SDI. In general, SDI measurements are taken by filtering a water sample through a 0.45-μm flat sheet filter with a 47-mm diameter at a pressure of 30 psi. The time required to collect two separate 500 mL volumes of filtrate is measured, and the resulting data become the inputs to a formula used to calculate SDI. Water samples that contain greater quantities of particulate matter require longer to filter and thus have higher SDI values.

As a general rule of thumb, spiral-wound NF and RO modules are not effective for treating water with a SDI of 5 or greater, as this quality of water contains too much particulate matter for the non-porous, semi-permeable membranes, which would foul at an unacceptably high rate. Thus, some form of pretreatment to remove particulate matter is generally required for SDI values exceeding 5 in NF/RO applications. NF/RO membrane module manufacturers can usually provide a rough estimate of the range of anticipated operating fluxes based on the type of source water, which is roughly associated with a corresponding range of SDI values. A summary of these estimates is included in Table 7.3.

Table 7.3 Estimated NF/RO Membrane Fluxes as a Function of SDI

SDI (dimensionless)	Estimated NF/RO Flux (gfd)	Characteristic Source (Typical)
2 - 4	8 - 14	Surface Water
< 2	14 - 18	Ground Water

However, SDI is only one measure of water quality, and there are a number of site- and system-specific water quality and operational factors that combine to dictate the flux for a given system. Thus, the ranges cited in Table 7.3 should only be used as a rough guideline. Caution should also be exercised when interpreting SDI results, as measurements can vary from test to test and with the analyst, as well as with both temperature and the specific type of membrane used. Consequently, it is important that the results are given for comparable conditions when evaluating SDI data. Note that because SDI determination is a batch process, it is not conducted continuously on-line and thus is not typically utilized as gauge of water quality or system performance during daily operation in the way turbidity or conductivity monitoring are often employed.

SDI is typically not used as a tool for estimating flux for MF, UF, and MCF systems. Since these systems are designed to filter particulate matter, parameters such turbidity that are more commonly associated with conventional drinking water filtration are used to assess system performance.

Turbidity

Turbidity is a measure of the scatter of incident light caused by particulate matter in water. Because turbidity is widely used as a performance gauge for conventional media filters, among the various types of membrane filtration systems turbidity is most often used as an assessment tool for MF, UF, and MCF, since these systems are specifically designed to remove particulate matter. Higher turbidity measurements are indicative of greater quantities of suspended solids, and thus the potential to cause more rapid membrane fouling. Therefore, water with higher turbidity is usually filtered at lower fluxes to minimize fouling and the consequent backwash and chemical cleaning frequency. In some cases when turbidity levels are extremely elevated, it may be more economical to provide pretreatment for a MF/UF system to reduce the solids loading to the membranes. In general, if the turbidity of the water normally exceeds 10 NTU on a sustained basis, some type of pretreatment (e.g., prefiltration or pre-settling) should be considered. MCF systems may be applied to untreated source waters; however, as turbidity levels increase, disposable MCF cartridges will foul more rapidly, requiring more frequent replacement. In some applications with higher turbidity source water, coarse bag or cartridge filters may be used as pretreatment for MCF systems.

Because spiral-wound NF and RO membrane modules are not designed to handle significant solids loading, these systems are typically not applied to treat water with turbidity levels exceeding approximately 1 NTU. For water turbidity levels greater than about 1 NTU, pretreatment would be necessary to reduce solids loading upstream of NF or RO.

Organic Carbon

Another water quality constituent that influences membrane flux is the organic carbon content, which is typically expressed in terms of either total or dissolved organic carbon (DOC). Organic carbon in the feed water can contribute to membrane fouling, either by adsorption of the dissolved fraction onto the membrane material or obstruction by the particulate fraction. Thus, lower fluxes may be necessary if membrane filtration is applied to treat a water with significant organic carbon content. The tendency for a membrane to be affected by TOC is partially influenced by the nature of the organic matter in the water. TOC can be characterized as either hydrophilic or hydrophobic in composition, and studies suggest that the hydrophobic fraction contributes more significantly to membrane fouling. The character of the organic carbon content can be roughly quantified by measuring the specific ultraviolet (light) absorbance (SUVA) of the water, as calculated using the following equation:

$$SUVA = \frac{UV_{254}}{DOC}$$
Equation 7.1

Where: $SUVA$ = Specific ultraviolet (light) absorbance (L/mg-m)
UV_{254} = ultraviolet (light) absorbance at 254 nm (1/m)
DOC = Dissolved organic carbon (mg/L)

Because TOC is more commonly measured than DOC in drinking water treatment, SUVA is sometimes estimating using values for TOC in place of those for DOC.

Higher SUVA values tend to indicate a greater fraction of hydrophobic organic material, thus suggesting a greater potential for membrane fouling. Generally SUVA values exceeding 4 L/mg-m are considered somewhat more difficult to treat. However, organic carbon (as well as turbidity) can often be removed effectively via coagulation and pre-settling, particularly if more hydrophobic in character, thus minimizing the potential for membrane fouling and facilitating operation at higher fluxes. Coagulation can also be conducted in-line (i.e., without pre-settling) with MF/UF systems. Pretreatment using the injection of powdered activated carbon (PAC) may also reduce DOC in the membrane feed; however, because spiral-wound membrane modules cannot be backwashed, PAC should not be used in conjunction with NF/RO systems unless provisions are made to remove the particles upstream.

Dissolved Solids

The total dissolved solids (TDS) and the particular species of dissolved solids present in the membrane feed are both critical considerations for NF/RO systems. Species such as silica, calcium, barium, and strontium, which can precipitate as sparingly soluble salts, can cause scaling and a consequent rapid decline in flux under certain conditions. Scaling is typically controlled using pretreatment chemicals such as an acid to lower the pH and/or a proprietary scale inhibitor, as discussed in section 7.2.1.2. However, the total quantity of dissolved solids of any species also influences system operation, as the net driving pressure required to achieve a target flux is related to the osmotic pressure of the system, which is directly proportional to the TDS (as discussed in section 2.2.2). Thus, as the TDS increases, so does the required feed pressure.

TDS is generally not a significant consideration for MF, UF, and MCF systems, since these processes do not remove dissolved solids. In some cases, however, the use of upstream oxidants may cause the precipitation of iron or manganese salts (either unintentionally or by design as a pretreatment process), which could accelerate membrane fouling.

7.3.3 Temperature Compensation

Like other water quality parameters such as turbidity and TDS (for NF/RO systems), the temperature of the feed water also affects the flux of a membrane filtration system. At lower temperatures water becomes increasingly viscous; thus, lower temperatures reduce the flux across the membrane at constant TMP or alternatively require an increase in pressure to maintain constant flux. The means of compensating for this phenomenon varies with the type of membrane filtration system used. General viscosity-based means of compensating for temperature fluctuations for both MF/UF and NF/RO systems are described below, although membrane manufacturers may have a preferred product-specific approach.

MF/UF Systems

MF/UF membrane systems usually operate within a relatively narrow range of TMPs, which may limit increasing the TMP in order to maintain constant flux as the water temperature decreases. Because the membrane modules can be damaged if the TMP exceeds an upper limit, as specified by the manufacturer, it may not be possible to operate the system at a TMP that is sufficient to meet the required treated water production during colder months if demand remains high. As a result, additional treatment capacity (i.e., increased membrane area or number of membrane modules) is incorporated into the design of the system such that the water treatment production requirements can be satisfied throughout the year.

For the microporous MF/UF membranes, the relationship between flux, TMP, and water viscosity is given by the following equation, with a cross-reference to the same equation as listed in Chapter 2 given in parentheses:

$$J = \frac{TMP}{R_t \bullet \mu_w}$$

Equation 7.2   (2.7)

Where: J   =   flux (gfd)
         TMP   =   transmembrane pressure (psi)
         R_t =   total membrane resistance (psi/gfd-cp)
         μ_w =   viscosity of water (cp)

If the system is operated at constant flux, then increases in viscosity require proportional increases in operating TMP (assuming constant membrane resistance). However, once the TMP approaches the rated maximum for the membranes, further increases in viscosity necessitate a reduction in flux. Thus, in order to maintain the required filtered water production flow (so as to satisfy customer demand), the membrane area must increase in proportion to the flux decrease, as shown in Equation 7.3 (also Equation 2.1):

$$J = \frac{Q_p}{A_m}$$

Equation 7.3   (2.1)

Where: J   =   flux (gfd)
         Q_p =   filtrate flow (gpd)
         A_m =   membrane surface area (ft^2)

Combining Equations 7.2 and 7.3 demonstrates that the additional membrane area required is directly proportional to the increase in water viscosity (for constant flow, TMP, and membrane resistance), as shown in Equation 7.4:

$$\frac{Q_p}{A_m} = \frac{TMP}{R_t \bullet \mu_w}$$

Equation 7.4

Where: Q_p =   filtrate flow (gpd)
         A_m =   membrane surface area (ft^2)
         TMP   =   transmembrane pressure (psi)
         R_t =   total membrane resistance (psi/gfd-cp)
         μ_w =   viscosity of water (cp)

Membrane filtration systems are commonly designed to operate at a particular flux (e.g., as determined via pilot testing or mandated by the State) to produce a specific flow (i.e., rated system capacity) at a given reference temperature. Thus, the required membrane area at the reference temperature can be calculated using Equation 7.3. The increased membrane area required to compensate for cold weather flow can be determined by multiplying this area by the ratio of the viscosity at the coldest anticipated temperature (e.g., the coldest average monthly temperature) to that at the reference temperature. Values for water viscosity can be found in the literature or approximated using Equation 7.5 (also Equation 2.8):

$$\mu_T = 1.784 - (0.0575 \bullet T) + (0.0011 \bullet T^2) - (10^{-5} \bullet T^3) \qquad \text{Equation 7.5 \quad (2.8)}$$

Where: $\quad \mu_T \quad = \quad$ viscosity of water at temperature T (cp)

$\qquad\qquad\quad$ T $\quad\; = \quad$ water temperature ($^\circ$C)

After the appropriate values for water viscosity have been determined for both the reference temperature (commonly 20 $^\circ$C for MF/UF systems) and the coldest anticipated temperature, then the design membrane area, as compensated for seasonal temperature variation, can be calculated as shown in Equation 7.6:

$$A_d = A_{20} \bullet \left(\frac{\mu_T}{\mu_{20}} \right) \qquad \text{Equation 7.6}$$

Where: $\quad A_d \quad = \quad$ design membrane area (adjusted to temperature T) (ft^2)

$\qquad\qquad\; A_{20} \quad = \quad$ membrane area required at 20 $^\circ$C reference temperature (ft^2)

$\qquad\qquad\; \mu_T \quad\; = \quad$ viscosity of water at temperature T (cp)

$\qquad\qquad\; \mu_{20} \quad = \quad$ viscosity of water at 20 $^\circ$C (cp)

Equation 7.4 can also be utilized to calculate the required membrane area using a less conservative approach that accounts for seasonal fluctuations in demand. First, if the information is available, the average daily flow and temperature over each calendar month can be tabulated, and the temperature data converted into associated values for water viscosity using Equation 7.5. Then, the 12 sets of paired flow (Q_p) and viscosity (μ) data can be applied to Equation 7.4 to generate 12 values of membrane area. The largest of these values for the required membrane area over each of the 12 calendar months, along with its corresponding flow, is applied to Equation 7.3 to generate an associated flux. If this flux is less than the maximum permitted value, then this largest of the 12 calculated values for membrane area represents the design value for the system. However, if the resulting flux exceeds this threshold, the design area must be increased to lower the flux to the maximum permitted value. A somewhat more conservative variant of this approach described in the literature utilizes the maximum daily flow and minimum daily temperature recorded over each month of the year to generate 12 sets of paired flow and viscosity data; these data can then used to calculate 12 values for membrane area and flux, as described above (Schideman et al. 2004).

Note that that use of Equation 7.4 in this approach requires values for both the TMP and the total membrane resistance, R_t, both of which should be considered constants for the purposes of calculating the various values for membrane area. While a reasonable TMP can be easily identified, appropriate values of R_t are more difficult to determine. The total membrane resistance represents the sum of the intrinsic resistance of the membrane (which may be considered a constant and can generally be obtained from the manufacturer) and the resistance attributable to fouling at any given point during operation, as shown in Equation 7.7 (also Equation 2.6):

$$R_t = R_m + R_f \qquad\qquad \text{Equation 7.7 \quad (2.6)}$$

Where:
R_t = total membrane resistance (psi/gfd-cp)
R_m = intrinsic membrane resistance (psi/gfd-cp)
R_f = resistance of the foulant layer (psi/gfd-cp)

Because it difficult to both identify and justify a single, specific value for the fouling resistance for use with this approach, the contribution attributable to fouling may be ignored for practical purposes. Thus, for the purposes of calculating membrane area, the total resistance used in Equation 7.4 may be approximated by the membrane's intrinsic resistance. This approximation may be reasonable for the membrane system at the start of a filtration cycle when the fouling resistance is minimal. Since the minimum value for membrane resistance expected over a filtration cycle is used, the minimum TMP anticipated over a filtration cycle should also be used. This minimum TMP occurs at the beginning of a filtration cycle before gradually increasing until the MF/UF system must be backwashed. Use of the intrinsic membrane resistance and minimum TMP should result in a reasonable and not excessively conservative estimate of the membrane area requirements. In rare cases in which the membrane is experiencing significant fouling under conditions of peak demand such that the flux and/or TMP are approaching their maximum values, the backwash and/or chemical cleaning frequencies can be increased temporarily to compensate and keep fouling to a minimum. Alternatively, standby membrane units may be used when necessary, as described in section 7.3.5.

The MF/UF membrane manufacturer may have an alternate preferred method of determining the design membrane area for a particular application based on temperature, membrane material, or other site- or system-specific factors. It is recommended that the utility collaborate with the State, the membrane manufacturer, and its consulting engineer (if applicable) to select the most appropriate method for determining the required area. Note that the addition of membrane area to compensate for low-temperature flow will also help the system to meet higher flow demands during warm weather without operating at an exceedingly high membrane flux. Temperature compensation for MCF systems, if necessary, can be determined using the methodology for MF/UF systems.

NF/RO Systems

NF/RO membrane modules are designed to operate over a larger range of TMPs than MF/UF modules, and thus the TMP is simply increased to maintain constant flux as either the temperature of the feed water decreases or the membrane fouling increases. The required increase in TMP at the temperature of interest relative to that at a given reference temperature (typically 25 °C for NF/RO systems) is dependent on the specific proprietary membrane used and can be calculated by means of a temperature correction factor (TCF), as shown in Equation 7.8 (also Equation 2.19):

$$TCF = \exp\left[U \bullet \left(\frac{1}{T + 273} - \frac{1}{298}\right)\right]$$
Equation 7.8   (2.19)

Where: TCF   =   temperature correction factor (dimensionless)
            T   =   water temperature (°C)
            U   =   membrane-specific manufacturer-supplied constant

Alternatively, many NF/RO membrane manufacturers may supply tables specifying TCF values over a range of temperatures for a given membrane. Once the appropriate TCF is known, the required TMP at temperature T of interest can be calculated by dividing the TMP at the reference (i.e., design) temperature (commonly 25 °C for NF/RO systems) by the TCF value, as shown in Equation 7.9:

$$TMP_T = \frac{TMP_{25}}{TCF}$$
Equation 7.9

Where: TMP_T =   transmembrane pressure (adjusted to temperature T) (psi)
           TMP_{25} =   transmembrane pressure at 25 °C reference temperature (psi)
           TCF   =   temperature correction factor (dimensionless)

Note that Equations 7.8 and 7.9 also allow the calculation of the TMP reduction resulting from decreased water viscosity at higher feed water temperatures during warmer months of the year.

7.3.4 Cross-Connection Control

In the context of membrane filtration systems, cross-connection control measures are implemented to prevent chemicals from the cleaning process from contaminating the feed or filtrate streams. States may have particular requirements for cross-connection control, although in general there are two strategies that are commonly used: a double block and bleed valving arrangement or a removable spool. These strategies are applicable to both MF/UF and NF/RO systems; however, because MCF systems typically utilize modules that are disposable cartridges, these systems are not usually subject to chemical cleaning and thus cross-connection control measures are generally unnecessary for these systems.

The double block and bleed valve arrangement is the most common method of cross-connection control for large membrane filtration systems. A schematic illustrating this method is shown in Figure 7.1. In summary, two isolation valves (V-1A and V-2A in Figure 7.1) are placed in the feed line to isolate it from the cleaning chemicals. These are the block and bleed valves, respectively. During the cleaning process, valve V-4A and V-6A are opened to bring cleaning solution(s) into Membrane Rack (i.e., unit) A. Although valve V-3A is kept closed during this operation, if it were to leak it would allow chemicals to pass into the common feed manifold, causing contamination. In order to prevent this potential contamination, valve V-1A is also closed and valve V-2A is opened. With this configuration, if valve V-3A leaks, the cleaning solution(s) simply flows through valve V-2A to waste. A similar block and bleed valve arrangement is utilized to prevent the cleaning chemicals from contaminating the filtrate manifold as they are recirculated, which gives rise to the term "double" block and bleed. If the cleaning process starts automatically, valves V-1A and V-2A (i.e., the block and bleed valves, respectively) must be actuated automatically.

As an alternative to the double block and bleed valving arrangement, a removable spool (i.e., a short section of pipe) can be placed between valves V-1A and V-3A in lieu of utilizing valve V-2A. The spool is then removed during the chemical cleaning. As with the block and bleed valving arrangement, a removable spool configuration would be used to prevent the cleaning chemicals from potentially contaminating the filtrate in the event of an isolation valve leak. A removal spool should not be used for cross-connection control if chemical cleaning is automated, since the process could potentially be initiated with the spool still in place. Air gaps could also be used as cross-connection control measures, where applicable.

Figure 7.1 Double Block and Bleed Valving Arrangement

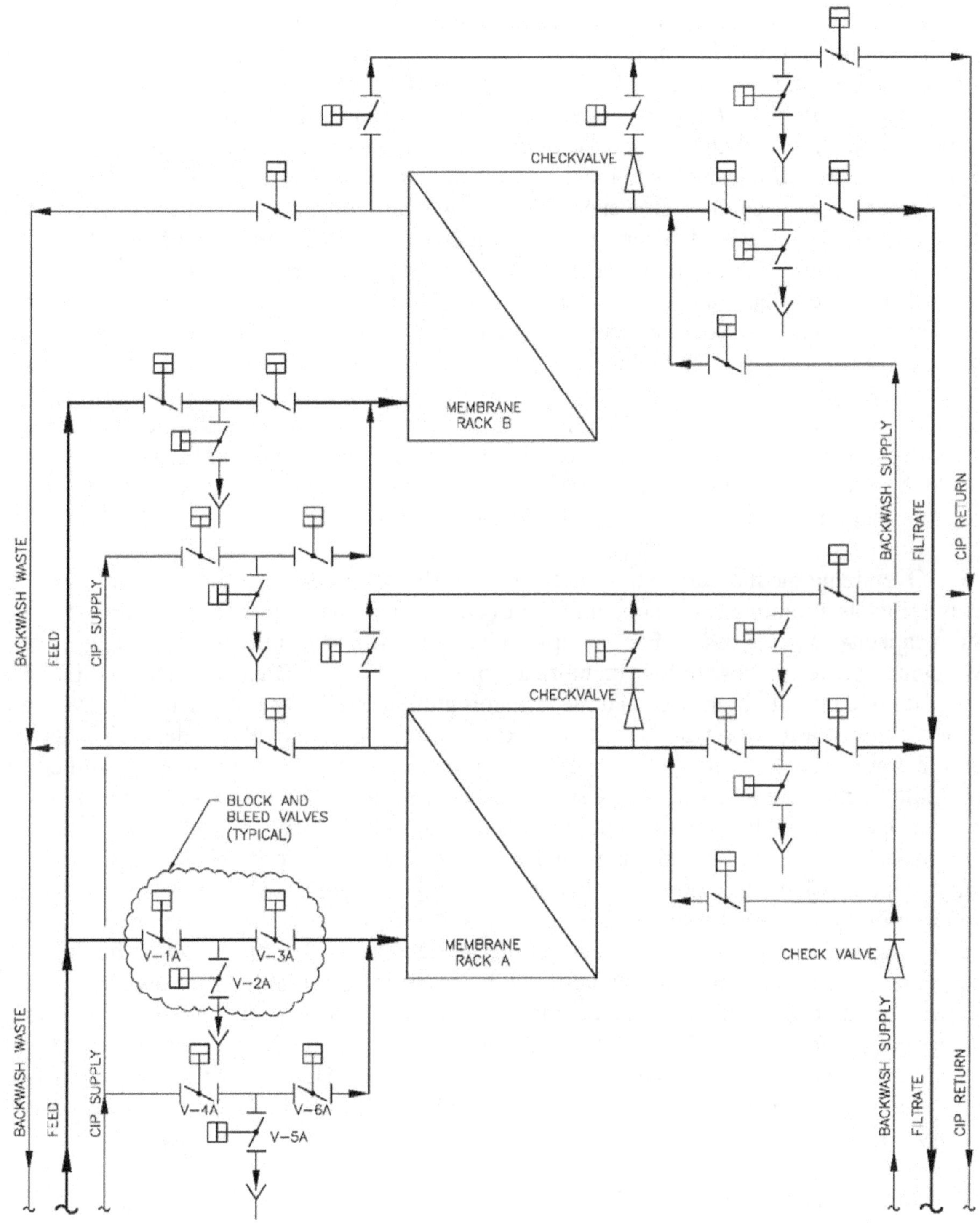

7.3.5 System Reliability

System reliability is an important consideration in the design of a membrane filtration system. Design capacities are typically maximum values and may not necessarily account for the possibility of one or more units being taken out of service for repair or routine maintenance. Even standard operational unit processes such as backwashing, chemical cleaning, and integrity testing that are normally accounted for in accurately sizing the facility and determining system capacity may be problematic if it becomes necessary to conduct these processes more frequently than was planned. For example, most membrane filtration systems are designed with sufficient storage and equalization capacity to meet average demand even when a unit is taken out of service for chemical cleaning. However, this storage would generally not allow the system to operate at capacity for extended periods (e.g., longer than about 24 hours) with a unit of out service for repair. Reliability issues such as this may be particularly pronounced for smaller systems with fewer membrane units, since one out-of-service unit may significantly impact overall system capacity. Note that any State sizing and redundancy requirements should be considered in the design of the facility.

In order to ensure system reliability, it is common for membrane filtration systems to incorporate some measure of redundancy. There are two strategies that utilities commonly use to provide this redundancy: 1) oversizing the membrane units at the design flux; and 2) providing a redundant membrane unit.

Oversizing the membrane units at the design flux generally allows for operation at average flow as a minimum with one unit out of service and may allow for operation at or near system capacity in some cases. For example, a water treatment plant with rated capacity of 7 MGD may be designed with four membrane units rated at 2 MGD each rather than 1.75 MGD. Thus, with one unit out of service and the remaining three operating at capacity, the system can produce as much as 6 MGD (at the design flux) consistently. Although the design in this example would not allow the system to operate at maximum capacity, it is likely that the average flow could be met or exceeded. A system could also be designed with greater excess capacity to allow for operation at the maximum output with one unit out of service (in this example, using four units at 2.33 MGD each), although this approach may be cost prohibitive, particularly for smaller systems, since the percentage that each unit must be oversized increases as the total number of units is reduced. Note that because this method of providing system redundancy is based on the assumption that the rated capacities are all specified with respect to the same design flux, the rate of fouling would not be increased by operating three of the four oversized units at capacity. In addition, continuing this example of a treatment plant with 7 MGD of permitted capacity consisting of four 2.33-MGD membrane filtration units, with all four units in operation the system could operate at reduced flux and still produce 7 MGD in accordance with its permit. This operation at reduced flux would decrease the rate of fouling and lower operating costs to partially offset the increase in capital cost associated with the extra unit capacity.

The second common method of adding system redundancy is by providing an additional membrane unit. Thus, using the same example of a water treatment plant permitted for a maximum rated capacity of 7 MGD, the membrane filtration system would consist of five 1.75-MGD units, rather than four. This design would always allow for operation at capacity with one unit either out of service or in standby mode. In this case the standby unit can be rotated such that any time a unit is taken offline for chemical cleaning the standby unit is activated. The newly cleaned unit then reverts to standby mode until the next unit is due for chemical cleaning. This method of adding redundancy is more cost effective for larger facilities with a greater number of membrane units, such that the addition of an extra unit does not represent a large percent increase in capital cost. This approach is recommended when the treatment facility utilizing membrane filtration is the primary source of drinking water for the public water system, such that operating the membrane system at capacity may be critical to the ability of the utility to satisfy customer demand. Alternatively, under this scenario all the units may be operated at lower flux to extend the interval between chemical cleanings.

Independent of the particular strategy for providing some redundant capacity for the membrane filtration system, utilities should also provide some redundancy for any major ancillary mechanical equipment that may be utilized, such as pumps, compressors, and blowers. Utilities must comply with any State-mandated redundancy requirements for either membrane capacity or ancillary equipment.

Membrane filtration systems may also have some inherent redundancy if the system is normally operated significantly below the maximum flux permitted by the State. In this case, if one membrane unit is taken out of the service, the utility can increase the temperature- and pressure-normalized flux through the remaining units to partially or fully compensate for the loss of production attributable to the out-of-service unit. However, if a unit is out of service for a prolonged period, the increase in flux may accelerate fouling in the remaining in-service units to an unacceptable rate.

Timely membrane replacement is another consideration for maintaining system performance and reliability. Membrane replacement is usually conducted on an as-needed basis, typically either in cases in which the membranes have been damaged or the flux has declined to an unacceptable level as a result of irreversible fouling. Although the useful life of membranes is commonly cited in the range of 5 to 10 years (a period generally consistent with manufacturer warranties), the use of membrane technology – particularly MF and UF – has increased almost exponentially in the decade preceding this guidance manual, and thus most membrane filtration systems have not been in continuous operation for more than 5 to 10 years. Consequently, there is limited field data available to document the typical useful life of membrane filtration modules. Nevertheless, it is recommended that utilities keep a small number of surplus membrane modules on site in the event that emergency replacement becomes necessary.

7.4   Residuals Treatment and Disposal

As with many water treatment processes, membrane filtration systems may generate several different types of residuals that must be treated and/or disposed of, including concentrate and both backwash and chemical cleaning residuals. The types of residuals that are generated vary with both the type of membrane filtration system and the hydraulic configuration in which the system is operated. For example, NF/RO systems produce a continuous concentrate stream and periodic chemical cleaning waste, but because these spiral-wound membrane systems are not backwashed, no backwash residuals are generated. MF/UF systems are regularly backwashed and undergo periodic chemical cleanings, thus producing residuals from both these operational unit processes. However, only MF/UF systems that are operated in a crossflow hydraulic configuration and waste (rather than recirculate or recycle to the plant influent) the unfiltered flow – the so-called "feed and bleed" mode – will generate a concentrate stream. Because cartridge filters operate in a deposition mode hydraulic configuration (see section 2.5) and are designed to be disposable, there are typically no residuals streams associated with MCF systems. Nevertheless, the spent filter cartridges (as well as all membrane filtration residuals) must be properly disposed of in accordance with any applicable State and local regulations, particularly if the cartridges have been used to filter any potentially hazardous materials.

The various potential residuals streams from membrane filtration systems – backwash residuals, chemical cleaning residuals, and concentrate – are discussed in the following subsections. Note that these discussions are not meant to represent a comprehensive review of residuals treatment and disposal, but rather a general overview of some of the primary considerations that should be taken into account when planning and designing a membrane filtration system. For additional information, the reader is referred to the following references:

- ***Water Treatment Plant Waste Management*** (1987)
  (Prepared by Cornwell et al. of Environmental Engineering & Technology for the American Water Works Association Research Foundation)

- ***Treatment of MF Residuals for Contaminant Removal Prior to Recycle*** (2002)
  (Prepared by MacPhee et al. of Environmental Engineering & Technology for the American Water Works Association Research Foundation)

- ***Current Management of Membrane Plant Concentrate*** (2000)
  (Prepared by Kenna et al. of Clarkson University for the American Water Works Association Research Foundation)

- ***Major Ion Toxicity in Membrane Concentrate*** (2000)
  (Prepared by Mickley of Mickley & Associates for the American Water Works Association Research Foundation)

- ***Membrane Concentrate Disposal: Practices and Regulation*** (2001)
  (Prepared by Mickley of Mickley & Associates for the United States Bureau of Reclamation)

- ***Membrane Concentrate Disposal*** (1993)
  (Prepared by Mickley et al. for the American Water Works Association Research Foundation)

- ***The Desalting and Water Treatment Membrane Manual: A Guide to Membranes for Municipal Water Treatment (2nd Ed.)*** (1998)
  (Prepared by the United States Bureau of Reclamation)

- ***Reverse Osmosis and Nanofiltration (M46)*** (1999)
  (Prepared by the American Water Works Association)

- ***Microfiltration and Ultrafiltration Membranes (M53)*** (2005)
  (Prepared by the American Water Works Association)

7.4.1 Backwash Residuals

Among the various types of membrane filtration, only MF/UF systems employ backwashing, thus generating backwash residuals. Although the frequency varies on a site- and system-specific basis, backwashing is typically conducted every 15 to 60 minutes. Under normal operating conditions the backwash frequency should remain relatively consistent, allowing for the quantity of residuals generated to be estimated fairly accurately.

As a general rule of thumb, the residual stream produced from backwashing MF/UF membranes has a concentration of suspended solids that is approximately 10 to 20 times greater than that of the feed water. Although MF/UF systems remove approximately the same types of feed water constituents as conventional media filters, the volume and characteristics of the residuals may be significantly different. In many current applications of MF/UF for municipal water treatment, filter aids such as coagulants and polymers are not necessary. In these cases, the amount of solids removed in the backwash process may be significantly less than that for a comparable conventional filtration plant. In addition, disposal of these coagulant- and polymer-free MF/UF backwash residuals may be less problematic. However, in some applications of MF/UF, coagulants may be added in-line (i.e., without pre-settling) to help facilitate the removal of TOC, which is not normally removed to a significant degree by MF or UF. In such applications the MF/UF backwash residuals characteristics will generally be more similar to those for conventional media filtration.

Disposal options for MF/UF backwash residuals are similar to those for conventional water treatment plants, and typically include the following:

- Discharge to a suitable surface water body

- Discharge to the sanitary sewer

- Treatment with supernatant recycle and solids disposal

The discharge of backwash residuals to surface water bodies or the sanitary sewer is likely to be subject to State and/or local regulations and, in the case of surface water discharge, to require a permit. Moreover, the potential to utilize one of these options may be complicated if the residuals include chemical wastes. In addition to the use of coagulants added to the feed, some backwash procedures utilize chlorine or other chemicals, as described in section 7.2.2. Small amounts of chlorine may be quenched, and acids or bases can be neutralized prior to discharge, although larger amounts or other types of chemicals added to the backwashing processes may require additional treatment or preclude discharge to a surface water body or sanitary sewer.

On-site treatment options for MF/UF backwash residuals are also similar to those that might be used with conventional media filtration and include clarification, sedimentation lagoons, gravity thickening, centrifuging, belt filter presses, or a combination of these processes. A second stage of MF or UF may also be utilized to further concentrate residuals and increase process recovery. If a sedimentation process is used to treat MF/UF backwash residuals, the addition of a coagulant may be necessary to improve the settling characteristics of the solids if coagulant is not already applied in the MF/UF pretreatment process. With on-site treatment, the supernatant is generally recycled to the treatment plant influent while the concentrated solids are transported off site for landfilling or other means of disposal. As with discharging, the addition of chlorine or other chemicals to the backwash process may complicate residuals treatment.

It is important to note that backwash residuals concentrate any pathogenic organisms that are present in the feed water, as well as other suspended solids. The potential treatment of this stream should be taken into consideration if these residuals are to be discharged into a surface water receiving body.

7.4.2 Chemical Cleaning Residuals

Both MF/UF and NF/RO membranes undergo periodic chemical cleaning, and thus both types of systems generate spent chemical waste as a byproduct of these processes. As with the backwashing process for MF/UF, the frequency of chemical cleaning varies on both a site- and system-specific basis. Although chemical cleaning is conducted much less frequently than backwashing, the frequency is also more difficult to predict. Generally, MF/UF systems are cleaned no more frequently than once every month for efficient operation and the minimization of system downtime, although it is not uncommon for these systems to operate for much longer without requiring chemical cleaning. The cleaning frequency for NF/RO may vary from 3 months to 1 year or longer, depending on the feed water quality and the effectiveness of feed water pretreatment for minimizing fouling. However, because chemical cleaning is a relatively infrequent batch process, estimating the quantity of residuals generated is not as critical for day-to-day operation, as is the case with backwashing.

Chemical cleaning residuals are generally treated on-site and discharged to either a suitable surface water body or sanitary sewer, subject to State and/or local regulations. Oxidants such as chlorine used in the chemical cleaning process can be quenched prior to discharge, and acids and bases can be neutralized. The use of other chemicals, such as surfactants or proprietary cleaning agents, may complicate the process of obtaining regulatory approval for discharge and could require additional treatment.

Note that the rinse water applied to the membranes after the cleaning process may also represent a chemical waste and thus could require treatment prior to discharge. Although the rinse water increases the volume of the chemical cleaning residuals, this increase can be balanced somewhat by the recovery and reuse of a significant portion of the cleaning solutions.

In some cases, as much as 90 percent of the applied cleaning solutions can be reused, reducing residuals treatment and disposal costs, as well as chemical usage.

7.4.3 Concentrate

The term "concentrate" is usually associated with the continuous waste stream of concentrated dissolved solids produced by NF and RO processes. This waste stream is typically 4 to 10 times more concentrated than the feed with respect to suspended and dissolved constituents and represents about 15 to 25 percent of the total feed flow, although it can exceed 50 percent or more in some cases. As a result, concentrate disposal is a significant logistical and regulatory concern for utilities and is often a critical factor in the planning and design of an NF or RO facility.

There are a number of methods for concentrate treatment and disposal, including the following options:

- Surface water discharge

- Sanitary sewer discharge

- Land application / irrigation

- Deep well injection

- Evaporation

Because there are complicating factors associated with each of these options, no single option is ideal or most appropriate for every application. In many cases surface water discharge is the least expensive option, although the permitting process may be difficult, since there are potential environmental impacts if the salinity of the concentrate is significantly higher than that of the receiving body. Discharge to the sanitary sewer may have similar issues, since the wastewater treatment process does not typically affect dissolved solids concentrations to a significant degree, and the treatment plant effluent may ultimately be discharged to a surface water receiving body. High salinity or major ion toxicity may also preclude land application of the concentrate if levels exceed the threshold tolerances of the irrigated crops. Bioaccumulation of metals has also been cited as a potential concern for land application of NF/RO concentrate. Deep well injection is an effective and commonly used technique for concentrate disposal, although this method risks the escape of brackish water into less saline or freshwater aquifers and may have unknown long-term environmental affects. The use of evaporation ponds is generally limited to areas with low precipitation and high evaporation rates, as well as an abundance of inexpensive and available land.

Another option for dealing with NF/RO concentrate is the concept of "zero liquid discharge," which involves sufficiently concentrating the residual stream through the use of such technologies as crystalizers and evaporators to allow remaining solids to be landfilled. While zero liquid discharge is typically an expensive option, is does offer a number of advantages, including avoiding the discharge permitting process and the ability to be utilized at any location independent of factors such as the proximity to a suitable surface water body or available land for evaporation ponds. In addition, zero liquid discharge maximizes facility recovery and has minimal environmental impact. Although still fairly uncommon, this option has become increasingly feasible relative to other methods as environmental and discharge regulations have become more stringent.

The considerations noted above represent only a cursory discussion of the issues associated with the various options for concentrate treatment and disposal. More detailed information is available in several of the references cited in section 7.4. Note that all of these options are subject to applicable federal, State, and local regulations.

Some MF/UF systems that are operated in a suspension mode hydraulic configuration and waste the unfiltered flow generate a continuous concentrate stream. This stream differs from that associated with NF/RO systems in two significant ways: MF/UF systems concentrate suspended rather than dissolved solids; and the concentrate stream represents only a small fraction of total feed flow. (Note that NF/RO membranes represent a barrier to particulate matter, and thus these systems will also concentrate suspended solids; however, because suspended solids rapidly foul the semi-permeable membranes (which cannot be backwashed), most particulate matter is typically removed with pre-filters.) An MF/UF concentrate stream has characteristics somewhat similar to those of backwash residuals, and therefore can be considered comparable for the purposes of treatment and disposal. Alternatively, these two residuals streams may be blended together.

8.0 Initial Start-Up

8.1   Introduction

The initial start-up phase is an important step in the successful installation of a full-scale membrane filtration system and thus is an essential consideration in the facility planning and design process. This period includes such tasks as initial system flushing and disinfection, system diagnostic checks, membrane module installation, subjecting new equipment to integrity testing, and operator training, all of which must be completed prior to placing the system into service. Note that the Long Term 2 Enhanced Surface Water Treatment Rule (LT2ESWTR) does not contain any requirements for the initial start-up of a membrane filtration system. Consequently, this chapter is included in the *Membrane Filtration Guidance Manual* as a useful reference for utilities that summarizes widely recognized industry best practices and important considerations for the start-up process. Any applicable State requirements would supersede the guidance provided in this chapter.

In general, the primary objectives of the initial start-up phase are to ensure that the installation is successfully completed and that the equipment is in proper working order and ready to produce potable water that achieves all target quality standards. A well-planned initial membrane filtration start-up phase can create a smooth transition from testing to drinking water production. This chapter discusses important general start-up considerations, pointing out any significant differences between nanofiltration (NF)/reverse osmosis (RO) and microfiltration (MF)/ultrafiltration (UF) systems. For the purposes of this discussion, membrane cartridge filtration (MCF) systems are considered to be similar to MF/UF systems, except as otherwise noted.

This chapter is divided into the following sections.

Section 8.2:   Temporary System Interconnections
This section describes the provisional requirements that may be necessary during start-up of a membrane filtration system.

Section 8.3:   Flushing and Testing Without Membranes
This section describes the general procedures that are associated with flushing a membrane filtration system and confirming that each unit is operating properly without the membrane modules installed.

Section 8.4:   Membrane Installation
This section reviews considerations associated with the installation of membrane modules.

Section 8.5:   System Disinfection
This section discusses the initial disinfection process for systems using both chlorine-tolerant and chlorine-intolerant membranes.

Section 8.6:   Initial Direct Integrity Testing
This section outlines some considerations for conducting initial direct integrity testing on the newly installed membrane modules.

Section 8.7:   Acceptance Testing
This section describes some of the typical practices associated with acceptance testing.

Section 8.8:   Operator Training
This section describes the operator training that the equipment supplier should provide during the start-up phase.

8.2   Temporary System Interconnections

During the start-up process, the filtered water produced from the membrane units may not be acceptable for distribution. Therefore, facility design should include provisions for the recycle and/or temporary disposal of feed and filtrate water. Provisions for disposal usually consist of a removable pipe spool or a tee with a "dump" valve placed in the common feed or filtrate line. This water can usually be diverted to the sanitary sewer. If the feed and filtrate piping are interconnected, the connection is removed and replaced with blind flanges after start-up has been completed. Some facilities simply divert and recycle the treated water to the inlet structure until the system is fully commissioned. Any State requirements governing the disposal and potential recycle of the water produced during the start-up phase should be incorporated into the planning process.

8.3   Flushing and Testing Without Membranes

Prior to installing the membrane modules, any debris introduced during construction should be flushed from the system. This flushing is usually conducted by running the appropriate pump(s) at high velocity and low pressure through the piping and discharging the water to a suitable drain. Typically, the flush water can be discharged to the sanitary sewer, although the utility should comply with any State requirements for disposal. Because the intent of the flushing process is to remove debris from the system, this water should not be recycled unless pretreatment processes would be expected to remove contaminants flushed from the system.

After the piping system has been flushed, the operational sequences and chemical addition systems should be tested to ensure that they are operating properly before installing the membrane modules. Because of the complexity of the equipment involved, it is suggested that a plan for conducting these diagnostic checks be developed in advance. In general, the recommended testing can be divided into the following categories:

- **Mechanical Equipment** – Inspect automated mechanical equipment to ensure that it is properly installed and that there are no leaks in the piping system. Operate mechanical equipment in manual mode and then in automatic mode to verify that it is working as designed.

- **Instrumentation** – Verify that the instrumentation is properly installed and calibrated. Confirm that the instruments are operating correctly and are responsive to the control system.

- **Control System** – Verify control inputs and outputs (both digital and analog), instrumentation alarm limits, programming logic, instrumentation loops, and operational sequences.

- **Membrane Units and Related Systems** – Verify that the membrane filtration system and each respective membrane unit can be both started and shut down smoothly. Operate the chemical feed systems (if any) to verify that they are each in proper working condition and that all chemicals are delivered at the proper dosages.

For NF/RO systems, the procedure for testing the operating sequences is commonly known as the "48-hour test," as it may take two or more days to complete. Unlike the initial flushing, the 48-hour test should be conducted at actual operating parameters, including both typical pressures and flows. Therefore, it is common to insert flow restriction devices in the system in place of the membrane modules during this operation to simulate the anticipated system backpressure. These devices typically consist of orifice plugs placed in the permeate (i.e., filtrate) port inside the pressure vessel. The orifice plugs are sized to simulate the flow and pressure parameters of the operating membrane filtration system. Alternatively, an isolation valve on the permeate piping could be throttled to generate the appropriate backpressure. Although the term "48-hour test" is less commonly used in association with MF and UF, these systems are also generally tested either with "dummy" modules that have similarly designed orifices or by throttling filtrate effluent valves to create sufficient backpressure during the testing sequences.

8.4   Membrane Installation

Membrane module installation should be conducted according to the instructions obtained from the manufacturer. Care should be taken not to damage the membranes during the installation process. It is recommended that the location of each individual membrane in the system be recorded according to its serial number. Factory test data are often shipped with the membranes, and this information should be collected and filed.

Membrane modules are typically shipped "wet" with a liquid preservative solution. The particular preservative depends upon the type of membrane. Many membranes are preserved with a 1 percent solution of sodium bisulfite, a reducing agent that acts as a biocide to control microbial growth. Prior to installation, the membrane modules should be stored in an appropriate manner such that they are not subjected to freezing conditions that could damage the membranes. Regardless of the membrane type, the preservative should be flushed to waste after the membranes are installed. Note that the disposal of any preservative solutions should be conducted in accordance with any applicable regulations.

Some membrane modules are shipped with more problematic storage solutions. For example, glycerin solutions may pose a waste treatment issue because of the biochemical oxygen demand (BOD) that may result from the solution being flushed down the drain. The use of formaldehyde, once a common membrane preservative, is generally no longer acceptable, and it may pose a significant disposal problem. Both State and local regulations regarding chemical disposal may apply to membrane preservative solutions, and these may dictate whether it is permissible to discharge these preservatives to the sanitary sewer or if collection and alternate disposal and/or treatment is required.

8.5   System Disinfection

Initial system disinfection typically involves the application of a disinfectant, such as a chlorine solution, throughout the entire system, including both the feed and filtrate piping. Although both soaking and recirculating procedures are used, recirculating the solution through the system will generally provide more effective disinfection. This disinfection step may be required by the State, and in any case is recommended to inactivate bacteria or other pathogenic organisms that may contaminate the membrane filtration system and associated piping. Because some membranes have limited compatibility with disinfecting chemicals, the manufacturer may have specific requirements for the disinfection process, and approval will be necessary if manufacturer recommendations differ from State requirements. General procedures for the disinfection of both chlorine-tolerant and chlorine-intolerant membranes are described in the following sections. In either case, once the system disinfection is complete, the entire system (including the membranes) should be flushed prior placing the membrane unit(s) into continuous service. This final flush can be conducted according same guidelines outlined in section 8.3 and should continue until the desired filtered water quality is attained as measured by site- or membrane system-specific water quality parameters.

8.5.1   Chlorine-Tolerant Membranes

The American Water Works Association (AWWA) and the American National Standards Institute (ANSI) have developed standards pertaining to the disinfection of water treatment facilities. Specifically, ANSI/AWWA Standards C651 and C653 provide guidance for the disinfection of water treatment plants and associated piping systems. The procedure involves surface contact with a high strength chlorine solution for a specific time period. The disinfection is complete when bacteriological sampling and testing indicate the absence of coliform organisms. For systems that use chlorine-tolerant membranes, the membrane modules should be installed prior to initiating the disinfection procedure. If the disinfection process is conducted while the membranes are in place, sufficient pressure should be applied to ensure adequate flow across the membranes such the filtrate side piping is thoroughly disinfected. Some MF/UF membranes and most MCF membrane are chlorine-tolerant.

8.5.2 Chlorine-Intolerant Membranes

Most NF and RO membranes, as well as some MF and UF membranes, are not chlorine-tolerant. For systems that use these types of membranes, there are two options that are generally available for system disinfection. One option consists of conducting the system disinfection with chlorine as described in section 8.5.1 but prior to membrane installation. Subsequently, the membranes are installed and flushed, and then another disinfection specific to the chlorine-intolerant membrane is conducted. This second disinfection typically utilizes a high pH (i.e., ~10 or greater) solution of caustic or other alkaline chemical that is very effective for bacteriological inactivation. The membrane manufacturer should be consulted to ensure that this second disinfection is conducted within the pH tolerance limits of the membrane. A second option, if permissible under State regulatory requirements, involves the elimination of chlorine or other oxidants and using caustic (or other alkaline chemical) to disinfect the entire system, including the membranes. The State should be consulted prior to performing this type of alternative disinfection procedure. If an alternate disinfection procedure is used, additional microbiological monitoring may be warranted to ensure the efficacy of the disinfection process.

8.6 Initial Direct Integrity Testing

Once the system is thoroughly flushed and disinfected, a direct integrity test should be conducted on each membrane unit. (Both general guidance for direct integrity testing and specific requirements for compliance with the LT2ESWTR are provided in Chapter 4.) If the direct integrity test utilized requires that the membrane be fully wetted, the system may need to be operated for a period of time prior to conducting the test. After the direct integrity test has been completed, any defective membrane modules or leaks in o-rings, pipe connections, or valve seals should be repaired or replaced, as necessary. Until the direct integrity test has been successfully completed, the filtrate should be discharged to waste or recycled, as appropriate.

For most membrane systems it is advisable to conduct relatively frequent direct integrity testing during the initial start-up phase, as there may be a higher incidence of integrity failure observed during the initial stages of facility operation. As a general rule, most membrane manufacturing related defects will manifest within 72 hours of continuous operation. However, the propensity to exhibit such defects varies significantly among different products. In order to ensure that all initial manufacturing defects are detected, it is recommended that direct integrity testing be conducted 2 to 6 times per day during start-up until the test results are stable.

8.7   Acceptance Testing

The purpose of acceptance testing is to demonstrate equipment performance as a condition for transferring responsibility over the membrane filtration equipment from the manufacturer or contractor to the utility. Acceptance testing is conducted after other phases of the start-up and commissioning process (i.e., flushing, disinfection, and integrity testing) have been completed. Typically, acceptance testing is the final phase of the commissioning process and consists of the following two criteria:

- **Operation** – The entire system, including all membrane units, is continuously operated for a pre-determined period of time, usually from 3 to 30 days. Any interruptions in continuous operation as a result of faulty equipment or controls during this period of time may result in restarting or temporarily suspending the test. Design parameters (e.g., flux and backwash frequency (if applicable)) are also typically verified as a component of the operational criterion.

- **Water Quality Testing** – Periodic sampling and analysis is conducted to ensure that the treated water quality objectives are continuously satisfied. The required sampling can vary widely in terms of the number of constituents of interest and is very site-specific. The sampling frequency for each parameter is typically once per day. If the system is unable to produce the required water quality, the manufacturer or contractor should be obligated to correct the problem before the acceptance testing is determined to be successfully completed.

Utilities that apply membrane filtration for compliance with the LT2ESWTR may use the acceptance testing period to establish the required control limits for direct integrity testing and continuous indirect integrity monitoring, as discussed in Chapters 4 and 5, respectively, subject to any particular State requirements.

8.8   Operator Training

Because operators may not be familiar with the various types of membrane filtration systems, operator training is an important program element to include in the initial start-up phase. Even if some experience was gained during pilot testing, this experience should serve as a supplement to training on the completed full-scale facility, rather than a substitute. This training can also help facilitate a smooth transition of responsibility from the equipment supplier to the utility. A sample operator training schedule is shown in Table 8.1.

Table 8.1 Schedule of Training Events

Time Period	Topic	Duration[1] (hours)
During Construction and Commissioning	System Overview / Review of Unit Operations	1 - 4
	Facility Walk Through	1 - 2
	Principles of Membrane Operation	1 - 4
	Control System	4 - 24
	Pretreatment Unit Operations	1 - 4
	Operation / Pumps / Instrumentation	4 - 16
	Post-Treatment	1 - 2
	Chemical Cleaning (Including Demonstration)	4 - 16
	Integrity Testing / Module Isolation and Repair	4 - 8
	Monitoring / Troubleshooting / Data Normalization	2 - 4
	Details of Control System / Remote Monitoring	1 - 2
	Total	*24 - 86*
After Commissioning	Open Discussion with Membrane Manufacturer	1 - 3 days
After Three Months	Open Discussion with Membrane Manufacturer	1 - 3 days

1 Higher ends of the ranges cited are associated with more complex systems and less operator familiarity

References

Abbaszadegan, M., M.N. Hansan, C.P. Gerba, P.F. Roessler, B.R. Wilson, R. Kuennen, and E. Van Dellen. 1997. The disinfection efficacy of a point-of-use water treatment system against bacterial, viral, and protozoan waterborne pathogens. *Water Research*. 31(3):574-582.

Adams, B.H. 1959. Methods of study of bacterial viruses. In *Bacteriophages*. pp. 443-522. New York: Interscience Publishers.

Adamson, A.W. 1990. *Physical chemistry of surfaces*. 5th Edition. New York, NY: John Wiley & Sons, Inc.

Adham, S.S., J.G. Jacangelo, and J-M. Laine. 1995. Low-pressure membranes: assessing integrity. *J. AWWA*, 87:3:62.

American Public Health Association, American Water Works Association, and Water Environment Federation. 1998. *Standard methods for the examination of water and wastewater*. 20th Edition. Baltimore, MD.

American Society for Testing and Materials. 2000. F 658-00a – Standard practice for calibration of a liquid-borne particle counter using an optical system based upon light extinction. West Conshohocken, PA

American Society for Testing and Materials. 1995. D 4189-95 – Standard testing method for silt density index (SDI) of water. West Conshohocken, PA

American Society for Testing and Materials. 1998. D 3923-94 – Standard practices for detecting leaks in reverse osmosis devices. West Conshohocken, PA

American Society for Testing and Materials. 1993. F 838-83 – Standard test method for determining bacterial retention of membrane filters utilized for liquid filtration. West Conshohocken, PA

American Water Works Association. 1999. *Reverse osmosis and nanofiltration* (M46). Denver, CO.

American Water Works Association Research Foundation, Lyonnaise des Eaux, and Water Research Commission of South Africa. 1996. *Water treatment membrane processes*. New York: McGraw-Hill.

Banerjee, A., F. Hanson, E. Paoli, C. Korbe, R. Kolman, D. Nelson, K. Smith, and M. Lambertson. 1999a. Ultra low range instrument increases turbidimetric sensitivity by two orders of magnitude. *AWWA Water Quality Technology Conference Proceedings, October 31 – November 4, 1999.* Tampa, FL.

Banerjee, A., J. Lozier, and K. Carlson. 2001. An on-line, multi-sensor, membrane filtration permeate water quality monitoring system. *AWWA Membrane Technology Conference Proceedings, March 4-7, 2001.* San Antonio, TX.

Banerjee, A., K. Carlson, and J. Lozier. 2000. Monitoring membrane integrity using ultra high sensitive laser light. *AWWA Water Quality Technology Conference Proceedings, November 5-9, 2000.* Salt Lake City, UT.

Banerjee, A., M. Lambertson, and K. Carlson. 1999b. Sub-micron particles in drinking water and their role in monitoring the performance of filtration processes. *AWWA Water Quality Technology Conference Proceedings, October 31 – November 4, 1999.* Tampa, FL.

Barbeau, B., L. Boulos, R. Desjardins, J. Coallier, M. Prévost, and D. Duchesne. 1997. A modified method for the enumeration of aerobic spore-forming bacteria. Canadian Journal of Microbiology. 43:976-980.

Childress, A. and M. Elimelech. 1996. Effect of solution chemistry on the surface charge of polymeric reverse osmosis and nanofiltration membranes. *J. Membrane Science* 119:253-268.

Chowdhury, Z., A. Van Gelder, D. Lawler, and M. Moran. 2000. *Particle count method development for concentration standards and sample stabilization.* Denver, CO: American Water Works Association Research Foundation.

Colvin, C., R. Brauer, N. DiNatale, and T. Scribner. 2001. Comparing laser turbidimetry with conventional methods for monitoring MF and UF membrane integrity. *AWWA Membrane Technology Conference Proceedings, March 4-7, 2001.* San Antonio, TX.

Côté, P., I. Sutherland, N. Adams, and J. Cadera. 2003. Validation of membrane integrity methods in a challenge test with *Bacillus subtillis. AWWA Membrane Technology Conference Proceedings, March 2-5, 2003.* Atlanta, GA.

Cornwell, D., M. Bishop, R. Gould, and C. Vandermeyden. 1987. *Water treatment plant waste management.* Denver, CO: American Water Works Association Research Foundation.

Crane Co. 1988. *Flow of fluids through valves, fittings, and pipe.* Technical Paper No. 410. Stamford, CT.

Dwyer, P.L., M.R. Collins, A.B. Margolin, and S.B. Hogan. 1995. Assessment of MS2 bacteriophage *Giardia* cysts *Cryptosporidium* oocysts and organic carbon removals by hollow fiber ultrafiltration. *AWWA Membrane Technology Conference Proceedings, August 13-16, 1995.* Reno, NV.

Farahbakhsh, K., and D. Smith. 2003. Estimating air diffusion contribution to pressure decay during membrane integrity tests. *AWWA Membrane Technology Conference Proceedings, March 2-5, 2003.* Atlanta, GA.

Glucina, K., J-M Laine, C. Anselme, M. Chamant, and P. Simonie. 1999. Acoustic sensor: a novel technique for low pressure membrane integrity monitoring. *AWWA Membrane Technology Conference Proceedings, February 28 – March 3, 1999.* Long Beach, CA.

Hargesheimer, E., C. Lewis, and N. McTigue. 2000. *Fundamentals of drinking water particle counting.* Denver, CO: American Water Works Association Research Foundation.

International Organization for Standardization. 1999. 11943 – Hydraulic fluid power – on-line automatic particle counting systems for liquids – methods of calibration and validation. Geneva, Switzerland.

Jacangelo, J., S. Adham, and J-M Laine. 1997. *Membrane filtration for microbial removal.* Denver, CO: American Water Works Association Research Foundation.

Johnson, W.T. 1997. Predicting log removal performance of membrane systems using in-situ integrity testing. *AWWA Annual Conference Proceedings, June 15-19, 1997.* Atlanta, GA.

Jucker, C. and M. Clark. 1994. Adsorption of aquatic humic substances on hydrophobic ultrafiltration membranes. *J. Membrane Science.* 97:253-268.

Kenna, E.N., and A.K. Zander. 2000. *Current management of membrane plant concentrate.* Denver, CO: American Water Works Association Research Foundation.

Kruithof, J.C., P. Hiemstra, P. Kamp, J. van der Hoek, and J.C. Schippers. 1999. Disinfection by integrated membrane systems for water treatment. *AWWA Membrane Technology Conference Proceedings, February 28 – March 3, 1999.* Long Beach, CA.

Landsness, L.B. 2001. Accepting MF/UF technology – making the final cut. *AWWA Membrane Technology Conference Proceedings, March 4-7, 2001.* San Antonio, TX.

Li, S.Y., J.A. Goodrich, J.H. Owens, G.E. Willeke, F.W. Schaefer III, and R.M. Clark. 1997. Reliability of non-hazardous surrogates for determining *Cryptosporidium* removal in bag filters. *Journal AWWA.* 89(5):90-99.

Lozier, J., C. Colvin, J-H. Kim, M. Kitis, B. Mariñas, and B. Mi. 2003. *Microbial removal and integrity monitoring of high-pressure membranes.* Denver, CO: American Water Works Association Research Foundation.

MacPhee, M. J., Y. LeGouellec, and D. Cornwell. 2002. *Treatment of MF residuals for contaminant removal prior to recycle.* Denver, CO: American Water Works Association Research Foundation.

Meltzer, T.H. 1997. *High-purity water preparation for the semiconductor, pharmaceutical, and power industries.* Littleton, CO: Tall Oaks Publishing, Inc.

Meltzer, T.H. 1987. *Filtration in the pharmaceutical industry.* New York, NY: Marcel Dekker, Inc.

Mickley, M.M. 2001. *Membrane concentrate disposal: practices and regulation.* Denver, CO: United States Bureau of Reclamation Technical Service Center, Water Treatment Engineering and Research.

Mickley, M.M. 2000. *Major ion toxicity in membrane concentrate.* Denver, CO: American Water Works Association Research Foundation.

Mickley, M.M., R. Hamilton, L. Gallegos, and J. Truesdall. 1993. *Membrane concentrate disposal.* Denver, CO: American Water Works Association Research Foundation.

NSF International. 2005. EPA/NSF ETV equipment verification testing plan for the removal of microbiological and particulate contaminants by membrane filtration. Ann Arbor, MI.

Schideman, L., M. White, L. Landsness, M. Kosterman, and L. Rago. 2004. Challenges, opportunities, and solutions in regulating the design of low-pressure membrane water treatment plants. *AWWA Annual Conference Proceedings, June 13-17, 2004.* Orlando, FL.

Schneider, O.D., E. Acs, S. Leggerio, and D. Nickols. 1999. The use of microfiltration for backwash water treatment. *AWWA Annual Conference Proceedings, June 20-24, 1999.* Chicago, IL.

Schulze, J.C. 2001. The Texas approach to regulating MF/UF drinking water applications. *AWWA Membrane Technology Conference Proceedings, March 4-7, 2001.* San Antonio, TX.

Sethi, S., G. Crozes, D. Hugaboom, B. Mi, J. Curl, and B. Mariñas. 2004. *Assessment and development of low-pressure membrane integrity monitoring tools.* Denver, CO: American Water Works Association Research Foundation.

Trussell, R., P. Gagliardo, S. Adham, and A. Olivieri. 1998. Membranes as an alternate to disinfection. *Microfiltration II Conference of the National Water Research Institute Proceedings, November 12-13, 1998.* San Diego, CA.

United States Bureau of Reclamation. 1998. *The desalting and water treatment membrane manual: a guide to membranes for municipal water treatment (2nd Ed.).* Technical Service Center, Water Treatment Engineering and Research, Denver, CO.

United States Department of Health and Human Services, Food and Drug Administration. 2002. Code of Federal Regulations Title 21 (Food and Drugs), Part 820 – Quality System Regulation (21 CFR 820).

USEPA. 2001. *Low-pressure membrane filtration for pathogen removal: application, implementation, and regulatory issues.* EPA 815-C-01-001.

USEPA. 1999. *Guidance manual for compliance with the Interim Enhanced Surface Water Treatment Rule: turbidity provisions.* EPA 815-R-99-010.

USEPA. 1993. *Methods for the determination of inorganic substances in environmental samples.* EPA 600-R-93-100.

USEPA. 1991. *Guidance manual for compliance with the filtration and disinfection requirements for public water systems using surface water sources.* EPA 68-01-6989.

Vickers, J. 1993. Aspects of integrity testing and module construction for microporous membrane filters. Technical Paper – Memtec America Corporation.

Willinghan, G.A., J.C. Vickers, and C. McGahey. 1993. Microfiltration as tertiary treatment – eighteen month trial. *AWWA Membrane Technology Conference Proceedings, August 1-4, 1993.* Baltimore, MD.

Appendix A

Development of a Comprehensive Integrity Verification Program

Appendix A:

Development of a Comprehensive Integrity Verification Program

A.1   Introduction

The ability to maintain system integrity is one of the most important operational concerns associated with any membrane filtration facility, whether applied for compliance with the Long Term 2 Enhanced Surface Water Treatment Rule (LT2ESWTR) or for any other treatment objective. Because a membrane represents a physical barrier to pathogens and other drinking water contaminants, the means to ensure that this barrier remains uncompromised is critical for the ongoing protection of public health. Moreover, the number and variety of integrity-related compliance requirements for membrane filtration under the LT2ESWTR, ranging from various forms of testing to repair to data collection and reporting, as specified in Chapters 4 and 5, illustrate the potential complexity of the process of verifying and maintaining system integrity. As a result, this appendix has been prepared to serve as a tool to guide utilities in the development of a comprehensive, effective, and efficient integrity verification program (IVP). The various sections of this appendix are organized into a series of introductory questions that would be posed in the preparation of an IVP and corresponding discussions to elaborate on how these questions would be addressed in the context of an IVP.

What is a comprehensive IVP?

A comprehensive IVP is a customized site- and system-specific program that details all operating procedures associated with maintaining membrane filtration system integrity, including both federal and State requirements and any additional practices that are voluntarily implemented at the discretion of the utility. In the broadest terms, an IVP should serve as a master plan for preserving system integrity.

What is the purpose of an IVP?

The primary purpose of an IVP is to provide a utility with rational and systematic blueprint for applying appropriate tools and techniques to efficiently conduct the following procedures:

1. Verifying integrity on an ongoing basis

2. Identifying and correcting any integrity problems

3. Recording and analyzing integrity test data

4. Preparing any required compliance reporting

The successful execution of these procedures, in turn, allows a utility to track system performance and determine whether or not it is consistent with either that established by challenge testing or other requirements that may be applicable.

Why is an IVP important?

Because the process of maintaining system integrity has many aspects, an IVP is critical for organizing all of the various operating procedures relating to system integrity into a comprehensive plan. As an organizational tool, an IVP is also important for its ability to provide a framework to assist operators with conducting the proper procedures in the correct sequence under both normal operating conditions and when an integrity breach is either suspected or confirmed. As a whole, these IVP functions help to both ensure the production of safe drinking water and facilitate regulatory compliance.

What are the regulatory requirements associated with an IVP?

Although an IVP is not required for membrane filtration systems under the LT2ESWTR, the development of IVP can facilitate regulatory compliance by organizing the rule requirements into a comprehensive operational program. However, whether membrane filtration is applied for LT2ESWTR compliance or to meet any other treatment objectives, any IVP should be consistent with all USEPA and State requirements governing the operation of membrane treatment facilities. In addition, any requirements relating to maintaining system integrity should be incorporated into an IVP. Thus, while the development of an IVP is not required under the LT2ESWTR, it is strongly recommended for all utilities that utilize membrane filtration, particularly for disinfection applications.

What are the components of an IVP?

An IVP should incorporate all of the operational procedures associated with maintaining system integrity, including as a minimum the following major program elements:

- Direct integrity testing

- Continuous indirect integrity monitoring

- Diagnostic testing

- Membrane repair and replacement

- Data collection and analysis

- Reporting

Other site-specific procedures or compliance requirements applicable to system integrity, but not specifically addressed under any of the above program elements, should also be included as components of an IVP.

How is IVP guidance presented in this appendix?

This appendix is organized into sections according to the major program elements described above. Each of these sections provides an overview of what the IVP should address with respect to that particular element and the various associated considerations that should be taken into account in the process of developing and implementing an IVP. Because compliance with the LT2ESWTR is critically related to maintaining system integrity, requirements from the rule are used as examples to illustrate program development, as well as how the various IVP components fit together into an integrated program.

The order of the sections in this appendix also illustrates the tiered approached to an IVP, ranging from the fundamental direct integrity test to successive levels of monitoring, testing, and repair that may be necessary. Note that this appendix is applicable to all types of membrane filtration addressed under the LT2ESWTR and consequently covered in this *Membrane Filtration Guidance Manual*, including microfiltration (MF), ultrafiltration (UF), nanofiltration (NF), reverse osmosis (RO), and membrane cartridge filtration (MCF). Significant technology-specific nuances associated with one or more particular types of membrane filtration are addressed in context whenever practical.

A.2   Direct Integrity Testing

Direct integrity testing is the primary means of verifying integrity in membrane filtration systems and thus represents a fundamental component of an IVP. The series of questions into which this section is organized represents important aspects of direct integrity testing that should be addressed in an IVP. In addition, the questions are presented in a logical progression designed to parallel the step-by-step process of formulating a direct integrity testing strategy.

What is the purpose of direct integrity testing?

Under the LT2ESWTR, a direct integrity test is defined as a physical test applied to a membrane unit in order to identify and isolate integrity breaches. Because direct integrity testing is the most accurate and precise means of determining whether or not a breach has occurred, it has historically been the primary means used to assess membrane integrity in potable water treatment applications in which pathogen removal is a principal concern. In addition, the test parameters and results can be correlated to the desired treatment objectives (e.g., log removal values) to yield a quantifiable assessment of system performance. In terms of LT2ESWTR compliance, as discussed in Chapter 4, requirements for direct integrity test *resolution* and *sensitivity* are specified for ensuring the necessary level of *Cryptosporidium* removal at a particular facility.

For example, the LT2ESWTR *resolution* requirement specifies that the direct integrity test parameters must be fixed such that a breach on the order of the smallest *Cryptosporidium* oocyst (i.e., 3 μm) is physically capable of contributing to a test response (40 CFR 141.719(b)(3)(ii). Thus, for pressure-based direct integrity tests, the applied pressure (or vacuum) must be great enough to overcome the capillary static forces that hold water in a breach (i.e., the bubble point) of 3 μm in diameter in a fully-wetted membrane, thereby allowing air to escape through a *Cryptosporidium*-sized hole and consequently enabling this loss of air to be potentially detected. If the applied pressure (or vacuum) were insufficient to overcome the bubble point, then the direct integrity test would be physically incapable of detecting any number of *Cryptosporidium*-sized breaches that would allow the passage of pathogens to the filtrate. Similarly, with marker-based tests, the marker used must be smaller than 3 μm to ensure that it would pass through a *Cryptosporidium*-sized hole, thus enabling a breach of this size to be potentially detected by the instrumentation that measures the concentration of the marker in the filtrate. Note that the 3 μm resolution requirement applies to all facilities utilizing membrane filtration for compliance with the LT2ESWTR.

However, unlike the resolution, the required *sensitivity* – the maximum log removal value (LRV) that can be reliably verified by the direct integrity test – may vary among different facilities under the LT2ESWTR, as the rule stipulates only that the test used have a sensitivity that exceeds the *Cryptosporidium* log removal credit awarded by the State (40 CFR 141.719(b)(3)(iii). In some cases, the sensitivity of the test may be determined from the threshold test result that signifies the smallest detectable integrity breach, information that can be provided by the membrane filtration system supplier. This result, which represents the test sensitivity, may then be converted into a LRV using the methodology described in Chapter 4 or an alternative methodology approved by the State.

Although the compliance framework developed for direct integrity testing under the LT2ESWTR applies only to those utilities that are subject to the *Cryptosporidium* removal requirements of the rule, the methodology could be applied for other pathogens either by mandate of the State or at the discretion of the utility. If this methodology is applied to more than one different pathogen simultaneously at the same treatment facility, then the direct integrity test used would be required to have a resolution corresponding to the smallest of the pathogens; in addition, the test would need to have a sensitivity greater than the removal credit awarded by the State for each pathogen to which the methodology was applied. Even if a compliance framework similar to that for the LT2ESWTR is not applicable to a particular utility's membrane filtration system, a direct integrity test is still the most reliable means of determining whether or not a breach has occurred.

For any system to which a compliance framework similar to that for the LT2ESWTR is applied, the resolution and sensitivity requirements for direct integrity testing should be established prior to placing the facility into service and specified in the IVP. If a different method is used to determine threshold integrity test results for applications other than for LT2ESWTR compliance (e.g., fiber-cutting studies), these critical values should also be incorporated into the IVP. As a minimum, the IVP should include the direct integrity test result threshold specified by the manufacturer as indicative of a potential integrity breach.

What type of direct integrity test should be used?

Currently, there are two general types of direct integrity tests that are commercially available for use with membrane filtration facilities: 1) pressure-based tests; and 2) marker-based tests. The various specific tests that fall under each of these two general categories are described in Chapter 4. The particular test utilized in conjunction with a given system may depend on regulatory requirements, the type of membrane filtration system, target organism (i.e., in terms of the required test resolution), test sensitivity, or utility preference.

The LT2ESWTR does not mandate the use of a particular direct integrity test for regulatory compliance; any test utilized must simply meet the criteria specified under the rule for resolution, sensitivity, and frequency. However, some States do stipulate the use of a specific test. In some cases, even if no particular test is specified, the required sensitivity relative to that of each potential direct integrity test method may govern which test(s) may be used. If no specific test is required by regulatory mandate, then any type of direct integrity test approved for use by the State may be used at the utility's discretion.

The direct integrity test used also depends to some extent on the type of membrane filtration system utilized, as some tests are not compatible with certain types of systems. For example, while a particulate marker test may be used with a MCF, MF, or UF system, a molecular marker test would not be a feasible means of assessing MCF, MF, or UF integrity, since the molecular marker may not be sufficiently removed by these membrane systems to demonstrate a reasonable LRV (e.g., 3 log). Conversely, a particulate marker test would probably not be used with NF or RO systems, since the particles could not easily be flushed from a spiral-wound membrane module and would potentially foul or otherwise damage the membranes. Thus, for NF/RO systems, a molecular marker would be a more appropriate marker-based direct integrity test.

Pressure (or vacuum) decay tests are compatible with all the various types of membrane filtration as defined under the LT2ESWTR (i.e., MF, UF, NF, RO, and MCF), and the equipment necessary to conduct this type of test is typically included with most of the currently available proprietary MF/UF systems. Similarly, some types of direct integrity tests may not be available from a proprietary membrane filtration system supplier. If several types of tests are available and the utility is not otherwise constrained by regulatory requirements, the selection of a direct integrity test should take into account any site- or system-specific considerations that would either favor or preclude certain tests.

The type of direct integrity test used for a particular system and the justifying rationale should both be included in the facility IVP, as well as specific procedures for conducting the test. If the test is automated (as is common), the procedures specified in the IVP should include how the test works in terms of the automatic sequencing that the system undergoes. In addition, the procedure for manually conducting the test whenever necessary should also be specified, as well as any other responsibilities that system operators may have with respect to direct integrity testing. One particular consideration for systems with automated direct integrity testing is whether or not an operator must be present during the test. Although the presence of an operator may be beneficial, particularly if an integrity breach is detected, the ability of an operator to respond quickly to an automated alarm and notification system may render this unnecessary. If the State does not require direct operator supervision during direct integrity testing, the utility should exercise its own discretion in determining whether the presence of an operator is critical based on site- and system-specific considerations and degree of comfort with unsupervised testing.

How frequently should direct integrity testing be conducted?

Because none of the various types of direct integrity tests currently available can feasibly assess integrity on a continuous basis while the system is on-line and producing filtered water, a membrane unit to be tested must be taken off-line and out of production during the testing process. Thus, more frequent direct integrity testing results in increased downtime for each membrane unit and consequently decreased system productivity. As a result, the frequency of direct integrity testing is an important parameter that hinges on striking an acceptable balance between the competing desires to maximize both confirmation of system integrity and treated water production.

A number of States have established minimum test frequency requirements for membrane filtration systems; currently, these requirements range from as often as every four hours to as relatively infrequently as once per week, depending on the State. If a utility applies membrane filtration for compliance with the LT2ESWTR, the rule requires that direct integrity testing be conducted on each membrane unit once per day at a minimum, unless the State approves less frequent testing based on demonstrated process reliability, the use of multiple barriers effective for *Cryptosporidium*, or reliable process safeguards (40 CFR 141.719(b)(3)(vi)). The State may also require testing on a more frequent basis at its discretion.

Even if subject to minimum test frequency requirements by federal or State regulations, a utility could opt to conduct direct integrity testing more frequently. In addition, a utility that is not otherwise constrained by any regulatory requirements should use its discretion to determine an appropriate direct integrity test frequency. In establishing a test frequency (regulatory requirements permitting), careful consideration should be given to the factors that may influence this decision. For example, testing less frequently may increase overall facility production and minimize the mechanical stress on the membrane module(s) that repeated testing might induce. However, increased testing provides more frequent assurance of system integrity, or in the case of an integrity breach, less operating time under compromised conditions that may allow the passage of pathogens or other undesirable pre-filtered water constituents. The most important concern should be the maximum length of time that a utility feels comfortable potentially operating with an integrity breach of any magnitude, which would dictate the minimum acceptable integrity test frequency. Although the use of continuous indirect integrity monitoring (see Chapter 5 and section A.3) provides some measure of integrity confirmation between direct test applications, currently available indirect integrity monitoring techniques may not be able to detect a breach of potentially significant magnitude. It is important to note that the frequency of direct integrity testing should be based on public health considerations rather than the observed or anticipated frequency of the occurrence of integrity breaches.

When should direct integrity testing be conducted?

In the process of developing an IVP, some consideration should be given to the timing of conducting direct integrity testing relative to the normal operating cycle(s) of the system. For example, with systems that utilize some type of regular reverse flow process to remove foulants from the membrane surface (e.g., backwashing), the membrane will be the least obstructed immediately after this reverse flow process is complete. Implementing a direct integrity test at this point in the operational cycle would provide the most conservative estimate of system integrity, since the accumulated foulants that could plug any breaches, and thus effectively mask potential integrity problems, would be minimized. However, using this same example, for systems that employ a pressure-based direct integrity test, it is important that the test not be conducted too soon after the reverse flow processes (particularly if air is used), since the membrane must be fully-wetted for the integrity test to be effective.

It is also recommended that a direct integrity test be conducted after chemical cleaning or any other routine or emergency maintenance in order to ensure that the system integrity has not been compromised by the procedure(s). The affected membrane module(s) or unit(s) should be returned to service only after system integrity has been confirmed by a direct integrity test. In cases in which a membrane unit has been taken out of service for diagnostic testing and/or repair, systems applying membrane filtration for LT2ESWTR compliance are required to conduct direct integrity testing on the affected unit to demonstrate system integrity prior to returning the unit to service (40 CFR 141.719(b)(3)(v)).

Any special circumstances that might call for direct integrity testing (i.e., in addition to the regularly scheduled periodic testing), such as subsequent to chemical cleaning or membrane repair, whether in accordance with federal or State requirements or at the discretion of the utility, should be specified in the IVP.

How should the direct integrity test results be interpreted?

For a given resolution, a marker- or pressure-based direct integrity test indicates whether or not an integrity breach has occurred by comparing the results of the test to the threshold result known to represent a breach. This threshold represents the test sensitivity and can be determined either by information provided by the manufacturer or through an on-site, system-specific assessment such as a fiber-cutting study.

Under the compliance framework of the LT2ESWTR, the ongoing test results obtained during facility operation for each membrane unit (as well as the threshold result representing the test sensitivity) can be converted into LRVs using the methodology described in Chapter 4 or other method approved by the State. Each successive test result (or LRV, as converted) is then compared to the log removal credit allocated to the system by the State (either as per the requirements of the LT2ESWTR or for other treatment objectives) to determine compliance on an ongoing basis. If the LRV yielded by the direct integrity test is greater than the regulatory allocation, the system remains in compliance. However, if the LRV is below the required log removal, the membrane unit must be taken off-line for diagnostic testing and repair (see sections A.4 and A.5, respectively).

Thus, under the LT2ESWTR compliance framework, the direct integrity test results yield two important pieces of information: 1) whether or not an integrity breach has occurred; and 2) the maximum LRV that can be verified at the time of the test (when converted using the technique(s) described in Chapter 4 or alternate methodology approved by the State). Because the test sensitivity may often be greater (i.e., higher LRV) than the required log removal for any particular pathogen, it is possible for the direct integrity test to indicate that an integrity breach of some magnitude has occurred without the system being out of compliance. Thus, under the LT2ESWTR framework, a system could knowingly continue to operate with some level of integrity breach and still meet regulatory requirements. However, USEPA recommends (and the State may require) that a membrane unit with a detectable integrity breach of any magnitude be immediately taken out of service for diagnostic testing and repair. At a minimum, utilities should conduct diagnostic testing and potential repair when the compromised unit is taken off-line for chemical cleaning, previously scheduled maintenance, or other routine purpose.

The LT2ESWTR defines a control limit as an integrity test response, which, if exceeded, indicates a potential integrity problem and triggers subsequent action. For the purposes of LT2ESWTR compliance, a direct integrity test control limit would be established as the test result that translates into an LRV equal to the *Cryptosporidium* removal credit awarded by the State. As noted previously, any membrane unit for which the test results exceed the control limit (i.e., the LRV drops below the awarded removal credit) must be immediately taken off-line for diagnostic testing and repair (40 CFR 141.719(b)(3)(v)). Other control limits could also be established if the LT2ESWTR compliance framework is simultaneously applied for the removal of other pathogens with membrane filtration. However, because the system would be out of compliance if either of the two (or more) control limits were exceeded, the most stringent control limit (i.e., the lowest permissible integrity test response) would always represent the governing limit.

In addition to the critical or "upper" control limit (UCL) that governs regulatory compliance for a membrane filtration facility under the LT2ESWTR framework, "lower" control limits (LCLs) may be established as performance benchmarks either at the discretion of the utility or by mandate of the State. One or more LCLs may be identified between the integrity test result that indicates the smallest detectable breach and a breach that reduces the performance of the system such that it is just capable of meeting the require log removal (i.e., the UCL). LCLs may be useful for alerting system operators to the presence of an integrity breach, even if the detected breach is not sufficient to bring the system out of compliance. Rather than triggering unit shutdown and subsequent diagnostic testing, as with an UCL, exceeding a LCL might trigger increased operator attention or an investigation that can be conducted by operators while the system is still on-line in an attempt to determine the cause of the integrity problem and prevent the breach from expanding, if possible. Alternatively, in cases in which there is a particularly large gap between the test sensitivity and the UCL a utility may choose to voluntarily take a membrane unit off-line for diagnostic testing if a preferred LCL is exceeded in order to minimize the risk to public health via the potential for pathogens to pass through a barrier known to be compromised, even if only to a small degree that is within regulatory tolerances.

As an example, even with the smallest detectable breach the direct integrity test might still be able to verify 6 log removal (i.e., the test sensitivity), although the State only awards 2.5 log *Cryptosporidium* removal credit (i.e., the UCL) for the membrane filtration process. Under this scenario, a direct integrity test result yielding a LRV of 4 would indicate the presence of a significant integrity breach, even though the membrane filtration system would still be capable of achieving the required log removal. Thus, a utility might opt to establish a LCL at a LRV of 4 if an integrity breach of this magnitude represents an unacceptable public health risk independent of the ability of the system to maintain regulatory compliance.

Fiber-cutting studies that allow integrity test results to be quantified and correlated to certain number of fiber breaks or particular reductions in log removal capacity may be used to determine a single appropriate LCL or a series of tiered LCLs. Setting a LCL equal to the test sensitivity (i.e., the maximum log removal value that can be reliably verified by the direct integrity test) would represent the most conservative scenario, as in this case any detectable integrity breach would, at a minimum, alert an operator and potentially trigger some responsive action. All control limits and the action triggered by exceeding each limit should be clearly specified in the IVP.

Another issue that should be addressed in the IVP is the possibility of false positive and false negative direct integrity test results For example, if a false positive result (i.e., a result incorrectly indicating a breach in a fully-integral system) is suspected, the operator should check isolation valves and fittings on the system that are associated with the direct integrity test. In addition, a follow-up direct integrity test should be conducted both to confirm the results of the first test and to closely monitor the test for any system malfunctions. In any case, LT2ESWTR compliance requires that any membrane unit for which a direct integrity test result exceeds an UCL be taken off-line for diagnostic testing and repair, independent of whether or not a false positive result is suspected (40 CFR 141.719(b)(3)(v)).

False negative results (i.e., results that indicate either a fully integral system in the presence of an integrity breach or which significantly underestimate a substantial integrity breach) of direct integrity tests may be more difficult to identify; since the methods of continuous indirect integrity monitoring are less sensitive to integrity breaches, indirect monitoring data may not be capable of detecting a breach that is masked by a false negative direct test result. However, if the continuous indirect integrity monitoring data do suggest an integrity breach in contradiction of the direct test results, the membrane unit should be taken out of service to investigate the source of the discrepancy. One potential scenario that might result in a false negative result is an integrity breach that occurs in a membrane that is partially fouled. The accumulated foulants may obscure the breach, thus masking the integrity problem until after the next backwash or chemical cleaning. Nevertheless, in this case the false negative result may not represent a significant concern; if the direct integrity test is functioning properly, the false negative result would suggest that the system is functionally integral even with an integrity problem as a result of the breach being plugged. In this case, the breach would likely be detected after the next backwash or chemical cleaning that successfully removes the foulants from blocking the breach. A more problematic scenario involving false negative results might occur if the integrity test equipment is malfunctioning. Thus, is it important to incorporate a routine maintenance program for the integrity test system as part of the IVP.

Any particular strategies for minimizing the potential for false positive and false negative results should be specified in IVP documentation. A maintenance schedule for direct integrity monitoring equipment should also be specified. It is recommended that the direct integrity monitoring system receive a thorough check-up on at least an annual basis.

A.3   Continuous Indirect Integrity Monitoring

Continuous indirect integrity monitoring is a secondary means of verifying membrane filtration system integrity intended to detect significant breaches between direct test applications. Thus, in the absence of a continuously applied direct test, continuous indirect integrity monitoring is critical to an IVP. As with section A.2, this section is organized into a series of questions that parallel the step-by-step process of formulating strategy for continuous indirect integrity monitoring. Each of these questions represents an important aspect of that strategy that should be included to some extent in an IVP.

What is the purpose of indirect integrity monitoring?

For the purposes of LT2ESWTR compliance, continuous indirect integrity monitoring is defined as monitoring some aspect of filtrate water quality that is indicative of the removal of particulate matter at a frequency of at least once every 15 minutes. Although the various indirect monitoring methods are less sensitive techniques for assessing membrane integrity than the direct integrity tests, the value of utilizing the indirect methods is that they can be applied to assess integrity continuously while the system is on-line and producing water. In fact, since by definition indirect monitoring is applied to the filtrate, these techniques require that the membrane unit be in continuous production to assess membrane integrity.

Because currently available methods of direct integrity testing cannot be applied continuously, a successful direct test only indicates that no breach has occurred since the previous application of the test. Consequently, if the system were to become compromised immediately after a successful direct integrity test had been conducted, the breach might not be detected until the next regularly scheduled direct test, which is typically as long as one day for the purposes of LT2ESWTR compliance. During this interval, a potentially significant breach could have occurred, allowing pathogens or other particulate matter to contaminate the filtrate for a period as long as an entire day. For applications other than LT2ESWTR compliance, this interval may be as short as four hours or as long as one week or more, depending on State-specific regulatory policy. Thus, although continuous indirect integrity monitoring may not be able to detect small compromises in integrity, these techniques do provide the ability to identify larger breaches on an ongoing basis during production. As a result, periodic direct integrity testing and continuous indirect integrity monitoring are complementary tools for assessing system integrity, and both are critical for a comprehensive IVP.

Under the LT2ESWTR, continuous indirect integrity monitoring is required in the absence of a direct integrity test that can be applied continuously and which meets the resolution and sensitivity requirements of the rule (40 CFR 141.719(b)(4)). States may also have regulations requiring some form of continuous indirect integrity monitoring. In the absence of any applicable requirements, a utility may opt to employ some form of continuous indirect integrity monitoring at its discretion; however, note that turbidity monitoring, which is required under the various federal surface water treatment regulations as a measure of overall system performance, may also serve the dual purpose of a means of continuous indirect integrity monitoring.

Unlike direct testing, there are no specific resolution or sensitivity requirements for continuous indirect integrity monitoring under the LT2ESWTR. However, these concepts, where applicable, may be useful tools for optimizing the ability of the various continuous indirect integrity monitoring methods to yield meaningful information about potential integrity breaches, as described in Chapter 5.

What type of indirect integrity monitoring method should be used?

There are a number of different methods and associated devices that may be used for continuous indirect integrity monitoring, including particle counting, particle monitoring, turbidimetry, laser turbidimetry, and conductivity monitoring, among others. In general, any method that measures particulate matter in the filtrate as an indirect means of assessing integrity (such as particle counting, turbidimetry, etc.) is applicable to any of the various types of membrane filtration systems. Other methods that may measure dissolved constituents in the filtrate, such as conductivity monitoring, would only be applicable to NF or RO systems. The particular method of continuous indirect integrity monitoring employed by a utility for its system may be a function of regulatory requirements, test resolution or sensitivity, cost, confidence in the technology, or simply preference based on prior experience or other subjective criteria.

The LT2ESWTR requires the use of turbidity monitoring on each membrane unit as the default method of continuous indirect integrity monitoring for compliance, unless an alternate method is approved by the State (40 CFR 141.719(b)(4)(i)). Because the various federal surface water treatment regulations (i.e., the Surface Water Treatment Rule (SWTR), the Long-Term 1 Enhanced Surface Water Treatment Rule (LT1ESWTR), and the Interim Enhanced Surface Water Treatment Rule (IESWTR)) require turbidity monitoring as a means of assessing filtration performance, surface water facilities implementing membrane filtration for LT2ESWTR compliance may use turbidity monitoring to satisfy both requirements. States may have other specific requirements for continuous indirect integrity monitoring independent of the LT2ESWTR or may approve other methods for LT2ESWTR compliance. All federal and State requirements must be satisfied in a utility's IVP.

If not otherwise constrained by regulatory requirements, a utility's decision to use a specific type of indirect integrity monitoring technique may be influenced by the ability of a particular method to provide sufficient resolution or sensitivity. For example, because particle counters have been demonstrated to be more sensitive to breaches than particle monitors or conventional turbidimeters, a utility may choose to use particle counting to maximize the ability to detect compromises in system integrity between periodic direct integrity test events (Jacangelo et al. 1997). Other utilities may select laser turbidimetry, which has been shown in some studies to perform comparably to particle counting in terms of sensitivity to integrity breaches (Banerjee et al. 2000; Colvin et al. 2001). In this case, if laser turbidimetry is approved by the State for both purposes, a utility may use laser turbidimeters for compliance with the applicable surface water treatment rules with the additional benefit of improved sensitivity for detecting integrity breaches over conventional turbidimeters.

Sensitivity may also be improved for any given continuous indirect integrity monitoring method by applying instrumentation to smaller groupings of membrane modules, such that any integrity breach would have a greater impact on filtrate quality. In this case, the benefits of the increased sensitivity should be weighed against the cost of additional instruments. This approach may be advantageous if the gain in sensitivity by using a greater number of less expensive instruments is justified by the comparable cost of fewer more expensive instruments. Note that for LT2ESWTR compliance, an instrument for continuous indirect integrity monitoring must be applied to each membrane unit (40 CFR 141.719(b)(4)). A utility may apply instruments to smaller groupings of membrane modules at its discretion. For other applications, State requirements for monitoring various groupings of membrane modules in an overall system must be satisfied in a utility's IVP.

If test resolution is an important criterion, than a utility might strongly consider particle counting. Because particle counting is the only method that assesses the size of particulate matter, it is the only method of indirect integrity monitoring to which the concept of resolution applies. For example, if membranes are applied specifically to remove *Cryptosporidium*, the particle counters should be well-calibrated to accurately detect particles approximately 3 μm in size or larger. The target resolution may vary depending on the particular contaminant of concern.

What constitutes "continuous" indirect integrity monitoring?

Under the LT2ESWTR, "continuous" monitoring is defined as monitoring conducted at a frequency of no less than once every 15 minutes. However, because the instrumentation used for the various methods of indirect integrity monitoring may allow data collection at much more frequent intervals, the State may have more stringent data collection requirements. In the absence of specific regulatory requirements, a utility may collect data at a frequency it determines to be appropriate. Nevertheless, since data acquisition can be automated, it is recommended that data be collected at interval no less than every 15 minutes, even if no other regulatory requirements apply.

While more frequent data collection both provides increased integrity monitoring and additional data to track system performance, there may be some potential complications from collecting data too frequently. For example, data could be collected frequently enough such that a backwash (for applicable systems) cannot be completed between readings. In this case, a utility must program the system to cease data collection during the backwash cycle and for any length of time afterwards that the data remain artificially high, such that a direct integrity test is not triggered. Data collection and analysis are further discussed in section A.6. The implications of collecting indirect integrity monitoring data at different intervals should be considered in developing an IVP, and the frequency and any other associated qualifying guidelines or restrictions should be specified in the facility IVP documentation.

How should indirect integrity monitoring results be interpreted?

Continuous indirect integrity monitoring is primarily intended to provide some indication of system integrity between direct integrity test applications. The indirect monitoring results are continuously compared to an established performance threshold that is known to represent a potential integrity breach. If this threshold level is exceeded, some type of response is triggered to further investigate the problem.

Under the LT2ESWTR, this performance threshold represents the UCL for continuous indirect integrity monitoring. If the UCL is exceeded, direct integrity testing is automatically triggered as a means to assess system integrity using a more sensitive technique. Unlike that for direct testing, the UCL for continuous indirect integrity monitoring does not necessarily correspond to a specific and quantifiable integrity loss. For example, the UCL established by the LT2ESWTR for turbidity monitoring (i.e., the default method in the absence of another technique specified by the State) is 0.15 NTU independent of site- or system-specific considerations; filtrate turbidity readings exceeding 0.15 NTU for a period of 15 minutes (or two consecutive 15-minute readings exceeding 0.15 NTU) would immediately trigger direct testing (40 CFR 141.719(b)(4)(iv). The 0.15 NTU threshold was selected because it is significantly below the 0.3 NTU threshold for filter performance required by the IESWTR for 95 percent of all turbidity samples, yet because membrane filtration systems are well-documented to consistently produce filtered water below 0.05 NTU, a sustained filtrate turbidity exceeding 0.15 NTU strongly suggests a potential integrity problem. Note that if turbidity monitoring with the default UCL of 0.15 NTU is used as a means of continuous indirect integrity monitoring, a utility could simultaneously be in compliance with the IESWTR but not with the LT2ESWTR.

Although the LT2ESWTR specifies a UCL of 0.15 NTU with the default method of turbidity monitoring, the State may establish a more stringent standard at its discretion. In addition, for any approved method of continuous indirect integrity monitoring, the State may require a site- or system-specific performance-based UCL that is linked to a certain level of integrity loss (in terms of a specific number of broken fibers or LRV) as determined by a fiber cutting study. These studies may also serve as the basis for establishing LCLs that trigger a particular response at a threshold prior to the point at which direct integrity testing would be required, either mandated by the State or voluntarily implemented by the utility. Voluntary LCLs could also be implemented after the membrane filtration system has been in operation for a certain period of time; this staggered implementation would allow sufficient baseline data to be collected such that operators could identify threshold levels that represent elevated or otherwise unusual results that are still below the UCL and which do not necessarily indicate an integrity breach, but nevertheless warrant observation. Some examples of actions associated with LCLs might be increased operator attention or diagnostic checks of the continuous indirect integrity monitoring instrumentation. All control limits (CLs) both mandated by the State and voluntarily implemented by the utility should be documented in an IVP, along with the rationale supporting the establishment of the CLs, as well as the respective subsequent action associated with the exceedance of each.

As with direct integrity tests, false negative and false positive results are also possible with indirect integrity monitoring. For example, some indirect integrity monitoring instruments may indicate elevated levels of a parameter (e.g., turbidity, particle counts, etc.) after a routine maintenance event such as a backwash, particularly if air is employed in the process to scour or pulse the membrane surface. If significant, this air entrainment error could cause a CL to be exceeded, generating a false positive result (i.e., a result incorrectly suggesting a breach in a fully integral system); in the case of the UCL, this exceedance would inappropriately trigger direct integrity testing (and, under the LT2ESWTR, consequent reporting). This type of false positive result can be minimized by first characterizing typical system performance under a variety of operating conditions (such as after a backwash) and subsequently programming the data acquisition system to account for regularly occurring data aberrations of previously quantified magnitude and duration which are known not to represent an integrity problem, even if CLs are exceeded. (The importance of collecting and analyzing baseline data for IVP optimization is further discussed in section A.6.) In some cases, devices such as bubble traps (e.g., in the case of air entrainment) may be utilized to minimize modes of error that might generate false positive results.

False negative results (i.e., results that indicate either a fully integral system in the presence of an integrity breach or which significantly underestimate a substantial integrity breach) may be more common with indirect integrity monitoring methods. Because currently available indirect integrity monitoring techniques are less sensitive to breaches than direct integrity tests, it is possible that small breaches may occur that could be detectable via direct but not indirect methods. This potential may be minimized by utilizing a more sensitive method of continuous indirect integrity monitoring (such as the use of laser vs. conventional turbidimeters), if a utility is permitted such flexibility under State regulations. Alternatively, a utility could increase indirect method sensitivity by utilizing a greater number of instruments (i.e., decreasing the number of membrane modules monitored per instrument). However, a utility considering these options should evaluate whether the cost of increasing indirect method sensitivity (i.e., via purchasing a greater number of instruments or more sensitive instruments or both) is justified by the consequent level of heightened integrity monitoring ability between direct test events.

Any particular strategies for minimizing the potential for false positive and false negative results should be specified in IVP documentation. A calibration schedule for continuous indirect integrity monitoring instrumentation should also be specified. It is recommended that this instrumentation be calibrated on at least an annual basis.

A.4   Diagnostic Testing

Diagnostic testing is a process of identifying and isolating integrity breaches that have already been detected and confirmed using other methods, and thus a critical component of an IVP. As with previous sections in this appendix, this section is organized into a series of questions that parallel the step-by-step process of formulating a diagnostic testing strategy for an IVP. Each of the questions presented in this section represents some aspect of diagnostic testing that should be clearly addressed and documented in an IVP.

What is the purpose of diagnostic testing?

The purpose of diagnostic testing is to identify and isolate integrity breaches in a membrane module that are detected via other methods. Because direct integrity testing and continuous indirect integrity monitoring techniques are only intended to detect the existence of a breach, diagnostic testing complements these methods, serving as a tool to pinpoint the exact location of a breach. In this way, diagnostic testing also serves as a critical link between identifying an integrity problem during the course of operation and repairing the breach. (Membrane repair and replacement is further discussed in section A.5). Thus, if an integrity breach is known or suspected in a membrane unit, that unit should be taken out of service for diagnostic testing in order to facilitate repairs. IVP documentation should clearly note the purpose of diagnostic testing so as to distinguish it from other forms of testing.

Under what circumstances should diagnostic testing be applied?

The LT2ESWTR requires that a membrane unit be taken out of service if the established direct integrity test UCL (i.e., that associated with the log removal credit awarded to the membrane process) is exceeded (40 CFR 141.719(b)(3)(v)). Under these conditions, diagnostic testing may be used to identify the location of an integrity breach. The use of diagnostic testing under these circumstances is also suggested for those utilities that do not use their respective membrane filtration systems for LT2ESWTR compliance. Thus, in essence diagnostic testing is recommended any time an integrity breach is detected from the results of a direct integrity test.

Note that the LT2ESWTR does not link diagnostic testing directly to continuous indirect integrity monitoring. Even if indirect monitoring results clearly indicate an integrity problem, it is advisable to confirm the existence of a breach using more sensitive direct integrity testing methods. In any case, under the compliance framework for the LT2ESWTR, any continuous indirect integrity monitoring results that would clearly indicate an integrity breach would almost certainly exceed the indirect integrity monitoring UCL and thus trigger direct integrity testing, as required. However, a utility that is not otherwise constrained by regulatory requirements could voluntarily take a membrane unit out of service for diagnostic testing based on indirect monitoring results alone.

An IVP should clearly identify the specific circumstances under which diagnostic testing should be applied, including both those conditions that require diagnostic testing under a regulatory framework and those that might trigger the use of diagnostic testing on a voluntary basis by the utility.

What type(s) of diagnostic testing should be used?

By definition, most diagnostic tests are categorized as types of direct integrity tests; however, diagnostic tests are distinguished from other types of direct tests by their ability to not just detect an integrity breach in a membrane unit, but to help locate the specific module or fiber containing the breach, as well. In addition, most methods considered to be diagnostic tests are designed to be applied to specific membrane units on an as-needed basis, and thus are impractical for implementation on a scale that would satisfy the direct integrity testing requirements of the LT2ESWTR. For example, sonic testing – one type of diagnostic test – requires an operator to manually apply an accelerometer to various locations on the membrane module to listen for vibrations caused by leaking air. While this technique fits the definition of direct integrity test, it would be infeasible to use such a test to check every module in a membrane filtration system for integrity breaches every day.

In addition to the sonic test described above, other methods of diagnostic testing include bubble testing, conductivity profiling, and simple visual inspection (where applicable). Each of these methods is described in further detail in section 4.8. A pressure (or vacuum) decay test applied to a smaller number of modules (i.e., a subset of a full membrane unit) in order to identify a particular breached module may also be used as a form of diagnostic test. Single module testing represents the smallest form of incremental membrane unit testing. This type of diagnostic test generally involves removing the individual modules from the membrane unit and testing each on a specially designed single module apparatus. Other types of direct integrity tests may also have the potential to serve as diagnostic tests in a scaled-down form.

In some cases, a battery of diagnostic tests may be used to identify an integrity breach. For example, if a MF membrane unit fails a direct integrity test (i.e., the results exceed the associated UCL), after the unit is taken out of service a sonic test might be applied to each module in the unit in turn to identify the affected module(s). (Note that although the membrane unit is taken out of service, it must still remain in operation in order to facilitate some types of diagnostic tests, such as the sonic test referenced in this example and described in section 4.8.3. Therefore, in such cases, the unit must be operating in filter-to-waste mode.) The module(s) may then be removed from the unit to conduct a bubble test (see section 4.8.2) in order to isolate particular fibers that may be subsequently removed from service permanently by pinning or sealing (as described in section A.5). Thus, just as diagnostic testing complements direct integrity testing or continuous indirect integrity monitoring, different diagnostic tests can also complement each other.

A utility should develop its own system-specific protocol for diagnostic testing and document the procedures in its IVP. The IVP should specify which diagnostic tests are prescribed, the particular purpose of each test (e.g., to isolate a module or identify a particular fiber), the circumstances under which each test should be conducted, a list of necessary testing equipment, and detailed instructions for conducting the test. Some membrane filtration system manufacturers may provide guidance in developing an appropriate diagnostic testing plan. Although diagnostic testing may not be commonly employed, a utility should still keep all the equipment necessary to conduct each diagnostic test specified in its system IVP both on site and in proper working order. Note that some diagnostic tests (such as sonic testing) require a greater degree of skill in both conducting the test and interpreting the results; operators that will have responsibility for conducting these tests should be designated and trained in advance in order to minimize membrane unit downtime.

A.5 Membrane Repair and Replacement

In the context of this guidance, "membrane repair and replacement" does not necessarily apply to just the membranes themselves, but to any component of a membrane filtration system that might allow an integrity breach if it were to fail. This section is organized into a series of questions presented in a logical order intended to parallel the step-by-step process of considering membrane repair and replacement in the context of an IVP. Each of the questions presented in this section should be addressed in a utility's IVP to an appropriate degree.

What is the purpose of membrane repair / replacement?

The purpose of conducting repairs on a membrane filtration system or replacing irreparably damaged components is to correct any integrity breaches that have been detected, thus restoring and maintaining a fully-integral system. In cases in which membrane filtration is applied for the removal of one or more specific pathogens of interest (e.g., for compliance with the LT2ESWTR), a more specific objective is to enable the system to maintain the full removal credit allocated by the State. As previously noted, although other system components may be essential for overall system operation, in this context repair and replacement are discussed in regard to only those system components that are critical to system integrity.

When should membrane repair / replacement be conducted?

In the simplest terms, repair or replacement should be conducted whenever an integrity breach is detected (e.g., using direct integrity testing or continuous indirect integrity monitoring). After the source of the breach has been isolated (e.g., via diagnostic testing), the problem should be corrected via component repair or replacement, as appropriate.

Under the LT2ESWTR, if the results of a direct integrity test exceed the upper (i.e., mandated) control limit, the affected membrane unit(s) must be immediately taken out of service (40 CFR 141.719(b)(3)(v)). Because the membrane unit(s) cannot be returned to service until a direct integrity test confirms that the UCL is no longer exceeded, some type of repair must be conducted to correct the integrity problem. If a utility has voluntarily implemented one or more tiered LCLs, it may in some cases be able to detect an integrity problem without exceeding the UCL. In this case the utility could opt to take corrective action, conducting diagnostic testing and subsequent repair immediately, or instead keep the affected membrane unit(s) in continued operation under increased observation until the next scheduled maintenance event. Although repair of any breach is recommended as soon as possible, a utility may exercise discretion in determining whether or not to implement any repairs based on severity of the breach (assuming the UCL has not been exceeded). For example, if a utility is required to achieve only 2 log *Cryptosporidium* removal credit for LT2ESWTR compliance, it is possible that a significant integrity breach could occur without jeopardizing the ability of the membrane filtration system to obtain this credit. However, continued operation with a known integrity breach of any magnitude may not be permitted by a State. Even if operation under some compromised conditions is not explicitly prohibited, a utility should carefully consider the risk associated with the potential for pathogen passage before engaging in continued operation.

If a utility does not apply its membrane filtration system for compliance with the LT2ESWTR, it may have more flexibility with respect to the timing and necessity of conducting immediate repairs if an integrity breach is detected, unless otherwise constrained by State requirements. In the absence of any regulations prescribing system repair requirements, it is recommended that utilities adopt a conservative approach in order to help ensure the integrity of the membrane barrier against pathogens. The system IVP should clearly specify any regulatory requirements relating to membrane system repair, as well as other circumstances under which the utility would conduct repairs.

Membrane module repair – as opposed to replacement – is often advisable, if possible, since new membranes are typically expensive. However, if a membrane module is subject to repeated integrity breaches, a utility should consider replacing the module. Chronic repairs adversely affect treated water production and prevent operators from carrying out routine responsibilities. Also, if it is critical that an integrity breach be repaired as quickly as possible, for some types of membrane filtration systems it may be more efficient to replace the module with a new one from the utility's supply of spare components maintained on-site. Thus, in some cases the most expedient and cost-effective system "repair" could actually be membrane replacement.

Although the LT2ESWTR requires the use of direct integrity testing to confirm the success of any system repairs concerning integrity problems before returning the affected membrane unit(s) to service, this practice is recommended for all utilities, even if membrane filtration is not conducted for LT2ESWTR compliance (40 CFR 141.719(b)(3)(v)). This post-repair direct integrity testing not only confirms the success of the repair process, but also whether any module(s) that may have been removed for repair are properly re-installed. After any repair or replacement measures have been completed, both direct integrity test and continuous indirect integrity monitoring results for the repair unit(s) should be closely tracked for an extended period to gauge the long-term success of the repair and perhaps whether the problem would be likely to recur.

What are some common modes of integrity breaches?

The types of integrity breaches to which a particular membrane filtration system is most susceptible can depend on both the type of system (i.e., MCF, MF/UF, or NF/RO) and the manufacturer. For example, while many NF/RO membranes are subject to chemical degradation by oxidants such as chlorine, only some MF/UF membranes are vulnerable to oxidation, depending on the membrane material used. This same example of chemical oxidation illustrates potential causes of integrity breaches that are specific to a particular treatment application of membrane filtration, as well. A treatment process that utilizes chlorine as a disinfectant upstream of RO membranes has an inherent potential source of chemical degradation, even though dechlorination may be implemented prior to the membranes. If the dechlorination system fails or is miscalibrated, the membranes could be subject to chlorine exposure. By contrast, an identical RO system used for a different application in which upstream disinfection is not required to meet treatment objectives would not have this additional element of risk. It is important to note that this example should not be interpreted as a recommendation against utilizing oxidants upstream of oxidant intolerant membranes. It is not uncommon to use upstream oxidants effectively, particularly for disinfection and biofouling control. However, a utility should be aware of the potential for chemical degradation of membrane material and take appropriate measures to prevent integrity breaches and protect treatment equipment.

In addition to chemical degradation, the most common causes of integrity breaches with NF and RO membranes are associated with o-rings and seals, which can be cracked, rolled, and/or improperly sized. Each of these defects can result in an integrity breach, as can foreign matter such as hairs or other fibrous material underneath o-rings. Other mechanisms for integrity breaches may be related to membrane defects, such as failures that may occur along glue lines or at weak spots in the membrane (i.e., creases or thin areas). Fiber breaks and potting problems are the most common types of integrity breaches associated with MF/UF membranes, although chemical degradation may also result in membrane failure. MCF systems are most likely to be compromised by improperly seated/sealed membrane cartridges or tears or punctures in the membrane material. MCF membranes may also exhibit integrity breaches along folds and creases, depending on the construction of the filter.

Although it is possible for breaches to occur at any time during operation, it is most common for integrity problems to occur during system start-up. These breaches typically result from either manufacturing defects or improper installation. As a result, it is important that an initial shakedown period be included as a part of the start-up process to enable these initial problems to be identified and corrected prior to putting the system into service.

An IVP should identify and document the most probable types of integrity breaches for a utility's specific membrane filtration system. These modes may be identified initially by consultation with the membrane manufacturer and consideration of site-specific circumstances (e.g., the use of upstream oxidants). Subsequently, the shakedown period and ongoing operational experience may yield important information about the most frequently occurring types of integrity problems. Also, the experience of other utilities using the same or similar filtration equipment may yield valuable information about the types of integrity failures common to a particular system.

How should membrane repair / replacement be conducted?

The types of repairs that may be conducted to correct integrity breaches vary significantly with the type of membrane filtration system. For example, for MF/UF systems, fiber breaks (the most common mode of failure) are not technically repaired, but rather isolated by inserting small pins or epoxy in the end(s) of the broken fiber, effectively removing them from service permanently and thus eliminating the system integrity breach. By contrast, it is generally not possible to repair comprised NF/RO membranes, although problems with o-rings and other seals may often be corrected by replacing the seals and making sure the membrane is properly seated in its housing. Likewise, because MCF systems utilize cartridges that are designed to be disposable, unless an integrity breach is the result of a seal problem, a damaged membrane cartridge would generally not be repaired. Thus, any integrity problems associated directly with either NF/RO or MCF membranes themselves generally necessitate membrane replacement.

Prior to placing the system into operation, it is important to consult with the membrane manufacturer to determine what types of repairs can be made and which types of integrity breaches require membrane replacement. The manufacturer should also be able to provide both instructions and training for all applicable repair procedures. Because it is somewhat common for some types of integrity problems to occur during the system shakedown phase, it is recommended that this period be used for practicing repair procedures under the directions of a qualified manufacturer's representative.

When integrity problems occur during operation, it is important to identify the cause of the integrity problem, as well as its source. While some fiber breaks in MF/UF systems may be expected due to wear or mechanical stress over time, other breaches may have a specific cause that can be isolated and corrected to avoid further integrity problems. If an integrity breach occurs, it is important to check both the membrane filtration system as well as any upstream treatment processes to ensure that these are operating properly. For example, if a chemical incompatible with the membrane material is added upstream but is not being properly removed or quenched, some membrane damage and loss of integrity may occur.

Any necessary repair equipment or spare parts, including replacement modules that may be required in the event of an integrity breach should be kept on site. The system IVP should specify a list of these components, along with any applicable instructions. The IVP should also include suggestions for troubleshooting integrity problems.

A.6 Data Collection and Analysis

Diligent and rigorous collection and analysis of integrity testing and monitoring data are an important component of any IVP. Careful data collection and analysis can serve as useful tools for preventing integrity problems, as well as for optimizing system performance and troubleshooting problems. Like other sections in this appendix, the following discussion is organized by addressing a series of critical questions in regard to data collection and analysis, each of which should be encompassed in a comprehensive IVP.

What is the purpose of data collection and analysis?

Although the primary purpose of data collection and analysis may be to demonstrate regulatory compliance, a thorough and well-planned program can result in a number of other important benefits for preventing integrity problems and optimizing system performance. For example, a consistently maintained record of membrane unit performance during both direct integrity testing and continuous indirect integrity monitoring may help determine when some membranes are approaching the end of their useful lives. In addition, because the resistance or permeability of some membranes changes after an initial "setting" period, a careful record of integrity test results may indicate that either the UCL (subject to regulatory approval) or any voluntarily implemented LCLs should be adjusted.

Continuous indirect integrity monitoring data also have a number of advantages in addition to helping gauge membrane integrity. This data may be used to identify performance trends, such as those that may occur between backwashing or chemical cleaning events, or over the entire life of the membranes. A well-documented data record can also identify any systematic or periodic trends and potentially help isolate the cause(s). In addition, data records facilitate the comparison of performance trends among different membrane units. (Note that a normal amount of instrument variation should be taken into account when conducting a statistical analysis comparing data among different membrane units.) A substantial amount of data collected over time may also enable operators to identify aberrations that do not necessarily indicate integrity problems. For example, if the turbidity is consistently higher than normal after a backwash, operators may be able to attribute this consistent aberration to air-entrainment error associated with the instrumentation. It is important that these types of aberrations are identified such that the system can be programmed not to trigger direct integrity testing during these events. Direct integrity testing data can also be a valuable tool for identifying trends; however, because continuous indirect integrity monitoring data are typically collected much more frequently and while the unit is on-line, it may often be the most practical means of tracking performance trends.

What data should be collected?

In addition to any regulatory requirements regarding data collection during operation, a utility should collect baseline data for each membrane unit (both with respect to direct integrity testing and continuous indirect integrity monitoring) before putting the plant into service. This data will serve as a reference baseline against which to evaluate membrane unit performance and also help refine a utility's strategy for collecting continuous indirect integrity monitoring data during regular operation. For example, for MF/UF systems the baseline data should demonstrate how long after a backwash event turbidity or particle count data might remain elevated. Based on this information, a utility can account for such data spikes that are known to not represent integrity problems, such that direct integrity testing and consequent loss of production are not unnecessarily triggered. Throughout operation, as well as during integrity testing conducted during routine maintenance or repair, it is recommended that data be collected in a spreadsheet or with software that establishes a database and allows data to be readily plotted in order to identify trends occurring over time.

What are some methods for reducing continuous indirect integrity monitoring data?

As noted in section A.3, the LT2ESWTR defines "continuous" monitoring as a frequency of no less than once every 15 minutes. However, instruments used to collect integrity monitoring data – particle counters, particle monitors, turbidimeters, etc. – may allow data collection at much more frequency intervals. Thus, in the absence of more specific requirements from the State, a utility may collect data as often as possible, if desired, provided the minimum frequency is met or exceeded. Furthermore, in the absence of specific guidelines for how a large amount of data should be reduced for compliance and reporting purposes, a utility has the latitude to select a statistical method that it determines to be appropriate for its system. Some potential methods include:

- Maximum value

- 95th percentile

- Average value

- Singular timed measurement

Using the example of "continuous" monitoring as defined under the LT2ESWTR (i.e., minimum frequency of once every 15 minutes), each of the above methods are described in context as follows.

Maximum Value
Using this method, the maximum value that occurs over every 15-minute period represents the entire data set. Thus, direct integrity testing is triggered if even one measurement exceeds the UCL. This method is very conservative and could potentially result in excessive direct integrity testing and subsequent loss of filtered water production from any anomalous data spikes, which may be attributable to any number of factors aside from an integrity problem.

95th Percentile
With this method, the 95th percentile datum represents the entire 15-minute data set, effectively screening out the largest five percent of data spikes. Direct integrity testing would be triggered if this datum exceeded the UCL. This method is less conservative than the maximum value approach and is more likely to screen anomalous data spikes that are not indicative of an integrity problem. The rationale behind this method is that if an integrity breach occurs, it may be likely that more than five percent of the data collected exceed the UCL. The premise for this method may be used with any percentile, and it may be advantageous for a utility using this technique to conduct a statistical analysis to determine an appropriate percentile to eliminate anomalous spikes without screening data that might indicate an integrity breach.

Average Value
Using this method, the average value represents the entire 15-minute data set. This technique is roughly equivalent to a 50th percentile approach, and thus is less conservative than the 95th percentile method. However, the average value method lessens the need to artificially exclude

anticipated integrity spikes that are known not to indicate an integrity problem (e.g., data collected immediate after a backwash with MF/UF systems), as the effect of these spikes may be sufficiently dampened such that the average value is below the UCL.

Singular Timed Measurement

The singular timed measurement approach uses one reading collected exactly every 15 minutes for comparison to the UCL (i.e., for regulatory compliance), independent of how frequently data are collected between these compliance readings. Because many other non-compliance measurements collected during these 15-minute intervals could exceed the UCL without triggering direct integrity testing, this method is one of the least conservative approaches. This technique represents the minimum requirement for compliance with the LT2ESWTR.

Note that because the LT2ESWTR does not specify a statistical reduction technique for data collected more frequently that at 15-minute intervals, the methods described in this section are not specific compliance options under the rule. These methods are not meant to represent an exclusive or exhaustive list and are simply examples of approaches that utilities could employ if not otherwise constrained by State requirements. A utility's IVP should clearly specify its data collection and analysis practices for both continuous indirect integrity monitoring and direct integrity testing.

A.7   Reporting

Utilities that use membrane filtration for compliance with the LT2ESWTR are required to submit a monthly operating report to the State summarizing the UCL exceedances for both direct integrity testing and continuous indirect integrity monitoring, respectively, as well as any corrective action that is taken in response (40 CFR 141.721(f)(10)(ii)). Because these monthly reports are directly linked to integrity testing results, it is important that reporting be incorporated into a comprehensive IVP. As with other sections in this appendix, the following discussion on reporting is organized into a series of critical questions that should be addressed in an IVP.

What is the purpose of reporting?

As it relates to integrity verification, the primary purpose of reporting is to document the ability of a membrane filtration system to meet its required log removal or other performance-based objectives on an ongoing basis. Under the minimum requirements of the LT2ESWTR, this documentation generally consists of all direct integrity test and continuous indirect integrity monitoring results that exceed the respective UCL and the subsequent corrective action that was taken in each case (40 CFR 141.721(f)(10)(ii)). Note that collecting, recording, and storing an abundance of integrity test data may be very beneficial for optimizing membrane filtration system performance, as discussed in section A.6, even if the majority of the accumulated data are not necessary for complying with reporting requirements.

If membrane filtration is not applied specifically for LT2ESWTR compliance, reporting may not necessarily be directly related to integrity verification, depending on State requirements. For example, membrane filtration might be considered an alternative filtration technology (as provided for under the federal SWTR), in which case reporting requirements might simply include turbidity, similar to those for conventional media filters. In this case, including reporting requirements in an IVP may not be critical for a utility. Nevertheless, because verifying and preserving membrane integrity are critical to successful membrane filtration system operation, it is recommended that even turbidity, particle counts, or other required filtrate quality data always be considered continuous indirect integrity monitoring results, thus linking reporting requirements to membrane integrity and, subsequently, an IVP.

What should an IVP include with respect to reporting?

An IVP should specify any State requirements for reporting, including both the content and the reporting frequency. (Reporting requirements for utilities that employ membrane filtration for compliance with the LT2ESWTR are addressed in Chapters 4 and 5 for direct integrity testing and continuous indirect integrity monitoring, respectively.) Any utility-specific procedures for preparing the compliance report should also be included in an IVP, along with a sample report form. An IVP should also indicate the duration of time over which the utility is required to keep records related to reporting data. Under the LT2ESWTR, utilities must keep records of all treatment monitoring associated with membrane filtration and used for rule compliance (including both direct integrity testing and continuous indirect integrity monitoring results, as applicable) for a period of three years (40 CFR 141.422(c)).

A.8 Summary

Although not required under the LT2ESWTR, the development of a comprehensive IVP can be a valuable organizational tool to help a utility verify and maintain membrane system integrity. An IVP should essentially serve as a utility's system-specific "how-to" guide for all aspects of operation and maintenance that are related to system integrity, including (but not necessarily limited to) the following:

- Regulatory requirements

- Voluntarily implemented system-specific practices

- Clear objectives for all IVP procedures

- Instructions for all IVP procedures

- Equipment listing, description, and purpose

- System troubleshooting tips

- Guidelines for interpreting test results

- Sample calculations (where applicable)

- Membrane manufacturer contact information

A well-developed IVP containing these elements can make the process of integrity verification more effective and efficient, thus helping a utility maximize the benefit of its membrane filtration system for serving as a barrier to pathogens and other particulate matter.

Appendix B

Overview of Bubble Point Theory

Appendix B:
Overview of Bubble Point Theory

B.1   Introduction

The various methods of pressure-based direct integrity testing are predicated on capillary theory as described by the bubble point equation, which is derived from a balance of static forces on the meniscus in a capillary tube. The bubble point itself is defined as the threshold gas pressure required to displace liquid from the pores or capillary-like breaches of a fully-wetted membrane. In the context of porous membranes, bubble point theory was originally used as the basis for developing a test to characterize pore sizes. Because the bubble point equation describes an inverse relationship between the applied pressure and capillary (or pore) diameter, the pressure at which bubbles are first detected in a fully-wetted membrane can be used to calculate the diameter of the largest pore (see Equation B.1). Accordingly, larger threshold pressures are indicative of membranes with smaller pores. A diagram illustrating a membrane pore as a capillary tube is shown in Figure B.1.

Figure B.1 Diagram of a Membrane Pore Modeled as a Capillary Tube

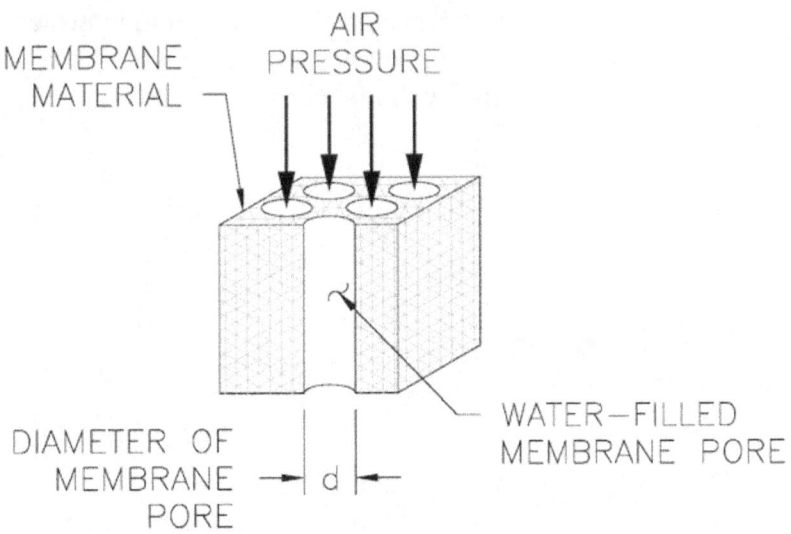

Bubble point theory has also been applied to the detection of integrity breaches in the form of the various pressure-based direct integrity tests, such as the pressure or vacuum decay tests (sections 4.7.1 and 4.7.2, respectively), the diffusive airflow test (section 4.7.3), and the water displacement test (section 4.7.4). Integrity breaches such as broken hollow fibers or holes in the surface of the membrane are analogous to pores, and larger test pressures enable the detection of smaller breaches. If the applied pressure is below the bubble point of the largest membrane pore and does not decay (pressure and vacuum decay tests), generate airflow (diffusive airflow test), or displace water (water displacement test) to a degree that exceeds normal tolerances over the duration of the direct integrity test, the membrane is determined to be integral at the level of the threshold pore or breach size corresponding to that applied pressure. In the context of the Long Term 2 Enhanced Surface Water Treatment Rule (LT2ESWTR), this threshold pore or breach size is called the test resolution, as described in section 4.2.

The purpose of this appendix is to provide a general overview of bubble point theory as it relates to direct integrity testing under the LT2ESWTR. In addition to the background provided in this introduction, subsequent sections of this appendix describe the bubble point equation and its parameters.

B.2   The Bubble Point Equation

The bubble point equation is derived from a balance of static forces on the meniscus in a capillary tube and is given as Equation B.1 (without specific units). A derivation of this equation is given in the literature by Meltzer (1987).

$$P_{bp} = \frac{4 \bullet \sigma \bullet \cos\theta}{d_{cap}}$$
Equation B.1

Where:

P_{bp}	=	bubble point pressure
σ	=	surface tension at the air-liquid interface
θ	=	liquid-membrane contact angle
d_{cap}	=	capillary diameter

Converting Equation B.1 to a form that utilizes convenient units for the various parameters yields Equation B.2:

$$P_{bp} = \frac{0.58 \bullet \sigma \bullet \cos\theta}{d_{cap}}$$
Equation B.2

Where:

P_{bp}	=	bubble point pressure (psi)
σ	=	surface tension at the air-liquid interface (dynes/cm)
θ	=	liquid-membrane contact angle (degrees)
d_{cap}	=	capillary diameter (μm)

Because the pore structure of most membranes cannot be accurately represented by a perfectly cylindrical capillary, a shape correction factor, κ, can be included in this equation to account for non-ideal conditions, as shown in Equation B.3:

$$P_{bp} = \frac{0.58 \bullet \kappa \bullet \sigma \bullet \cos\theta}{d_{cap}}$$

Equation B.3

Where:

P_{bp}	=	bubble point pressure (psi)
κ	=	pore shape correction factor (dimensionless)
σ	=	surface tension at the air-liquid interface (dynes/cm)
θ	=	liquid-membrane contact angle (degrees)
d_{cap}	=	capillary diameter (µm)

Equation B.3 serves as the general form of an expression for the resolution of a pressure-based direct integrity test, yielding the minimum required test pressure as a function of the capillary diameter. Equation B.4 represents this same expression, but incorporates variables more specific to the application of determining pressure-based direct integrity test resolution.

$$P_{test} = \frac{0.58 \bullet \kappa \bullet \sigma \bullet \cos\theta}{d_{res}}$$

Equation B.4

Where:

P_{test}	=	minimum test pressure (psi)
κ	=	pore shape correction factor (dimensionless)
σ	=	surface tension at the air-liquid interface (dynes/cm)
θ	=	liquid-membrane contact angle (degrees)
d_{res}	=	direct integrity test resolution requirement (µm)

Substituting the direct integrity test resolution requirement (d_{res}) of 3 µm for membrane filtration under the LT2ESWTR into Equation B.4 yields an expression specific to the rule, as shown in Equation B.5.

$$P_{test} = 0.193 \bullet \kappa \bullet \sigma \bullet \cos\theta$$

Equation B.5

Where:

P_{test}	=	minimum test pressure (psi)
κ	=	pore shape correction factor (dimensionless)
σ	=	surface tension at the air-liquid interface (dynes/cm)
θ	=	liquid-membrane contact angle (degrees)

If there is any backpressure in the membrane filtration system during direct integrity testing, this pressure must also be accounted for in calculating the minimum test pressure as a function of the required resolution. Because the backpressure acts in opposition to the applied test pressure, the test pressure must be increased by this same amount, as shown in Equation B.6. Note that this equation is also cited as Equation 4.1 in section 4.2.1.

$$P_{test} = (0.193 \bullet \kappa \bullet \sigma \bullet \cos\theta) + BP_{max} \qquad \text{Equation B.6 (4.1)}$$

Where:

P_{test}	=	minimum test pressure (psi)
κ	=	pore shape correction factor (dimensionless)
σ	=	surface tension at the air-liquid interface (dynes/cm)
θ	=	liquid-membrane contact angle (degrees)
BP_{max}	=	maximum backpressure on the system during the test (psi)

For the purposes of pressure-based direct integrity testing, Equation B.3 can be rearranged to yield an expression for the diameter of smallest integrity breach (i.e., defect diameter) that contributes to the response at a given applied pressure, as shown in Equation B.7:

$$d_{fect} = \frac{0.58 \bullet \kappa \bullet \sigma \bullet \cos\theta}{P_{test}} \qquad \text{Equation B.7}$$

Where:

d_{fect}	=	defect diameter (μm)
κ	=	pore shape correction factor (dimensionless)
σ	=	surface tension at the air-liquid interface (dynes/cm)
θ	=	liquid-membrane contact angle (degrees)
P_{test}	=	integrity test pressure (psi)

B.3   Bubble Point Equation Parameters

As shown in Equations B.4 - B.6, the pore shape correction factor (κ), liquid-membrane contact angle (θ), and the surface tension (σ) all affect the minimum direct integrity test pressure necessary to meet a given resolution requirement. Each of these parameters is discussed in the following subsections, including considerations for selecting appropriate values.

Liquid-Membrane Contact Angle (θ)

The liquid-membrane contact angle ranges from 0-90° and is primarily a function of the membrane hydrophilicity, which can be characterized in general terms as the affinity of the membrane material for water or the ability of the membrane to become wetted with water. For an ideally hydrophilic membrane, the liquid-membrane contact angle is 0 degrees. Although many membranes used for drinking water applications are manufactured using hydrophilic materials, an ideally hydrophilic membrane is purely theoretical. However, a value of 0 degrees yields the largest minimum integrity test pressure (as shown in Equations B.4 - B.6), and thus represents the most conservative value for the contact angle. In the absence of specific information from the membrane manufacturer regarding a more accurate value of θ for a particular material, a contact angle of 0 degrees should be assumed. The determination of values for θ using methods based on bubble point theory (Meltzer 1987), the Wilhelmy method (Adamson 1990), and other techniques is discussed in the literature. Because a less conservative contact angle can significantly reduce the minimum required integrity test pressure, any value for θ other than 0 degrees should be well-documented and approved by the State if used for the purposes of regulatory compliance, such as under the LT2ESWTR.

Pore Shape Correction Factor (κ)

The dimensionless pore shape correction factor ranges from 0 - 1 and is a function of the pore structure, accounting for deviations from perfectly cylindrical pores, as well as for the torturous flow path across the membrane. A correction factor of one represents a perfectly cylindrical pore, and as shown in Equations B.4 - B.6, maximizes the value of the minimum required test pressure. However, it is generally recognized that microporous membranes used for drinking water treatment do not have perfectly or approximately cylindrical pores, thus have associated correction factors less than one. Nevertheless, because the pore shape correction factor can significantly reduce the minimum required test pressure, it is important that any value less than one be well-documented and approved by the State if used for the purposes of regulatory compliance, such as under the LT2ESWTR. In the absence of data supporting the use of a non-ideal pore shape correction factor, a conservative value of one should be used.

Surface Tension

Although the surface tension of water does not vary significantly over the typical range of ambient water temperatures, it still can affect the minimum direct integrity test pressure necessary to meet a given resolution requirement, as shown in Equations B.4-B.6. Thus, because the surface tension varies inversely with temperature, the surface tension at the lowest anticipated water temperature should be used to calculate a conservative value for the minimum required test pressure. For reference, Table B.1 includes values of surface tension over a range of water temperatures.

Table B.1 Surface Tension of Water at Various Temperatures[1]

Temperature (°C)	Surface Tension (dynes/cm)
5	74.9
10	74.2
15	73.5
20	72.8
25	72.0
30	71.2
40	69.6

1 CRC Handbook, 66th ed., 1985-1986

Appendix C

Calculating the Air-Liquid Conversion Ratio

Appendix C:

Calculating the Air-Liquid Conversion Ratio

C.1   Introduction

As described in Chapter 4, the regulatory framework for the Long-Term 2 Enhanced Surface Water Treatment Rule (LT2ESWTR) requires that the flow through the smallest integrity breach that generates a measurable response from the direct integrity test (i.e., the critical breach size - Q_{breach}) be determined in order to establish the sensitivity of a pressure-based test method (40 CFR 141.719(b)(3)(iii)(A)). (Under the LT2ESWTR, sensitivity is defined as the maximum log removal value (LRV) that can be reliably verified by the direct integrity test (i.e., LRV_{DIT}).) However, because most pressure-based direct integrity tests yield results in terms of airflow or pressure decay, it may be necessary to convert these results to an equivalent value for the flow of water through the critical breach under typical filtration conditions. This conversion is necessary for calculating both the sensitivity and the upper control limit (UCL) for a pressure-based direct integrity test.

Although there are a number of methods for converting a direct integrity test response to a corresponding flow of water, each can be generally categorized as one of two types of approaches: mathematical modeling or experimental determination. This appendix describes a mathematical approach based on a parameter called the air-liquid conversion ratio (ALCR), which is defined as the ratio of air that would flow through a breach during a direct integrity test to the amount of water that would flow through the breach during filtration, as shown in Equation C.1 (also Equation 4.5):

$$ALCR = \left(\frac{Q_{air}}{Q_{breach}} \right)$$

Equation C.1   (4.5)

Where: ALCR =   air-liquid conversion ratio (dimensionless)
            Q_{air} =   flow of air through the critical breach during a pressure-based direct integrity test (volume / time)
            Q_{breach} =   flow of water through the critical breach during filtration (volume / time)

Because of the many variations in membrane configurations, breaches in the membrane may exhibit either turbulent or laminar flow characteristics depending upon the location and size of the defect, as well as the pressure differential between the feed and filtrate. In addition, there are fundamental differences between hollow-fiber and flat sheet membrane breaches, since the most common breaches associated with hollow-fiber modules exhibit pipe flow characteristics, while flat sheet breaches are best represented by an orifice model. Consequently, three different hydraulic models have been developed for determining the ALCR for a particular membrane system, depending on the configuration of the membrane material (i.e., hollow-fiber vs. flat sheet) and the type of flow (i.e., laminar vs. turbulent) that is expected through the critical breach. These three models include the Darcy pipe flow model (for breaches in a hollow-fiber (or hollow-fine-fiber) module under conditions of turbulent flow), the orifice model (for modules utilizing flat sheet membranes such as spiral-wound and membrane cartridge configurations under conditions of turbulent flow), and the Hagen-Poiseuille model (for any configuration under conditions of laminar flow). Table C.1 summarizes the various approaches for calculating the ALCR based on these three models and the conditions under which the use of each model is appropriate.

Table C.1 Approaches for Calculating the ALCR

Module Type	Defect Flow Regime	Model
Hollow-fiber[1]	Turbulent[2]	Darcy Pipe Flow
	Laminar	Hagen-Poiseuille
Flat sheet[3]	Turbulent	Orifice
	Laminar	Hagen-Poiseuille

1 Or hollow-fine-fiber
2 Typically characteristic of larger diameter fibers and higher differential pressures
3 Includes spiral-wound and cartridge configurations

Note that the various methods presented in this appendix for determining the ALCR implicitly assume that the flow regime for airflow through a breach during direct integrity testing is the same as that for liquid flow though a breach during filtration (i.e., either both laminar or both turbulent). If this assumption is determined to be inappropriate for a given membrane filtration system such that inaccurate and non-conservative estimates for sensitivity may result, then a hybrid approach may be considered. An example of such a hybrid approach is to assume laminar water flow and turbulent airflow, which could be modeled through the application of the Hagen-Poiseuille equation for water and the Darcy equation for air. The ALCR for such cases could then be derived using a similar methodology to that described in sections C.2, C.3, and C.4.

Procedures for calculating the ALCR and subsequently the sensitivity and UCL for applicable pressure-based direct integrity tests are given in Chapter 4, but the derivations of the various hydraulic models that form the basis for the respective ALCR equations are provided in the following sections of this appendix. Note that while the derivation of the ALCR equations relies on various hydraulic models that could be used to directly calculate the flow of air (Q_{air}) and water (Q_{breach}) through an integrity breach, direct application of these equations requires knowledge of the critical breach size, which is difficult and impractical to accurately quantify.

The advantage of the ALCR is that the terms relating to the size of the breach cancel out, yielding equations for the ALCR that are a function of either known and/or more easily determined parameters and independent of the critical breach size or geometry. Thus, although the ALCR equations in this appendix are derived for conditions assuming the flow of air (Q_{air}) and water (Q_{breach}) through the critical breach, the ALCR is independent of integrity breach size (i.e., physical dimensions of the breach) or magnitude (i.e., number of distinct breaches), and thus is a scalable parameter. An additional section is also included at the end of this appendix (C.5) that discusses cases in which the ALCR equations derived in sections C.2, C.3, and C.4 may not be applicable for some membrane filtration systems, as well as modifying the derivations to accommodate these systems.

For the derivation of the various models described in this appendix, airflow equations have been developed using a standard temperature and pressure of 68 °F (528 R, 293 K, or 20 °C) and 0 psi (1 atm or 14.7 psia). These standard conditions were selected to be consistent with the convention for airflow measurement devices. The equations can be modified to a different set of reference conditions through application of the ideal gas law expressed in terms of absolute temperature and pressure, if necessary. In addition, the temperature of the air used in a pressure-based direct integrity test is assumed to be the same as that for the water in the membrane filtration system, since these temperatures are expected to rapidly equilibrate.

Additional background on the hydraulic modeling developed in this appendix may be found in Crane's *Flow of Fluids Through Valves, Fittings, and Pipes* (1988). All of the basic hydraulic equations used in the derivation of the ALCR equations are included in the Crane text.

C.2 Darcy Pipe Flow Model

The Darcy pipe flow model is used to describe turbulent flow through an integrity breach with characteristics similar to a broken hollow-fiber. Generally, turbulent flow may be expected through larger diameter broken fibers and at higher differential pressures. The Darcy equations for the flow of air and water through a pipe are given in Equations C.2 and C.3, respectively:

$$Q_{air} = 11.3 \bullet Y \bullet d_{fiber}^{2} \bullet \sqrt{\frac{(P_{test} - BP) \bullet (P_{test} + P_{atm})}{(460 + T) \bullet K_{air}}} \qquad \text{Equation C.2}$$

Where:

	Q_{air}	=	flow of air at standard conditions (ft³/s)
	Y	=	net expansion factor for compressible flow through a pipe to a larger area (dimensionless)
	d_{fiber}	=	fiber diameter (in)
	P_{test}	=	integrity test pressure (psi)
	BP	=	backpressure on the system during the integrity test (psi)
	P_{atm}	=	atmospheric pressure (psia)
	T	=	water temperature (°F)
	K_{air}	=	resistance coefficient of air (dimensionless)

$$Q_{breach} = 0.525 \bullet d_{fiber}^{2} \bullet \sqrt{\frac{TMP}{K_{water} \bullet \rho_w}} \qquad \text{Equation C.3}$$

Where:

Q_{breach}	=	flow of water through the critical breach during filtration (ft^3/s)
d_{fiber}	=	fiber diameter (in)
TMP	=	transmembrane pressure (psi)
K_{water}	=	resistance coefficient of water (dimensionless)
ρ_w	=	density of water (lbs/ft^3)

Assuming that the resistance coefficients for air and water are similar (i.e., $K_{air} \approx K_{water}$) and applying a value of 62.4 lbs/ft^3 for the density of water, the ratio of Equation C.2 to Equation C.3 yields an expression for the ALCR, as given by Equation C.4:

$$ALCR = 170 \bullet Y \bullet \sqrt{\frac{(P_{test} - BP) \bullet (P_{test} + P_{atm})}{(460 + T) \bullet TMP}} \qquad \text{Equation C.4}$$

Where:

ALCR	=	air-liquid conversion ratio (dimensionless)
Y	=	net expansion factor for compressible flow through a pipe to a larger area (dimensionless)
P_{test}	=	direct integrity test pressure (psi)
BP	=	backpressure on the system during the integrity test (psi)
P_{atm}	=	atmospheric pressure (psia)
T	=	water temperature (^{o}F)
TMP	=	transmembrane pressure during normal operation (psi)

The ALCR is used in the equations for determining both the sensitivity and the UCL for pressure-based direct integrity tests, as described in Chapter 4. Consequently, the values of the parameters in Equation C.4 should be selected to yield a lower, more conservative value for the ALCR. For example, the transmembrane pressure (TMP) in Equation C.4 comes from the expression for Q_{breach} (Equation C.3) during filtration, and thus the most conservative ALCR result would be generated from using the maximum anticipated TMP during normal operation.

The net expansion factor for compressible flow (Y) may be obtained from charts in various hydraulics references, such as Crane (1988) (page A-22). Using the appropriate chart for airflow, values for Y are given as a function of pressure and the flow resistance coefficient, as shown in Equation C.5 (a non-specific expression illustrating the relationship between Y and its variables):

$$Y \propto \left[\frac{1}{\left(\dfrac{P_{test} - BP}{P_{test} + P_{atm}} \right)}, K \right]$$

Equation C.5

Where: | | | |
|---|---|---|
| Y | = | net expansion factor for compressible flow through a pipe to a larger area (dimensionless) |
| P_{test} | = | direct integrity test pressure (psi) |
| BP | = | backpressure on the system during the integrity test (psi) |
| P_{atm} | = | atmospheric pressure (psia) |
| K | = | flow resistance coefficient (dimensionless) |

The flow resistance coefficient (K) is a common fluid flow parameter described by most hydraulics texts and is defined as shown in Equation C.6:

$$K = f \bullet \frac{L}{d_{fiber}}$$

Equation C.6

Where: | | | |
|---|---|---|
| K | = | flow resistance coefficient (dimensionless) |
| f | = | friction factor (dimensionless) |
| L | = | length of the defect (in) |
| d_{fiber} | = | fiber diameter (in) |

Using the conservative scenario of a fiber break at the point where the fiber enters the pot, the length of the defect (L) is represented by the length of the lumen encasement into the membrane pot. The friction factor (f) may be estimated from a Moody diagram or the corresponding tabulated values, both of which are readily in available in most hydraulics references. The relative roughness (e/d_{fiber}) that is required to estimate the value for the friction faction may be calculated by either obtaining a product-specific value for the specific roughness (e) from the manufacturer or by using the membrane pore size as an estimate of the specific roughness.

Note that the net expansion factor (Y) should remain constant over time for practical purposes if appropriately conservative values are used to calculate this parameter. Thus, the determination of Y should represent a one-time, site-specific calculation. Also, because the ALCR is directly proportional to Y (as shown in Equation C.4), lower values for Y result in lower, more conservative values for the ALCR.

An iterative solution may be required to determine a value for the net expansion factor (Y). A general outline for one such iterative process is given as follows. The use of a spreadsheet may help facilitate the various calculations required.

1. Select a reasonable value for the friction factor (f).

2. Calculate the flow resistance coefficient (K) using Equation C.6.

3. Obtain a value for the Reynolds number (Re) from tabulated values for the friction factor (f) as a function of the Reynolds number (Re) and the relative roughness (e/d_{fiber}).

4. Calculate airflow (Q_{air}) from the equation for the Reynolds number (Re) as a function of equivalent diameter, air velocity, and dynamic viscosity (as referenced in fluid mechanics and fluid dynamics texts). For the purposes of determining the equivalent diameter and velocity (i.e., the flow (Q_{air}) divided by the cross-sectional area), assume that the applicable integrity breach may be represented by a pipe (e.g., a hollow fiber) flowing full with air. Use the maximum anticipated temperature and the minimum pressure applied over the duration of the direct integrity test (i.e., accounting for baseline decay) to generate a conservative (i.e., low) value for the dynamic viscosity and thus, in turn a conservative (i.e., low) value for both airflow (Q_{air}) and the ALCR.

5. Using tables available in hydraulics texts (e.g., page A-22 of Crane (1988)), apply the flow resistance coefficient (K) and the pressure ratio to determine a value for the net expansion factor (Y), as shown in Equation C.5.

6. Calculate airflow (Q_{air}) using Equation C.2. Assume that $K \approx K_{air}$.

7. If the airflow (Q_{air}) calculated in steps 4 and 6 (above) is approximately the same, then the net expansion factor (Y) determined in step 5 is correct. Otherwise, select a revised value for the friction factor (f) and repeat steps 1-7 in an iterative process until the two calculated values for airflow (Q_{air}) converge.

C.3   Orifice Model

The orifice flow model may be used to approximate turbulent flow through an integrity breach with characteristics similar to a hole in a flat sheet membrane that may be configured as a cartridge or a spiral-wound module. The representative equations for airflow and water flow through an orifice are given as Equations C.7 and C.8, respectively:

$$Q_{air} = 11.3 \bullet Y \bullet d_{fiber}^{2} \bullet C \bullet \sqrt{\frac{(P_{test} - BP) \bullet (P_{test} + P_{atm})}{460 + T}}$$ Equation C.7

Where:

Q_{air}	=	flow of air at standard conditions (ft^3/s)
Y	=	net expansion factor for compressible flow through a pipe to a larger area (dimensionless)
d_{fiber}	=	fiber diameter (in)
C	=	coefficient of discharge (dimensionless)
P_{test}	=	direct integrity test pressure (psi)
BP	=	backpressure on the system during the integrity test (psi)
P_{atm}	=	atmospheric pressure (psia)
T	=	water temperature (°F)

$$Q_{breach} = 0.525 \bullet d_{fiber}^{2} \bullet C \bullet \sqrt{\frac{TMP}{\rho_w}}$$ Equation C.8

Where:

Q_{breach}	=	flow of water through the critical breach during filtration (ft^3/s)
d_{fiber}	=	fiber diameter (in)
C	=	coefficient of discharge (dimensionless)
TMP	=	transmembrane pressure (psi)
ρ_w	=	density of water (lbs/ft^3)

The ratio of Equation C.7 to Equation C.8 yields an expression for the ALCR, as given in Equation C.9. Note that this equation incorporates a value of 62.4 lbs/ft^3 for density of water.

$$ALCR = 170 \bullet Y \bullet \sqrt{\frac{(P_{test} - BP) \bullet (P_{test} + P_{atm})}{(460 + T) \bullet TMP}}$$ Equation C.9

Where:

ALCR	=	air-liquid conversion ratio (dimensionless)
Y	=	net expansion factor for compressible flow through a pipe to a larger area (dimensionless)
P_{test}	=	direct integrity test pressure (psi)
BP	=	backpressure on the system during the integrity test (psi)
P_{atm}	=	atmospheric pressure (psia)
T	=	water temperature (°F)
TMP	=	transmembrane pressure during normal operation (psi)

Note that although derivations are slightly different, the resulting ALCR equations for the Darcy and orifice models (Equations C.4 and C.9, respectively) are identical. However, the two models utilize different methodologies for determining the net expansion factor for compressible flow (Y). As described in section C.2 for the Darcy model, the values of the parameters in Equation C.9 should be selected to yield a conservative value for the ALCR. For example, the TMP in Equation C.9 comes from the expression for Q_{breach} (Equation C.8) during filtration, and thus the most conservative ALCR result would be generated from using the maximum anticipated TMP during normal operation.

As with the Darcy model, the net expansion factor for compressible flow (Y) may be obtained from charts in various hydraulics references, such as Crane (1988) (page A-21). However, for the orifice model, Equation C.10 may also be used to calculate the net expansion factor, as follows:

$$Y = 1 - \left[0.293 \bullet \left(1 - \frac{BP + P_{atm}}{P_{test} + P_{atm}} \right) \right] \qquad \text{Equation C.10}$$

Where:

Y	=	net expansion factor for compressible flow through a pipe to a larger area (dimensionless)	
BP	=	backpressure on the system during the integrity test (psi)	
P_{atm}	=	atmospheric pressure (psia)	
P_{test}	=	direct integrity test pressure (psi)	

Because the ALCR is directly proportional to Y (as shown in Equation C.9), lower values for Y result in lower, more conservative values for the ALCR. Also, for practical purposes the net expansion factor (Y) should remain constant over time if appropriately conservative values are used to calculate this parameter. Thus, the determination of Y should represent a one-time, site-specific calculation.

C.4   Hagen-Poiseuille Model

The Hagen-Poiseuille model is appropriate for small integrity breaches (such as a pin hole or a broken, small-diameter hollow-fiber under low differential pressure) that would result in laminar flow. Using this model, the equation for airflow through a small defect under laminar flow conditions is given by Equation C.11:

$$Q_{air} = \frac{49.5 \bullet \pi \bullet d_{fect}^{4} \bullet \Delta P_{eff} \bullet g}{L \bullet \mu_{air} \bullet (460 + T)}$$

Equation C.11

Where:

Q_{air}	=	flow of air at standard conditions (ft^3/s)
d_{fect}	=	defect diameter (in)
ΔP_{eff}	=	effective integrity test pressure (psi)
g	=	gravitational constant $(32.2 \text{ lbm-ft/lbf-s}^2)$
L	=	length of the defect (in)
μ_{air}	=	viscosity of air (lbs/ft-s)
T	=	water temperature $(^{\circ}\text{F})$

Because air is a compressible fluid, the airflow is determined using the effective integrity test pressure, ΔP_{eff}, calculated according to Equation C.12.

$$\Delta P_{eff} = [(P_{test} - BP)] \bullet \left[\frac{(P_{test} + P_{atm}) + (BP + P_{atm})}{2 \bullet (BP + P_{atm})} \right] \bullet \left[\frac{(BP + P_{atm})}{P_{atm}} \right]$$

Equation C.12

Where:

ΔP_{eff}	=	effective air integrity test pressure (psi)
P_{test}	=	direct integrity test pressure (psi)
BP	=	backpressure on the system during the integrity test (psi)
P_{atm}	=	atmospheric pressure (psia)

The elements of the effective integrity test pressure include three primary terms, as individually bracketed in Equation C.12. These three terms are, respectively:

- the differential pressure across the membrane during the integrity test

- a term that accounts for the average velocity gradient of the compressed air as it passes across the membrane

- a multiplier that is necessary to convert the backpressure as it leaves the membrane to standard atmospheric conditions

The Hagan-Poiseuille equation for liquid flow through a breach under conditions of laminar flow is shown as Equation C.13:

$$Q_{breach} = \frac{0.094 \bullet \pi \bullet d_{fect}^{\,4} \bullet g \bullet TMP}{L \bullet \mu_w}$$

Equation C.13

Where:

Q_{breach}	=	flow of water through the critical breach during filtration (ft^3/s)
d_{fect}	=	defect diameter (in)
g	=	gravitational constant (32.2 lbm-ft/lbf-s^2)
TMP	=	transmembrane pressure (psi)
L	=	length of the defect (in)
μ_w	=	viscosity of water (lbs/ft-s)

The ratio of Equation C.11 to Equation C.13 yields an expression for the ALCR, as given by Equation C.14:

$$ALCR = \frac{527 \bullet \Delta P_{eff} \bullet \mu_w}{TMP \bullet \mu_{air} \bullet (460 + T)}$$

Equation C.14

Where:

ALCR	=	air-liquid conversion ratio (dimensionless)
ΔP_{eff}	=	effective integrity test pressure (psi)
μ_w	=	viscosity of water (lbs/ft-s)
TMP	=	transmembrane pressure during normal operation (psi)
μ_{air}	=	viscosity of air (lbs/ft-s)
T	=	water temperature (^oF)

The ratio of the viscosity of water to the viscosity of air (μ_w / μ_{air}) may be combined and expressed as a single function of the water temperature that is derived by fitting a curve to discrete data points for the viscosity ratio. Equation C.15 simplifies the calculation of the ALCR to an expression that is function of only measured pressures and the water temperature. Note that this form of the equation is only valid in the temperature range from 32 to 86 ^oF, in accordance with the limitations of the binomial fit for the viscosity ratio. If the temperature is outside of this range, the more general expression in Equation C.14 should be used.

$$ALCR = \frac{527 \bullet \Delta P_{eff} \bullet (175 - 2.71 \bullet T + 0.0137 \bullet T^2)}{TMP \bullet (460 + T)}$$

Equation C.15

Where:

ALCR	=	air-liquid conversion ratio (dimensionless)
ΔP_{eff}	=	effective integrity test pressure (psi)
T	=	water temperature (^oF)
TMP	=	transmembrane pressure (psi)

As with the Darcy and orifice models, the values for the parameters ΔP_{eff} and TMP used in Equation C.14 or C.15 should be selected to yield a lower, more conservative value for the ALCR. For example, the TMP in Equation C.15 comes from the expression for Q_{breach} (Equation C.13) during filtration, and thus the most conservative ALCR result would be generated from using the maximum anticipated TMP during normal operation.

C.5   Applicability of ALCR Equations

The various expressions for the ALCR developed in this appendix are each derived under the assumption that water and air pass through the same integrity breach during normal operation and pressure-based direct integrity testing, respectively. This consistency allows the terms relating to breach characteristics to cancel out in the ALCR derivations, resulting in equations that are independent of any specific knowledge of the integrity breach. Although this assumption is consistent with the operation of most membrane filtration systems, there may be some systems that operate in a manner that renders this assumption invalid. In these cases, the ALCR equations described in this appendix are not directly applicable, and the derivation of the ALCR must be modified to match the specific operation of a particular membrane filtration system.

One example of such a scenario might be a hollow-fiber membrane filtration system utilizing modules operating in an inside-out mode in which feed water flows into the fiber lumen from both ends. In this case, a single broken fiber represents two different pathways for water to flow through an integrity breach. If the Darcy model is applicable (i.e., turbulent flow conditions prevail), careful consideration must be given to the selection of the appropriate value for the net expansion factor for compressible flow (Y). This parameter is a function of the defect length (L), which will be different for each of the two flow pathways created by a single broken fiber (assuming the fiber is not severed precisely in the middle). In order to generate the most conservative (i.e., lowest) value for the ALCR (thus resulting in the most conservative direct integrity test sensitivity), the smallest potential value for L (i.e., the shortest length) should be used, which is generally represented by the length of the fiber encasement into the potting material. This value for L should always be utilized to yield a conservative ALCR, independent of the magnitude of the integrity breach. This approximation is reasonable since the flow through the short fiber length (i.e., the length of fiber embedded in the pot) will be substantially greater than the flow through the longer fiber length in most cases. If there is any question as to whether or not this approximation is appropriate for a particular system, the ALCR for different lengths could be calculated to evaluate the sensitivity of this important parameter to the length of the flow path.

A more complex scenario might involve a system similar to one described in the previous example, but which applies pressurized air from only one end of the fiber during direct integrity testing. In this case, a single broken fiber would represent two separate pathways for water to flow through an integrity breach during normal operation, but only a single path for air during integrity testing. This difference must be accounted for in the derivation of an expression for the ALCR. As shown in Equation C.1, the ALCR is the ratio of the air flowing through an integrity breach during direct integrity testing (Q_{air}) to the flow of water through the same breach during normal operation (Q_{breach}). For the system described in this example, any number of broken fibers would generate double that number of pathways for feed water to bypass the membrane filtration process via integrity breaches. Thus, the term Q_{breach} must also be doubled in the ALCR expression for both the Darcy (i.e., turbulent flow) and the Hagen-Poiseuille (i.e., laminar flow) models.

The two examples addressed in this section are just two possible scenarios in which the ALCR equations developed in this appendix may not be directly applicable to some membrane filtration systems. Because this guidance manual cannot anticipate and address every such potential case, it is recommended that each membrane filtration system be evaluated on a site-specific basis to determine whether the given ALCR equations may be used or if the derivation(s) must be modified to accommodate system-specific characteristics. Thus, any deviations from the assumptions used to derive the ALCR equations presented in this document may require a more complex treatment in which the flows of water and air through an integrity breach are determined separately using system-specific assumptions that are valid for each of these two respective flows.

Appendix D

Empirical Method for Determining the Air-Liquid Conversion Ratio for a Hollow-Fiber Membrane Filtration System

Appendix D:

Empirical Method for Determining the Air-Liquid Conversion Ratio for a Hollow-Fiber Membrane Filtration System

D.1   Introduction

As described in Chapter 4, the regulatory framework for the Long-Term 2 Enhanced Surface Water Treatment Rule (LT2ESWTR) requires that the flow through the smallest integrity breach that generates a measurable response from the direct integrity test (i.e., the critical breach size – Q_{breach}) be determined in order to establish the sensitivity of a pressure-based test method (40 CFR 141.719(b)(3)(iii)(A)). (Under the LT2ESWTR, sensitivity is defined as the maximum log removal value (LRV) that can be reliably verified by the direct integrity test (i.e., LRV_{DIT}).) However, because most pressure-based direct integrity tests yield results in terms of airflow or pressure decay, for systems that utilize such tests it may be necessary to convert these results to an equivalent value for the flow of water through the critical breach under typical filtration conditions. Although there are a number of potential methods for converting a direct integrity test response to a corresponding flow of water, each can be generally categorized as one of two types of approaches: mathematical modeling or experimental determination. This appendix describes an empirical approach based on a parameter called the air-liquid conversion ratio (ALCR), which is defined as the ratio of air that would flow through a breach during a direct integrity test to the amount of water that would flow through the breach during filtration, as shown in Equation D.1:

$$ALCR = \left(\frac{Q_{air}}{Q_{water}} \right) \qquad \text{Equation D.1}$$

Where: $ALCR$ =   air-liquid conversion ratio (dimensionless)

Q_{air} =   flow of air through an integrity breach during a pressure-based direct integrity test (volume / time)

Q_{water} =   flow of water through an integrity breach during filtration (volume / time)

While Appendix C describes the hydraulic models that could be used to calculate the ALCR for various types of membrane filtration systems and under different flow regimes, this appendix provides an example of an empirical method based on bubble point theory that can be used to determine the ALCR for a system using microporous hollow-fiber membranes – the correlated airflow measurement (CAM) technique. The CAM technique measures the flow of air (Q_{air}) and water (Q_{water}) through a fiber break scenario to empirically determine the ALCR of a membrane filtration unit for a system utilizing a pressure-based direct integrity test. This method is specific to hollow-fiber membrane processes in which the geometry of the fiber and associated module is known.

The CAM methodology involves first measuring the flow of water through a known integrity breach at various transmembrane pressures (TMPs) representative of normal operation, and subsequently measuring the flow of air through the same integrity breach during a pressure-based direct integrity test using a variety of potential test pressures. The data obtained from these measurements may be fitted with respective equations to establish empirical relationships between the applied pressures and the flow of water or air through an integrity breach. These functions can be used to determine the ALCR for any given operating transmembrane pressure (TMP) and direct integrity test pressure (P_{test}) (accounting for backpressure during the test, as per Equations 4.1 and 4.2). Note that the ALCR varies as a function of the TMP during operation, since a higher TMP results in a greater flow of water through the defect and thus a lower ALCR. In addition, these empirical relationships assume constant temperature, since changes in air or water temperature may change the functional relationships.

Although the CAM technique for determining the ALCR empirically is more labor-intensive than calculating the ALCR using a hydraulic model, the procedure does have several advantages. First, because the measurement is empirical, it is more accurate than calculations based on general hydraulic models. In addition, the CAM procedure does not rely on assumptions that are necessary to estimate the ALCR from hydraulic models, but instead facilitates direct determination of the ALCR based on measured air and water flows through a known defect. Another advantage of the CAM procedure is that it allows the ALCR to be easily recalculated for any operating TMP and direct integrity test pressure (assuming constant temperature).

D.2 Methodology

The following general procedure is provided as a guide for conducting the CAM technique for experimentally determining the ALCR. Note that although the CAM technique is described below for use with a single bench- or full-scale module for convenience in conducting the procedure, the resulting ALCR is scalable and independent of the size of the module or the integrity breach, as discussed in Appendix C (section C.1). Thus, the ALCR determined via this procedure would be applicable to an entire full-scale membrane unit.

1. Determine the baseline integrity test response for an integral bench- or full-scale membrane module. (See section 4.3.1.3 for a discussion of diffusive flow through the wetted pores of an integral membrane.)

2. For reference, measure the flow of water through the integral membrane module at various TMPs representative of the potential range of operating conditions.

3. Cut a known number of fibers (e.g., between 1 and 100). (Note that for many hollow-fiber membrane filtration systems, cutting a fiber at the point at which it enters the potting material represents the most conservative condition.)

4. Measure the water flow through the cut fiber(s) (i.e., Q_{water}) over the range of potential operating TMPs. One potential approach for determining Q_{water} is to compare the flow

through an integral membrane module (as determined in step 2, above) with that through the compromised module; the difference would represent the flow through the cut fiber(s) for each discrete TMP assessed. (Note that other approaches for determining Q_{water} may also be used.)

5. Develop an equation for a fitted curve that represents the water flow through the cut fiber(s) as a function of TMP.

6. Determine the minimum bubble point of the porous membrane material. (This information should generally be available from the manufacturer.)

7. Establish the direct integrity test pressure. As a general rule, the test pressure should be less than 80 percent of the bubble point pressure for the membrane and below the maximum TMP. However, the test pressure must be sufficient to meet the resolution requirement of the LT2ESWTR for the removal of *Cryptosporidium*, as expressed in Equations 4.1 and 4.2.

8. Measure the airflow from the cut fiber(s) at a variety of potential applied direct integrity test pressures. This may be particularly useful if the hydrostatic backpressure may vary between different direct integrity test applications.

9. If the diffusive flow (i.e., the baseline response, as determined in step 1) is significant (i.e., more than 5 percent of the total airflow) at the target test pressure to be utilized during normal operation (P_{test}), then a lower test pressure should be considered, if possible. (The test pressure must enable the direct integrity test to meet the resolution requirement.) Alternatively, the diffusive flow will have to be accounted for in determining the ALCR, as described in section 4.3.1.3.

10. Determine the ALCR using Equation D.1.

$$ALCR = \frac{Q_{air}}{Q_{water}}$$

Where: $ALCR$ = air-liquid conversion ratio (dimensionless)
Q_{air} = flow of air through the broken fiber(s) at the direct integrity test pressure (mL/min)
Q_{water} = flow of water through the broken fiber(s) at the reference TMP (mL/min)

Note that the reference TMP described in association with the variable Q_{water} above refers to the TMP that is used in the determination of the ALCR for the purpose of establishing the direct integrity test sensitivity for regulatory compliance, as described in section 4.3.1.2 and Appendix C. For example, the most conservative ALCR result would be generated from using the maximum anticipated TMP during normal operation.

11. Use Equation 4.7 to calculate the direct integrity test method sensitivity (i.e., LRV_{DIT}), incorporating the test result (Q_{air}) (either as directly measured with the diffusive airflow test or as converted from the pressure decay rate (ΔP_{test}) using Equation 4.8 if the pressure decay test is used) and the ALCR determined in the previous step, as described in Chapter 4. Note that the use of Equations 4.7 and 4.8 to calculate sensitivity require that the parameters Q_{air} and ΔP_{test}, respectively, represent the smallest integrity test response that can be reliably measured and associated with an integrity breach, as specified in section 4.3.1.2.

Appendix E

Application of Membrane Filtration for Virus Removal

Appendix E:

Application of Membrane Filtration for Virus Removal

E.1   Introduction

Although the Long Term 2 Enhanced Surface Water Treatment Rule (LT2ESWTR) only regulates the use of membrane filtration for the removal of *Cryptosporidium* for the purposes of compliance with the rule requirements, States could opt to apply the regulatory framework in a broader context to other applications of membrane filtration, at their discretion. For the removal of *Giardia* or other relatively large pathogens that are approximately the same size as *Cryptosporidium* in order-of-magnitude terms, the LT2ESWTR regulatory framework can be applied almost directly with only minor modifications, such as adjusting the direct integrity test resolution to coincide with the lower bound of the size range characterizing the pathogen of interest. However, for pathogens that are significantly smaller than *Cryptosporidium*, there are a number of important considerations that must be taken into account and which may limit the applicability of the LT2ESWTR regulatory framework in some cases. The purpose of this appendix is to summarize some of the critical issues associated with the application of membrane filtration for the removal of very small pathogens using the LT2ESWTR regulatory framework. Because viruses represent the smallest class of pathogen and are directly regulated under the Surface Water Treatment Rule (SWTR), this appendix is primarily focused on the application of membrane filtration for virus removal; however, it is important to note that these same issues may generally apply to other small pathogens of comparable size, as well.

Among the five classes of membrane filtration discussed in the *Membrane Filtration Guidance Manual* – microfiltration (MF), ultrafiltration (UF), nanofiltration (NF), reverse osmosis (RO), and membrane cartridge filtration (MCF) – only MF, UF, and MCF are typically applied specifically for the purpose of pathogen (or other particulate) removal. However, because the nominal pore sizes of the types of membranes associated with these processes vary widely (UF = 0.01 µm; MF = 0.1 µm; MCF = 1 µm or smaller), their removal characteristics are different. While all three processes can remove *Cryptosporidium*, with a size range lower bound of about 3 µm, only UF membranes have pores small enough to filter viruses, which generally range in size from about 0.01 to 0.1 µm. Thus, UF is the primary membrane filtration technology utilized for the objective of virus reduction. Accordingly, the discussion in this appendix is primarily presented in the context of virus removal using UF. Nevertheless, note that while MF cannot generally be used as an effective means of virus treatment, it may be applied to remove organisms such as some species of bacteria that are larger than 0.1 µm but still as much as an order of magnitude smaller than *Giardia* or *Cryptosporidium*. In these cases, some of the issues addressed in the appendix may be applicable to MF. It is important to note that although some virus removal by MF has been reported in the literature, it is generally attributed to formation of a cake layer on the membrane surface. Since this cake layer is dynamic and removed by the backwash process, the filtration of particulate matter via this mechanism varies during the course of an operational cycle and thus is not consistent with the LT2ESWTR framework, which considers only the removal efficiency of the membrane barrier itself.

While NF and RO utilize non-porous semi-permeable membranes that represent a barrier to viruses, the associated modules are generally not manufactured to be aseptic. In addition, because these membranes are not able to be backwashed, particulate matter can cause rapid irreversible fouling. As a result, NF and RO are not typically applied to directly treat raw water supplies with significant concentrations of suspended solids. However, NF or RO may be directly applied (i.e., without significant pretreatment for particulate removal) to remove dissolved solids from a ground water source that is subject to virus contamination but very low in suspended solids. In such cases, NF and RO may be used to obtain virus reduction credits under the Ground Water Rule (GWR). Under these circumstances, the issues addressed in this appendix may also be applicable to NF or RO.

In general, application of the LT2ESWTR regulatory framework for the removal of viruses necessitates that the membrane filtration process comply with appropriate pathogen-specific criteria for the three primary regulatory elements: challenge testing, direct integrity testing, and continuous indirect integrity monitoring. The subsequent sections of this appendix address some of the practical and regulatory issues with respect to virus removal for each of these three program elements. An additional preceding section summarizes current State regulatory policy for the use of UF for virus removal and highlights some associated considerations for the application of the LT2ESWTR regulatory framework.

E.2   Overview of Current Regulatory Policy

Because there is no general federal framework for the broad regulation of membrane filtration as a water treatment technology, States have developed policies that are widely varying in some cases. However, most States are consistent in allowing very little virus removal credit for membrane filtration. While numerous challenge studies have demonstrated a clear difference in the ability of MF and UF to remove viruses (as shown in Table E.1), a 2001 survey of State primacy agencies conducted by USEPA indicated that only 6 of 29 States with regulatory policies specific to membrane filtration technology differentiated between these two processes in terms of the virus removal credit awarded, as summarized in Table E.2 (USEPA 2001).

The most significant factor limiting the virus removal credit awarded to UF is the infeasibility of using current direct integrity test methods to detect a virus-sized breach (as discussed in section E.3). Thus, it is possible that a number of very small integrity breaches could allow the passage of viruses through the membrane barrier undetected, contaminating the filtrate. While this mode of failure may not be as common as a broken fiber, such a very small integrity breach may occur as the membranes age or as a result of degradation due to exposure to incompatible treatment chemicals.

Table E.1 Virus Removal Studies Using MF and UF

Researcher(s)	Year	Process	Log Removal
Jacangelo *et al.*	1997	MF	0 – 2.4
Wilinghan *et al.*	1992	MF	0.3 – 4
Schneider *et al.*	1999	MF	1.1
Trussel *et al.*	1998	MF	0.4 – 3.2
Jacangelo *et al.*	1997	UF	6.0 – 7.9
Dwyer *et al.*	1995	UF	6.2 – 6.8[1]
Trussel *et al.*	1998	UF	> 6.9[1]
Kruithof *et al.*	1999	UF	> 5.4

1 Removal to levels below detection limit

Table E.2 Summary of State Virus Removal Credit for MF and UF

Virus Log Removal Credit	Number of States	
	MF	UF
No standard credit[1]	8	8
0	18	14
0.5	3	1
1.0	-	1
1.5	-	-
2.0	-	-
2.5	-	-
3.0	-	1
3.5	-	-
4.0	-	4

1 Credit typically awarded on a case-by-case basis

Although this rigorous degree of testing and performance verification is generally not required of conventional treatment (e.g., daily testing to ensure the absence of short-circuiting in media filters), the combined processes of flocculation/sedimentation and media filtration represent a multiple barrier to particulate contaminants that provides some degree of insurance against difficult-to-detect treatment deficiencies (such as small integrity breaches or short-circuiting). The removal credits awarded to conventional treatment under the SWTR account for this multiple barrier treatment concept, while the log removal credits referenced for UF in Tables E.1 and E.2 represent membrane filtration as a stand-alone process. Consequently, it is possible that a multi-barrier treatment process scheme including flocculation, sedimentation, and UF (or other membrane filtration process) could be awarded equivalent or greater virus removal credit than conventional treatment by some States, at their discretion.

In addition, because viruses are easily inactivated by low chlorine doses and contact times that are unlikely to yield significant disinfection byproducts (DBPs), there is less of a necessity for States to award high virus removal credits to any alternative treatment process if some level of primary disinfection is required. By contrast, States generally award removal credit on par with that for conventional treatment for both MF and UF for larger pathogens such as *Giardia* and *Cryptosporidium,* which have more robust chlorine inactivation requirements relative to viruses and which are large enough such that similarly sized integrity breaches can be detected using common direct integrity test techniques (as discussed in section E.3). State-awarded log removal credits for both *Giardia* and *Cryptosporidium* using MF/UF (which are considered comparable technologies for *Giardia* and *Cryptosporidium* removal) are summarized in Table E.3 for the same 29 States with specific membrane filtration policies (USEPA 2001).

Table E.3 Summary of State *Giardia* and *Cryptosporidium* Removal Credit for MF/UF

Virus Log Removal Credit	Number of States	
	Giardia	*Cryptosporidium*
No standard credit[1]	8	22
0	-	-
0.5	-	-
1.0	1	-
1.5	1	-
2.0	2	4
2.5	7	-
3.0	9	2
3.5	-	-
4.0	1	1

1 Credit typically awarded on a case-by-case basis

In awarding virus removal credit to UF, States must balance the studies that consistently demonstrate the ability of UF to achieve 5 to 7 log virus removal with the practical problem of verifying virus-sized integrity breaches during operation (as discussed in section E.3) and the potential impact that such small integrity breaches might have on the ability of UF to achieve the awarded credit. Another consideration is the emphasis that many States place on multiple barrier protection. In cases in which chemical disinfection is utilized as a component of multiple barrier treatment, the level of virus inactivation generally negates the need to award virus removal credits to UF. As shown in Table E.2, most States currently consider the issues of integrity verification and/or multiple barrier protection more significant factors, since few States award UF any virus removal credit as a stand-alone process. However, even without significant State-awarded removal credit, utilities may still specifically opt to install UF as a barrier against viruses on the basis of the many studies indicating near complete virus removal, and that even with some very small and undetectable integrity breaches UF may remove viruses more efficiently than either MF or conventional treatment.

E.3   Direct Integrity Testing

In terms of direct integrity testing, a membrane process applied to obtain virus removal under the LT2ESWTR regulatory framework would need to meet applicable criteria for sensitivity, resolution, and frequency. While applying the LT2ESWTR framework for virus removal is similar to that for *Cryptosporidium* for sensitivity and frequency of direct integrity testing (since both of these concepts are independent of the target pathogen), the resolution that would be necessary for very small pathogens such as viruses is more problematic. Because viruses range in size from approximately $0.01 - 0.10$ µm, the resolution must be a very small 0.01 µm – two orders of magnitude smaller than that for *Cryptosporidium* – in order to provide an appropriate level of conservatism. A resolution requirement this small may be very difficult to achieve with currently available direct integrity tests.

A virus-sized resolution requirement is particularly problematic for pressure-based direct integrity tests, which are based on bubble point theory (as described in Appendix B). A form of the bubble point equation is shown as Equation E.1, which relates the required direct integrity test pressure to the size of an integrity breach.

$$P_{test} = \frac{0.58 \bullet \kappa \bullet \sigma \bullet \cos \theta}{d_{fect}}$$
Equation E.1

Where:
P_{test} = required test pressure (psi)
κ = pore shape correction factor (dimensionless)
σ = surface tension at the air-liquid interface (dynes/cm)
θ = liquid-membrane contact angle (degrees)
d_{fect} = defect diameter (µm)
0.58 = constant including applicable unit conversion factors

In the context of determining a minimum necessary test pressure, the defect diameter is equal to the required resolution (i.e., the lower bound on the size range of the pathogen of interest). Table E.4 summarizes the required direct integrity test pressure as a function of the target pathogen size (i.e., the test resolution) for two sets of conditions. In the most conservative case, the pore shape correction factor (κ) is equal to 1, and the liquid-membrane contact angle (θ)

is equal to zero (indicating a perfectly hydrophilic membrane). These respective values should be used as a default in the absence of more specific information about the true values of these two parameters. The second case utilizes less conservative values of $\kappa = 0.25$ and $\theta = 45°$ to illustrate the affect that these two parameters can have on the required test pressure. Both cases assume a conservative surface tension (σ) of 74.9 dynes/cm (i.e., at a temperature of 5 °C). Note that Table E.4 is logarithmic, citing pathogen sizes from 0.01 to 0.1 μm in 0.01 μm increments, and then continuing to 1.0 μm in 0.1 μm increments.

Table E.4 Required Direct Integrity Test Pressure as a Function of Target Pathogen Size (i.e., Resolution)

Target Pathogen Size (μm)	Required Test Pressure (psi)	
	Most Conservative $\kappa = 1$ $\theta = 0°$	Less Conservative $\kappa = 0.25$ $\theta = 45°$
0.01	4,344	768
0.02	2,172	384
0.03	1,448	256
0.04	1,086	192
0.05	869	154
0.06	724	128
0.07	621	110
0.08	543	96
0.09	483	85
0.1	434	77
0.2	217	38
0.3	145	26
0.4	109	19
0.5	87	15
0.6	72	13
0.7	62	11
0.8	54	10
0.9	48	9
1.0	43	8

As shown in Table E.4, for the most conservative case, the required test pressure for a virus-sized resolution of 0.01 μm is over 4,000 psi, a value far in excess of what any current, commercially available water treatment membrane could withstand without rupturing. Even the strongest polymeric hollow-fiber membrane can withstand a transmembrane pressure (TMP) of no more than 100 psi at most, limiting the smallest target pathogen (and thus the resolution) to about 0.4 μm in the most conservative case and perhaps just below 0.1 μm if more specific, less conservative values of κ and θ are known. Other small pathogens that are larger than viruses (e.g., bacteria in the 0.5 – 1.0 μm range) require lower test pressures and may allow more feasible application of LT2ESWTR regulatory framework. In addition, some manufacturers have experimented with stronger ceramic "membrane" processes with pore sizes similar to MF or UF, which may allow the application of direct integrity tests at higher pressures, thus facilitating the use of the LT2ESWTR regulatory framework for smaller pathogens than are possible with current polymeric membrane materials. Note that Table E.4 does not account for the system backpressure during the application of a pressure-based direct integrity test, which can increase the required test pressure to achieve a target resolution, as described in Equation 4.1 in Chapter 4.

Resolution requirements on the order of viruses also present some problems for the use of marker-based direct integrity tests. Suitable virus-sized challenge particulates may be prohibitively expensive, making daily testing infeasible. In addition, it may be difficult to manufacture these particulates to meet acceptable tolerances for size range variation. Thus, even if the practical problems associated with virus-sized direct integrity test resolution could be overcome and significant virus removal credit would be awarded by the State under the LT2ESWTR regulatory framework, it may be less expensive for a utility using UF to apply a small amount of chlorine to inactivate viruses. In this case, because the achievement of virus removal credit using UF is an issue of operational verification and not removal efficiency as demonstrated in challenge studies, the utility would have the benefit of a multiple barrier treatment process even if no removal credit is awarded to UF by the State.

E.4   Challenge Testing

As shown in Table E.1, a number of studies conducted in recent years have demonstrated the ability of UF to achieve high log removals of viruses. While it is not necessary to use the pathogen of interest itself for the purposes of conducting a challenge test in accordance with the LT2ESWTR regulatory framework, it is important to ensure that any challenge particulate selected is conservatively (i.e., equivalently or less efficiently) removed (as discussed in section 3.9).

One significant consideration in regard to challenge testing under the LT2ESWTR regulatory framework as applied to virus removal is the issue of representative performance testing to verify the removal efficiency of all membrane modules in a product line that are not subject to challenge testing. In order to demonstrate *Cryptosporidium* removal efficiency, all modules must be subjected to non-destructive performance testing that is consistent with the appropriate resolution requirement, as discussed in section 3.6. The results of this testing are compared to the results of similar testing conducted on the representative modules in the same product line that were subject to challenge testing as an indicator of performance. Common non-destructive performance tests (NDPTs) are types of direct integrity tests, such as the bubble test (see section 4.8.2). However, as discussed in section E.3, because pressure-based direct integrity tests may not be able to achieve the required 0.01 μm resolution requirement for viruses, non-destructive performance testing may not be possible. Thus, it may be necessary to develop different criteria and methods for validating the virus removal capability of modules in a product line that are not subjected to challenge testing, such as representative destructive performance testing on a statistically significant number of modules in each production lot. One example of such a destructive performance test might be a scanning electron microscopy (SEM) analysis of the membrane media to confirm the pore size distribution.

E.5   Continuous Indirect Integrity Monitoring

Although the LT2ESWTR regulatory framework does not specify any particular resolution or sensitivity criteria for continuous indirect integrity monitoring, related issues may impact the effectiveness of indirect integrity monitoring if the framework is applied to virus removal. For example, very small integrity breaches that permit the passage of viruses may not allow enough particulate matter across the membrane to be detected by indirect methods such as turbidity monitoring or particle counting. Thus, it may be difficult to establish a meaningful control limit for continuous indirect integrity monitoring that is specific to the application of membrane filtration for virus removal. As a result, continuous indirect integrity monitoring should be used as a general gauge of gross membrane integrity that is independent of the particular pathogen of concern.